Katarzyna Staszak, Karolina Wieszczycka, Bartosz Tylkowski (Eds.)
Chemical Technologies and Processes

I0131962

Also of interest

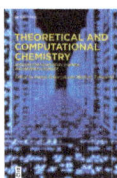

Theoretical and Computational Chemistry
Gulaczyk, Tylkowski (Eds.), 2020
ISBN 978-3-11-067815-4, e-ISBN 978-3-11-067821-5

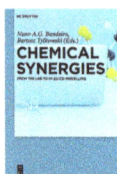

Chemical Synergies – From the Lab to In Silico Modelling
Bandeira, Tylkowski (Eds.), 2018
ISBN 978-3-11-048135-8, e-ISBN 978-3-11-048206-5

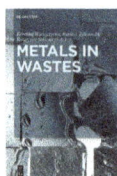

Metals in Wastes
Wieszczycka, Tylkowski, Staszak (Eds.), 2018
ISBN 978-3-11-054628-6, e-ISBN 978-3-11-054706-1

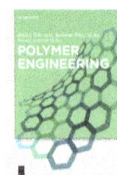

Polymer Engineering
Tylkowski, Wieszczycka, Jastrzab (Eds.), 2017
ISBN 978-3-11-046828-1, e-ISBN 978-3-11-046974-5

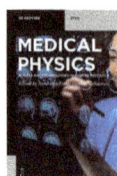

Medical Physics– Models and Technologies in Cancer Research
Bajek, Tylkowski (Eds.), 2021
ISBN 978-3-11-066229-0, e-ISBN 978-3-11-066230-6

Chemical Technologies and Processes

Edited by
Katarzyna Staszak, Karolina Wieszczycka
and Bartosz Tylkowski

DE GRUYTER

Editors

Dr. Katarzyna Staszak
Poznan University of Technology
Institute of Chemical Technology & Engineering
Berdychowo 4
60-965 Poznań
Poland
katarzyna.staszak@put.poznan.pl

Dr. Karolina Wieszczycka
Poznan University of Technology
Institute of Chemical Technology & Engineering
Berdychowo 4
60-965 Poznań
Poland
karolina.wieszczycka@put.poznan.pl

Dr. Bartosz Tylkowski
Eurecat, Centre Tecnològic de Catalunya
Marcel·lí Domingo s/n
43007, Tarragona, Spain
bartosz.tylkowski@eurecat.org

ISBN 978-3-11-065627-5
e-ISBN (PDF) 978-3-11-065636-7
e-ISBN (EPUB) 978-3-11-065643-5

Library of Congress Control Number: 2020938354

Bibliographic information published by the Deutsche Nationalbibliothek
The Deutsche Nationalbibliothek lists this publication in the Deutsche Nationalbibliografie;
detailed bibliographic data are available on the Internet at http://dnb.dnb.de.

© 2020 Walter de Gruyter GmbH, Berlin/Boston
Typesetting: Integra Software Services Pvt. Ltd.
Printing and binding: CPI books GmbH, Leck
Cover image: gremlin/E+/gettyimages

www.degruyter.com

Preface

One could say that there are a lot of books on chemical technology and why there should be another one. This need was due to several factors. First of all, the changes taking place in the chemical and petrochemical industry today are so dynamic that the information from several years ago may be out of date. This progress is the result of the introduction of advanced technologies, which must be in line with scientific progress and based on a sustainable economy. Nowadays environmental aspects are also very important. Technologies that may have been and have been widely used in the past must be modified and in many cases replaced by completely new ones. For example, until a few years ago the chlor-alkali industry relied mainly on mercury cell process, which are successively replaced by more environmentally friendly membrane cell process. On the other hand, these changes are also due to concerns about process safety. Challenges in the technology are therefore a necessary and indispensable part of the chemical industry. Information in this respect should therefore be constantly updated. This is not easy. It requires constant analysis of the latest literature, patent databases, technological innovations presented at exhibitions, conferences and symposia. It should also be realized that much of this information is closely guarded by the know-how of the chemical and petrochemical industries. Nevertheless, the authors' intention was to extract as much information as possible. And the most important goal was to write a book that would be an excellent compendium of modern knowledge for readers. We will leave it up to the readers to judge whether this was successful.

We would like to express our gratitude to the contributing authors in making this project a success, and to Vivien Schubert from DeGruyter Publisher, Germany for her assistance and encouragement in this venture.

Editors
Katarzyna Staszak
Karolina Wieszczycka
Bartosz Tylkowski

https://doi.org/10.1515/9783110656367-202

Contents

Maciej Staszak

Xavier Montané, Josep M. Montornes, Adrianna Nogalska, Magdalena Olkiewicz, Marta Giamberini, Ricard Garcia-Valls, Marina Badia-Fabregat, Irene Jubany and Bartosz Tylkowski

List of contributing authors

Francisco Jose Alguacil
Centro Nacional de Investigaciones
Metalurgicas (CSIC)
Avda. Gregorio del Amo 8
Madrid 28040, Spain
fjalgua@cenim.csic.es

Marina Badia-Fabregat
Sustainability Unit
Eurecat Centre Tecnològic de Catalunya
Manresa, Catalunya
Spain
marina.badia@eurecat.org

Filip Ciesielczyk
Faculty of Chemical Technology
Institute of Chemical Technology
and Engineering
Poznan University of Technology
Berdychowo
4, Poznan PL-60965, Poland,
filip.ciesielczyk@put.poznan.pl

Ricard Garcia-Valls
Department of Chemical Engineering,
Universitat Rovira i Virgili, Av. Països
Catalans 26, 43007 Tarragona, Spain,
ricard.garcia@urv.cat

Marta Giamberini
Department of Chemical Engineering,
Universitat Rovira i Virgili,
Av. Països Catalans 26,
43007 Tarragona,
Spain,
marta.giamberini@urv.cat

Irene Jubany
Sustainability Unit
Eurecat Centre Tecnològic de Catalunya
Manresa, Catalunya
Spain
irene.jubany@eurecat.org

Bartosz Tylkowski
Chemical Technologies Unit, Eurecat Centre
Tecnològic de Catalunya, Carrer Marcel·lí
Domingo, s/n, Tarragona 43007, Catalunya,
Spain
bartosz.tylkowski@eurecat.org

Dobrawa Kwaśniewska
Department of Technology and Instrumental
Analysis, Poznan University of Economics
and Business
al. Niepodległości 10
Poznań
Poland
dobrawa.kwasniewska@ue.poznan.pl

Katarzyna Materna
Institute of Chemical Technology
and Engineering, Poznań University
of Technology, ul. Berdychowo 4, Poznań
60-965, Poland
katarzyna.materna@put.poznan.pl

Xavier Montané
Departament de Química Analítica i Química
Orgànica, Universitat Rovira i Virgili, Carrer
Marcel·lí Domingo s/n, Campus Sescelades,
Tarragona 43007
Spain
xaviermontanemontanem@gmail.com

Josep M. Montornes
Chemical Technologies Unit, Eurecat Centre
Tecnològic de Catalunya, Carrer Marcel·lí
Domingo, s/n, Tarragona 43007, Catalunya,
Spain
josep.montornes@eurecat.org

Adrianna Nogalska
Chemical Technologies Unit, Eurecat Centre
Tecnològic de Catalunya, Carrer Marcel·lí
Domingo, s/n, Tarragona 43007, Catalunya,
Spain
adrianna.nogalska@eurecat.org

Magdalena Olkiewicz
Chemical Technologies Unit, Eurecat Centre
Tecnològic de Catalunya, Carrer Marcel·lí
Domingo, s/n, Tarragona 43007, Catalunya,
Spain
magda.olkiewicz@eurecat.org

Magdalena Regel-Rosocka
Institute of Chemical Technology
and Engineering, Poznań University of
Technology, ul. Berdychowo 4, Poznań
60-965, Poland
magdalena.regel-rosocka@put.poznan.pl

J. I. Robla
Centro Nacional de Investigaciones
Metalurgicas (CSIC)
Avda. Gregorio del Amo 8
Madrid 28040, Spain
jrobla@cenim.csic.es

Martyna Rzelewska-Piekut
Institute of Chemical Technology and
Engineering, Poznań University of
Technology, ul. Berdychowo 4, Poznań
60-965, Poland
martyna.rzelewska-piekut@put.poznan.pl

Katarzyna Staszak
Institute of Chemical Technology
and Engineering
Poznan University of Technology
Berdychowo St 4, Poznan 60-965,
Poland
katarzyna.staszak@put.poznan.pl

Maciej Staszak
Institute of Chemical Technology and
Engineering
Poznan University of Technology
Berdychowo St 4, Poznan 60-965,
Poland
maciej.staszak@put.poznan.pl

Daria Wieczorek
Department of Technology and Instrumental
Analysis, Poznan University of Economics
and Business
al. Niepodległości 10
Poznań
Poland
d.wieczorek@ue.poznan.pl

Karolina Wieszczycka
Institute of Chemical Technology and
Engineering, Poznań University of
Technology, ul. Berdychowo 4, Poznań
60-965, Poland
karolina.wieszczycka@put.poznan.pl

Filip Ciesielczyk

1 Inorganic acids – technology background and future perspectives

Abstract: The presented chapter deals with technological aspects concerning industrial production of major inorganic chemicals such as sulfuric, nitric and phosphorous acids. The idea was to highlight the main assumptions related to their fabrication as well as to point out novel aspects and perspectives concerning their industry. The main attention will be paid to raw materials and their transformation in order to obtain indirect precursors of inorganic acids – acid anhydrides. Characteristics of the individual stages of synthesis, with a special regard to the process conditions will be also of key importance. The environmental impact of this particular technology will be raised and discussed. Finally, the review over recently published scientific papers, concerning innovative solutions in inorganic acids production, will be performed.

Keywords: inorganic chemical technology, sulfuric acid, nitric acid, phosphorous acid, catalysts, heterogenous catalysis

1.1 Introduction

Inorganic acids, such as sulfuric, nitric or phosphorous ones, are commonly known inorganic bulk chemicals with a wide range of application. Generally, their production is based on the transformation of the naturally occurring raw materials e. g. sulfur, nitrogen or minerals such as apatite's, and its complexity includes various stages, numerous reactions, specific conditions and apparatus [1–5]. Historically, the evolution of production of inorganic acids has changed over the years. In case of sulfuric acid it started in early sixteenth century (production via heating of iron(II) sulfate heptahydrated), through eighteenth century when the so called "contact process" was established, which is actually used to this day with various modifications. The industrial production of nitric acid was developed in seventeenth century and concerned the distillation of potassium nitrate (alternatively sodium nitrate) and sulfuric acid mixture, but the technology used nowadays was firstly ran in 1906 in Germany, with a crucial step including catalytic oxidation of ammonia (based on the patent by Wilhelm Ostwald). The history of the phosphoric acid dates back to seventeenth century and concerns its fabrication via combustion of phosphorous with a burning glass, combined furthermore with water absorption. At the turn of the eighteenth and nineteenth centuries, first attempts were made in order to treat the

This article has previously been published in the journal Physical Sciences Reviews. Please cite as: Ciesielczyk, F. Inorganic acids – technology background and future perspectives *Physical Sciences Reviews* [Online] 2020, 5. DOI: 10.1515/psr-2019-0030.

https://doi.org/10.1515/9783110656367-001

phosphorous containing materials with sulfuric acid. The process was called "wet process" or "solvent extraction" and the first industrial plant for its realization ran in 1970s. Due to the economic aspect this technology is used up to date, not only for production of phosphoric acid, but mainly to fabricate phosphorous containing fertilizers [1–8]. Otherwise, all inorganic acids owe their interest to exhibited, specific properties and technological application. Figure 1.1 presents the key industries in which sulfuric, nitric and phosphorous acids are commonly used.

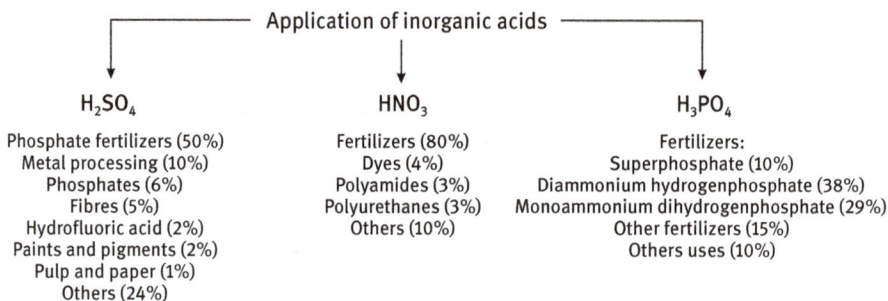

Application of inorganic acids

H_2SO_4	HNO_3	H_3PO_4
Phosphate fertilizers (50%)	Fertilizers (80%)	Fertilizers:
Metal processing (10%)	Dyes (4%)	Superphosphate (10%)
Phosphates (6%)	Polyamides (3%)	Diammonium hydrogenphosphate (38%)
Fibres (5%)	Polyurethanes (3%)	Monoammonium dihydrogenphosphate (29%)
Hydrofluoric acid (2%)	Others (10%)	Other fertilizers (15%)
Paints and pigments (2%)		Others uses (10%)
Pulp and paper (1%)		
Others (24%)		

Figure 1.1: Application of inorganic acids [1–5].

The largest amount of sulfuric acid is used for phosphoric acid production, which in turn, is applied for phosphate fertilizers fabrication. Additionally, it takes part also in production of ammonium sulfate, which is very important fertilizer in sulfur-deficient. On the other hand sulfuric acid is widely used in metal processing e. g. the manufacture of copper and zinc as well as in polymers technology – direct production of polyamide 6 via indirect caprolactam stage of manufacturing. In many cases it is used to treat (extract) raw minerals in order to obtain valuable inorganic compounds such as titania – white pigment commonly used in paint industry. The main application of nitric acid is the production of fertilizers, among which ammonium nitrate is of great importance. The same as in the case of sulfuric acid, it is also used in polymer technology, to fabricate some intermediates for e. g. adipic acid and polyurethanes production. Moreover, nitric acid finds its application in organic technology, where it is used for nitrobenzene and aniline synthesis, which are a key reagents for making dyes. Almost 90 % of the phosphoric acid produced is used to make fertilizers such as triple superphosphate, diamonium hydrogenphosphate and monoammonium dihydrogenphosphate. It is also added to soft drinks and as a supplement in feed. Among different applications of phosphoric acid we cannot forget about production of detergents, which decreases year by year due to law restrictions [1–5]. Wide range of potential application results in still increasing demand for inorganic acids. Their consumption worldwide is presented in Figure 1.2, which confirms that the main consumers of those particular chemical reagents are China and United States of America (data from 2016). It is related to the rapid pace of economic development of these two major powers [9].

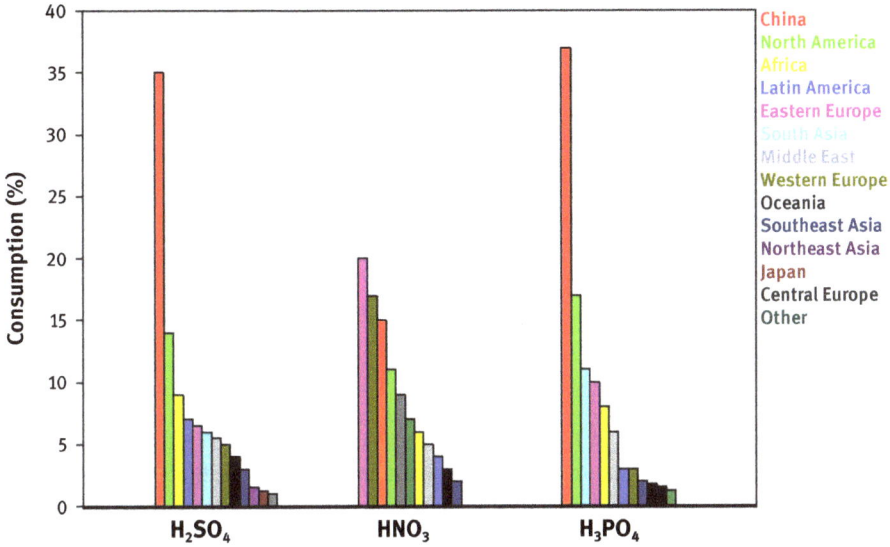

Figure 1.2: Worldwide consumption of commonly known inorganic acids [9].

Despite this dynamic economic development, the production of inorganic acids based on technologies known for years. This is related to the availability of natural resources and the significant efficiency of technological lines. As few literature reports indicate, the topic concerning the modification of industrial methods for the production of inorganic acids relates primarily to environmentally friendly aspects such as e. g. reduction of emissions of environmentally harmful gases, liquid or solid pollutants or searching for alternative catalysts affecting the course of individual stages of the production process [11–39]. Figure 1.3 presents the statistic of scientific papers published to this date for such keyword as *"sulfuric acid production"*, *"nitric acid production"* or

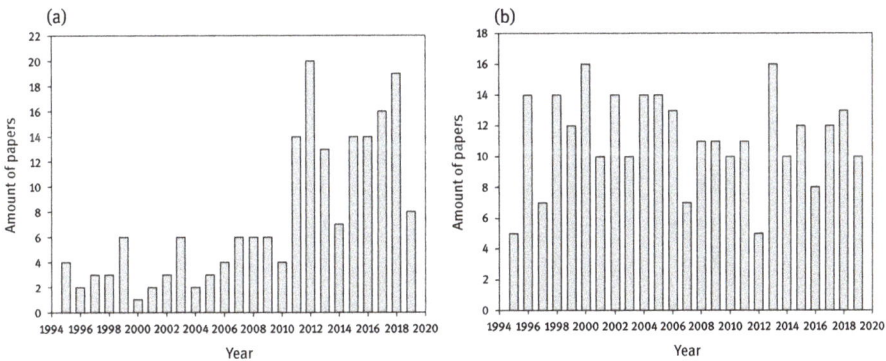

Figure 1.3: Web of science statistic concerning papers that deals with production of: (a) sulfuric acid and (b) nitric acid [10].

"phosphorous acid production" (source: *Web of Science*) as well as the disciplines in which they were ascribed to. As it can be seen, between 1995 and 2019 there only 330 papers concerning research on sulfuric acid production, 378 papers concerning nitric acid production and only 4 concerning phosphorous acid technology.

Taking into account all of above, an attempt was made to present the main technological assumptions of production of sulfuric, nitric and phosphorous acids. Additionally, the detailed literature review was done to highlight novel aspects and trends that are introduced to this specific industry, including especially their environmental impact.

1.2 Sulfuric acid

Since 1831, the technology of sulfuric acid production is based on so called "contact method" which is a great example of heterogeneous catalysis [1–6]. Since the beginning of industrial implementation its development included mostly the search for novel type of catalysts which would work efficiently without significant deactivation. In general, the contact method includes the transformation of pristine sulfur or sulfur dioxide. The process can be described with the Figure 1.4.

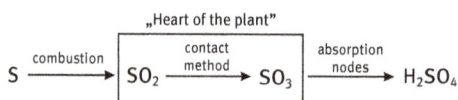

Figure 1.4: General visualization of contact method of production of sulfuric acid [1–6].

The process includes three separate stages: combustion of pristine sulfur, catalytic oxidation of sulfur dioxide toward SO_3 and its absorption in sulfuric acid solutions. There are various alternative sources of sulfur dioxide which include:

- combustion of H_2S

$$H_2S + 3/2O_2 \rightarrow SO_2 + H_2O + Q$$

- roasting of metal sulfides

$$4FeS_2 + 11O_2 \rightarrow 2Fe_2O_3 + 8SO_2 , \quad \Delta H = -803.0 \text{kcal}$$
$$3FeS_2 + 8O_2 \rightarrow Fe_3O_4 + 6SO_2 \quad \Delta H = -576.2 \text{kcal}$$
$$2ZnS + 3O_2 \rightarrow 2ZnO + 2SO_2 \quad \Delta H = -211.8 \text{ kcal}$$

- decomposition of sulfates

$$2CaSO_4 + C \rightarrow 2CaO + 2SO_2 + CO_2 \quad \Delta H = 138.0 \text{ kcal}$$

In these processes, a gas mixture consisting of SO_2, O_2, N_2 is obtained, while the SO_2 content in the exhaust gases depends on the type of raw material. Gases from sulfur combustion have the highest SO_2 content, and that is why in most cases it is its main source.

The pristine sulfur is in arrears at a depth of approx. 500 m and need to be extracted. The main process used for this purpose is a borehole method – hot water (120 °C) as well as compressed air are introduced into the exploitation field where sulfur is deposited, and sulfur is discharged with another pipe. The compressed air creates the effect of a "mammoth pump", which creates a vacuum causing sulfur to flow to the surface [1–6]. The realization of sulfur extraction from its deposits is presented in Figure 1.5.

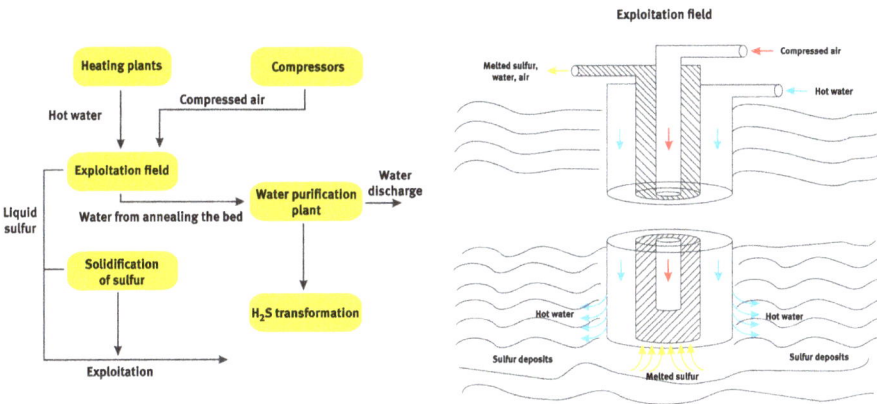

Figure 1.5: Extraction of sulfur from its deposits [1–6].

The next stage includes sulfur combustion (exothermic reaction, 800–1000 °C) which from the chemical point of view is the simplest method for SO_2 production.

$$S + O_2 + 3.76N_2 \rightarrow SO_2 + 3.76N_2 \quad \Delta H = -70.0 \text{ kcal}$$

To ensure complete combustion of sulfur, the process should be carried out with air excess, using good gas mixing. The result is the production of gases with a much lower SO_2 content than is theoretically possible (21 % SO_2) and the need to install secondary combustion chambers to complete the sulfur oxidation reaction. Combustion of very pure sulfur gives gas with a high SO_2 percentage. This gas is free of dust and As_2O_3, which is very important especially in the "contact method" because do not affect the catalyst efficiency. In case of using pyrite such impurities may occur and in most cases they cause the catalyst poisoning and deactivation.

(a)

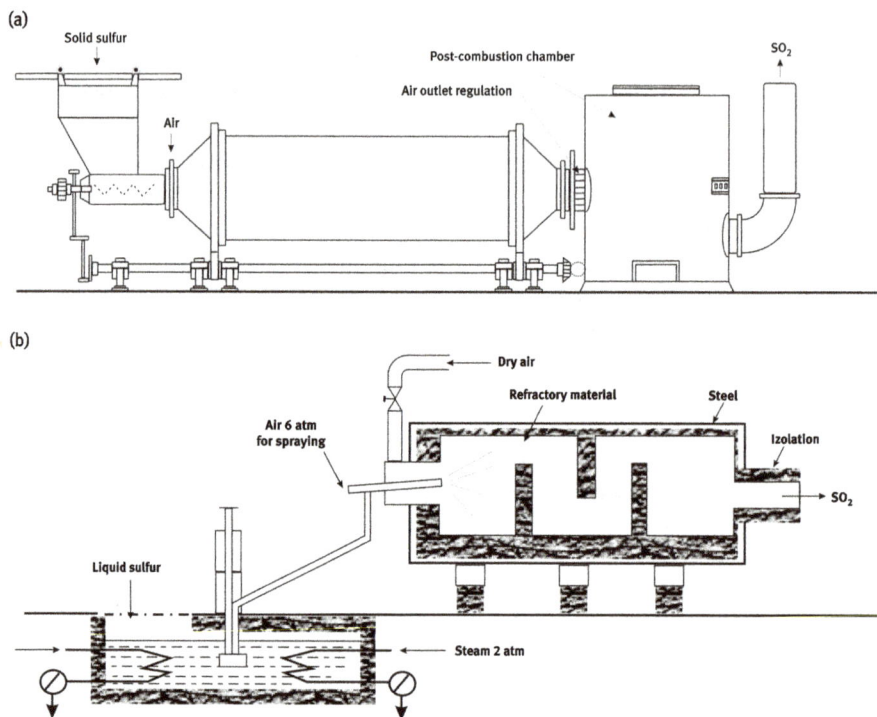

(b)

Figure 1.6: Installation used for combustion of (a) solid sulfur and (b) liquid sulfur.

Sulfur can be combusted as solid or in liquid form. Examples of chambers used for this purpose are presented in Figure 1.6.

Before sulfur dioxide will undergo oxidation, the gases raising from sulfur combustion (SO_2, O_2 and N_2) need to be purified from dust and humidity. About 98 % sulfuric acid is used for this purpose – the gases are washed with this acid in the drying towers in order to remove humidity. Schematically this technological step is presented in Figure 1.7.

The key part of sulfuric acid technological plant relates to the oxidation of sulfur dioxide toward SO_3 [1–6]. Process follows the reaction

$$2SO_2 + O_2 \leftrightarrow 2SO_3$$

which is exothermic and proceeds with a decrease in volume. The reaction equilibrium constant shifts to the right as the temperature decreases and pressure increases. The reaction is very slow and would not be of industrial importance without the use of a catalyst (Figure 1.8).

In order to increase the reaction rate it is necessary to use the catalyst, also named as a contact. Such a system is a typical example of heterogenous catalysis,

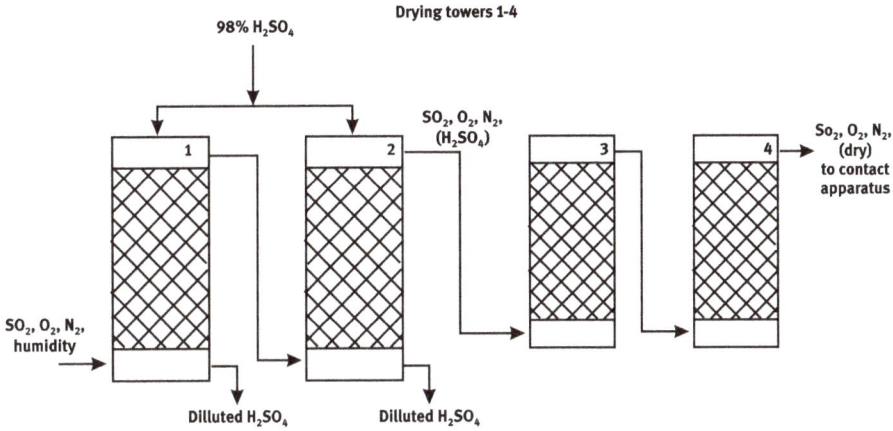

Figure 1.7: Purification of gases after sulfur combustion [1–6].

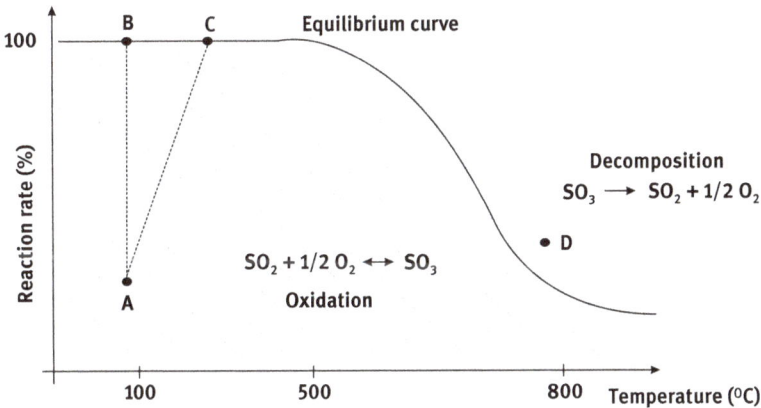

A - area of oxidation reaction (significant distance from point A to B and C means long
   reaction time - the need for a catalyst application)

B and C - 100% efficiency of oxidation

D - decomposition of SO_3

Figure 1.8: Equilibrium and reaction rate of SO_2 oxidation without a catalyst [1–6].

when the reagents are in a gas phase and the catalyst is in a form of solid. Each
catalyst is composed with an active phase (1–10 %), support (90 %) and activator
(1 %). The main criteria used for catalyst selection includes: selectivity, physicochemical parameters, cost, temperature of activation and susceptibility to poisoning. The most important is the temperature of activation because it affects the
whole process costs as well as it is related with the proper installation enabling
appropriate control of the reaction heat. Based on this, in Figure 1.9, different

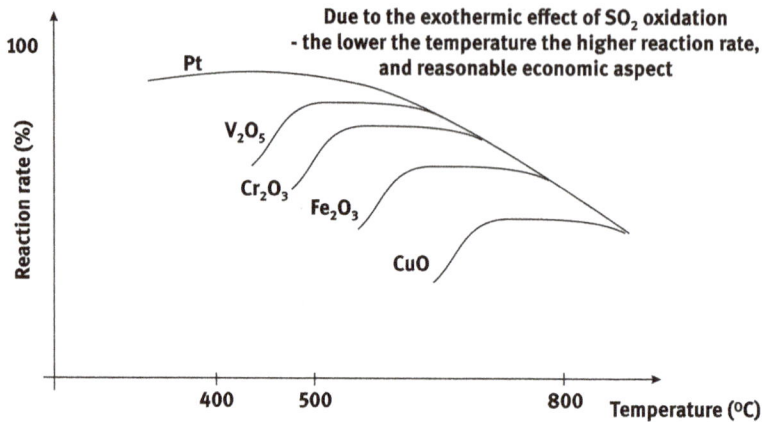

Figure 1.9: Working temperatures for different catalysts [1–6].

catalysts are presented together with corresponding temperature at which they may effectively work in mentioned oxidation reaction. At the beginning, the platinum was mainly used due to the lowest working temperature. Other metals from the platinum group also increase the SO_2 oxidation rate, but to a much lower extent Pt > Rh > Ir > Pd. Shortly after the discovery of the catalytic properties of platinum, mostly taking into account its high cost and susceptibility to poisoning with sulfur as well as problems with its regeneration, the catalytic activity of a number of metal oxides was discovered. Vanadium(V), iron(III) and copper(II) oxides at lower temperatures, under the influence of SO_2, form inactive sulfates, and thus their activity is manifested only after exceeding the dissociation temperature of the corresponding sulfates [1–6].

The SO_2 oxidation process on the catalyst is a complex process. In general it is based on adsorption theory which includes the following stages:

– oxygen binding through the catalyst surface (stage I);
– binding of molecules with oxygen atoms on the catalyst surface (stage II);
– rearrangement of atoms corresponding to the formation of SO_3 molecules on the catalyst surface (stage III);
– desorption of SO_3 molecules from the catalyst surface (stage IV).

Schematically, this process can be described as follows:

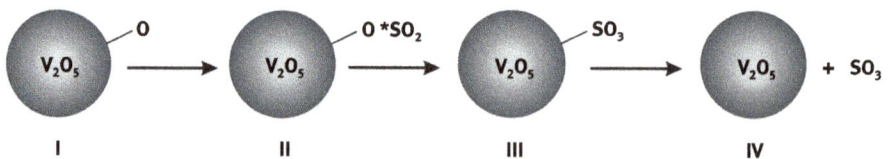

The disadvantage of heterogeneous catalysis process is the poisoning phenomenon of catalysts. This process is affected in most cases by the composition of gases going through the contact apparatus. The most common poisons of catalyst consist of water vapor, arsenic, selenium, tellurium, carbon oxide and dioxide, hydrogen sulfide and halogens. Contact poisoning causes a decrease in contact activity. Reasons

– chemical reaction of the poisoning substance with the catalyst, resulting in a new catalytically inactive compound;
– chemical transformation of the poisonous substance on the contact surface and the formation of a non-volatile, inactive product that covers the contact surface;
– activated poison adsorption on the contact surface.

Contact poisoning can be reversible or irreversible. Reversible poisoning can be removed directly in reactors, while in the case of irreversible poisonings it is not possible to recover catalytic mass. Regeneration of the catalysts is difficult because it include blowing the apparatus with a clean gas mixture at the highest possible temperature, chemical or thermal treatment. In most cases the effect of high temperature usually causes such significant changes in the contact structure, and on the other hand application of chemicals e. g. H_2SO_4 results in acid wastewaters emission, thus no satisfactory methods of regeneration are known till today. This aspect is very important because it affect the economical justification of the technological plant that produce sulfuric acid [1–6].

Contact node concerning the oxidation of SO_2–SO_3 step is presented schematically in Figure 1.10. Raw SO_2 together with dry air passes to the bottom of the contact apparatus and here oxidation occurs to the first degree of conversion. Then the gas leaves this lower part of the apparatus and goes to the exchanger where it gives off heat to a certain limit and cooled returns to the bottom of the oxidation apparatus. This is where the maximum oxidation occurs on the classic node 98.6–98.7 % (Figure 1.11).

The need to cool gas between contact layers is related with the exothermic nature of aforementioned oxidation process as well as the working temperature of the catalyst. This fact is explained in Figure 1.12. As gas flows through the first layer, the composition and temperature of the gas change in the manner defined by the bed 1. These changes are getting slower as the equilibrium curve approaches. Therefore, the process should be stopped at point D (Figure 1.12). The degree of transformation at this point, however, has a small value – only 40 %. To oxidize SO_2 the process continues, there is only a need to cool the gas, i. e. as far as possible from the equilibrium curve. Cooling is carried out to a temperature not lower than that corresponding to point F lying on the ignition line of catalyst. To achieve a high conversion (over 98.5 %), SO_2 oxidation must be carried out on several contact beds. It is necessary to cool the gas (in heat exchangers or by directly introducing cold air). Hence, a modern contact node was proposed (see Figure 1.13).

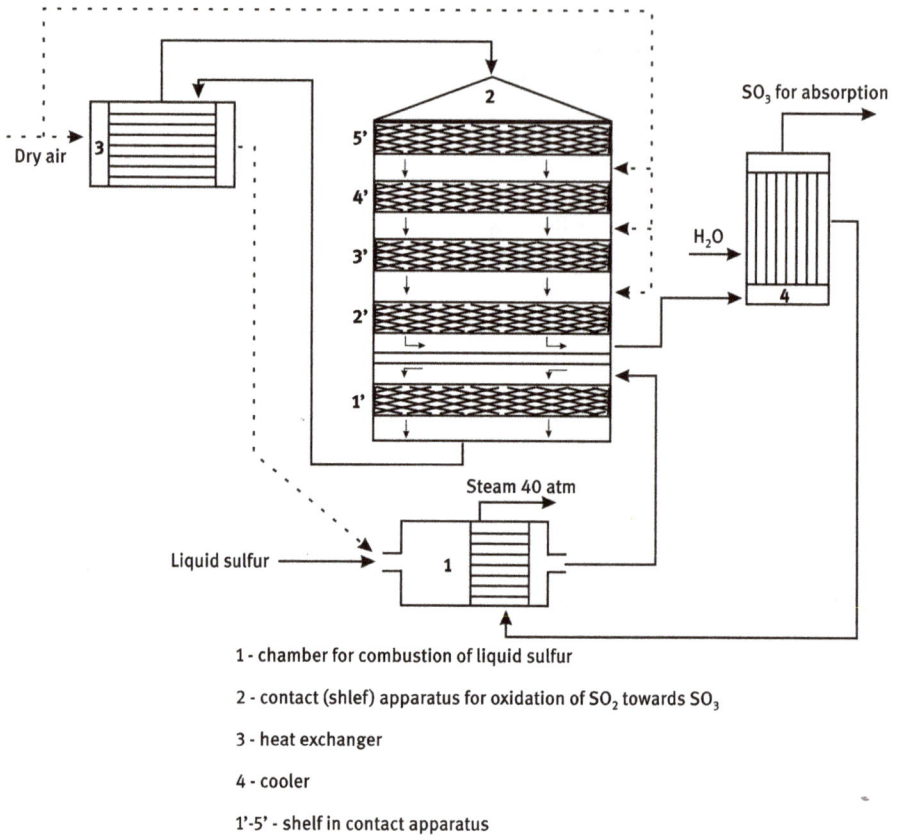

1 - chamber for combustion of liquid sulfur

2 - contact (shlef) apparatus for oxidation of SO_2 towards SO_3

3 - heat exchanger

4 - cooler

1'-5' - shelf in contact apparatus

Figure 1.10: Contact node concerning the oxidation of SO_2 to SO_3 [1–6].

The final stage of sulfuric acid production – absorption of SO_3 – is carried out in monohydrate (96 % H_2SO_4) or in oleum (concentration at least 20 %). In order to increase the absorption efficiency, absorbers with movable packing are used [1–6]. The absorber with a movable package ensures maximum mass and heat exchange (this is analogous to the fluidized phase). Absorption efficiency can be controlled by e. g. the appropriate diameter of the ceramic balls (laminar mode). Absorbers with turbulent mode have an advantage – the exchange of mass and heat is more intensive, and thus the absorption efficiency increases. Typical absorption node is presented in Figure 1.14.

Nowadays, huge amounts of H_2SO_4 are produced from the utilization of dilute H_2SO_4 and salt by concentration of dilute acids, regeneration from acid waste and regeneration from sulfates. In the past period, the most important improvement of sulfuric acid production technology by "contact method" was the introduction of two-stage conversion and absorption. In the meantime, great emphasis was placed

Gas containing 12% of SO_2

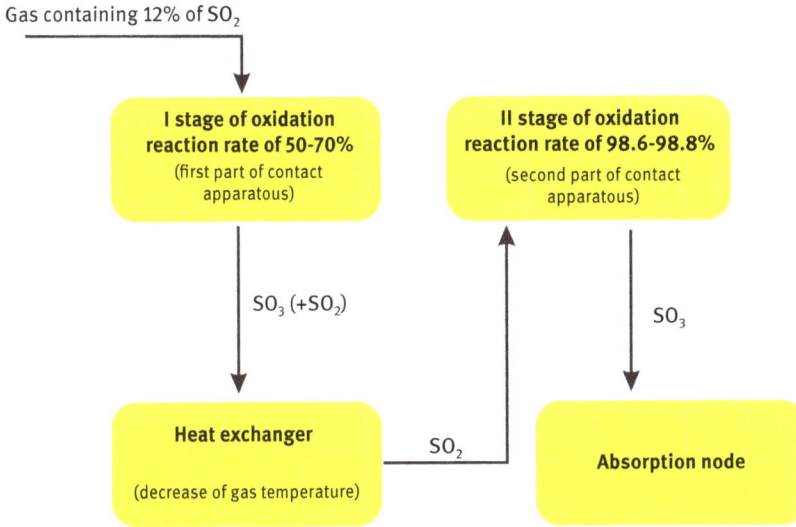

Figure 1.11: Classical contact node related to oxidation of SO_2 [1–6].

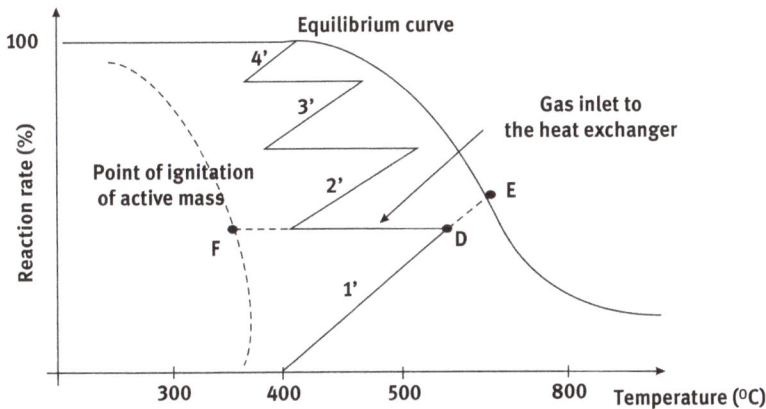

Figure 1.12: Equilibrium of oxidation of SO_2 to SO_3 including heat transfer between the catalyst beds [1–6].

on improving the catalyst and apparatus. Recently, improvements in the technological process have been mainly going toward increasing energy efficiency, including increasing the by-product of steam. For the sake of the environmental protection, the possibility of applying a third degree of conversion and absorption was considered, while the degree of sulfur utilization would increase up to 99.96 %. In this system, the use of the third degree of conversion and absorption was considered to be pointless, because such solutions would require additional external heat supply. Thus, the

Gas containing 12% of SO_2

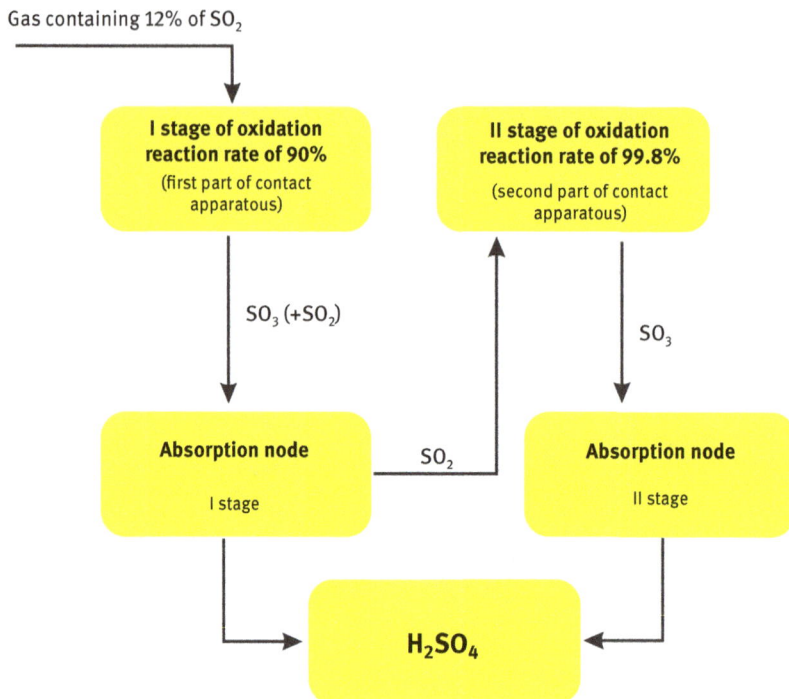

I stage of oxidation reaction rate of 90%
(first part of contact apparatous)

II stage of oxidation reaction rate of 99.8%
(second part of contact apparatous)

SO_3 $(+SO_2)$

SO_3

Absorption node
I stage

SO_2

Absorption node
II stage

H_2SO_4

Figure 1.13: Modern contact node related to the oxidation of SO_2 – double conversion with interstage absorption [1–6].

research concerning this particular technology are focused on development of novel type of catalyst, especially those which will work effectively at lower temperatures and will not be susceptible for poisoning [1–6].

Scientific literature concerning sulfuric acid production in most cases is related to the optimalization of main reactions conditions and whole plant efficiency [11–14]. They concern also the oxidation of sulfur dioxide toward SO_3, so important `sing the whole technological plant used for this acid production. In early 1990s many works have been undertaken in order to analyze the catalyst behavior in this process as well to estimate optimum conditions of the oxidation process [15–18]. The most important is also the enviornmental aspect cocnerning SO_2 emission. Thus, Vernikovskaya et al. [15] have analyzed the nonstationary state of the catalyst surface, in particular, the absorption properties of active component of vanadium catalysts in relation to SO_2. The proposed method of sulfur dioxide oxidation employs periodic flow of SO_2 and air through the catalyst bed, direction of mixture and air flow being opposite. According to the presented modeling Authors have achieved higher SO_2 conversions and, correspondingly, extremely low sulfur dioxide content in the outlet gases as compared to conventional continuous flow methods. Additionally, Authors have proved that the process may be

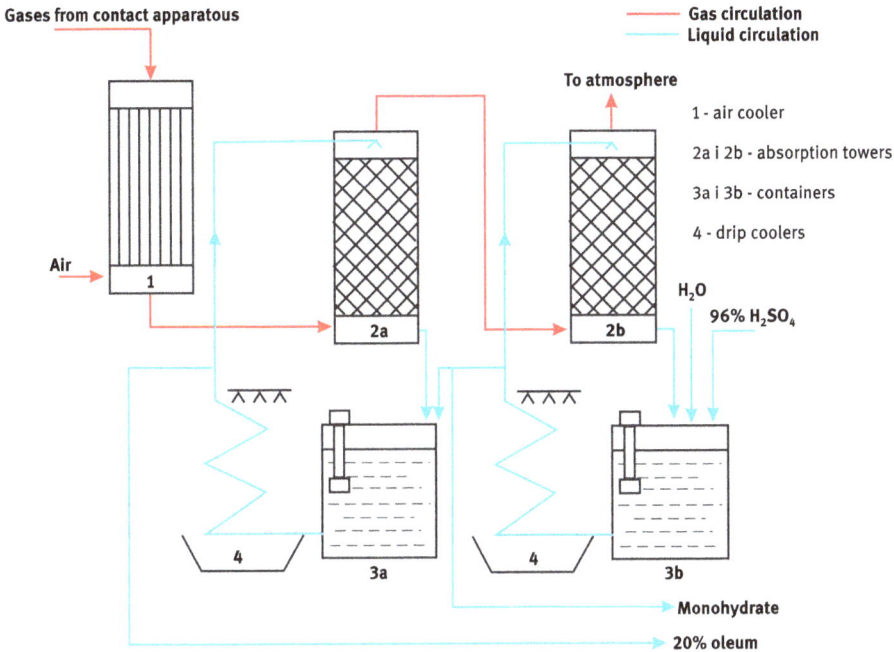

Figure 1.14: Absorption node for production of sulfuric acid [1–6].

carried out at reasonable and technologically appropriate values of parameters (duration, catalyst loading, inlet temperature, etc.), providing an extremely high degree of SO_2 purification from waste gases. Final remark was that the process may replace the conventional second stage of double contact/double absorption plants. Maximum estimated overall conversion in the plant may reach 99.9 %, which corresponds to an outlet SO_2 content as low as 50 ppm. This emission level satisfies the most severe environmental requirements. Quite similar research were conducted by Günther et al. [19] who present a new process concept, which uses the transient behavior of the reaction in two reactors operating under unsteady conditions. Besides several advantages, which can increase the efficiency of the whole sulfuric acid process drastically, the proposed reactor is a key component for an efficient operation of a sulfuric acid plant which reduces the emissions of SO_2 down to zero while keeping the necessary conditions for the hydrogenation unit installed downstream. On the other hand, Kiss et al. [20] proposed the model that enabled optimization of the operating parameters (feed flow rates or split fractions) in order to significantly reduce (by ~ 40 % or more) the SO_x emissions and increase in energy production.

Sulfuric acid plant reactors have been the subject of many studies, and thus there have been very limited work in the literature that have analyzed complete sulfuric acid plants. Tajeda-Iglesias et al. [21] have presented the process flowsheet for an industrial-scale single-absorption sulfuric acid plant with scrubbing tower that uses elemental sulfur as the feedstock. Detailed analysis of the process and proposed model enabled its use to understand the sensitivity of key response variables with respect to the operating parameters of the plant. Authors have proved that the air flow rate, gas strength and outlet temperatures of the heat exchangers following the second and third passes of the multistage reactor where the most sensitive variables in the plant. An optimization using those three variables as the decision variables was performed to identify the operating point that resulted in the maximum daily profit of the plant and the maximum H_2SO_4 production rate. Quite new approach was proposed by Zhou et al. [22] who conducted research onto reutilizing of semi-dry sintering flue gas desulfurized ash. In his work a new technology was proposed in which ash is added to a boiling furnace to obtain SO_2 during sulfuric acid production. Authors proved that pressure, temperature and oxygen content are the key factors affecting the decomposition of $CaSO_3$ in the ash. Based on the results Authors have proved the feasibility of preparing SO_2 by adding desulfurized ash during the production of sulfuric acid using a boiling furnace. This is very important factor that relates to the environmental protection. Nowadays there is a trend to use waste by-product in different technological approaches, which significantly decreases their amount and, as shown here, can effectively provide valuable components to the reaction system.

1.3 Nitric acid

Technological line for industrial production of nitric acid, which came on stream in 1906, is the same as complex as this one used for sulfuric acid fabrication. The part of the process concerns the heterogeneous catalytic oxidation of ammonia developed by Ostwald [1–7]. In general it can be presented as follows (Figure 1.15).

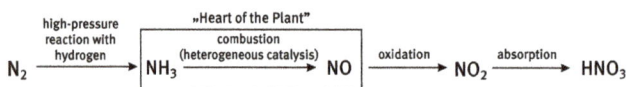

Figure 1.15: General visualization of production of nitric acid [1–7].

For such a technology, ammonia or nitrogen monoxide is needed. Nitrogen monoxide can be produced in direct process via reacting of atmospheric nitrogen and oxygen at least at 2000 °C, in a so called "electric arc". It is a reversible, extremely endothermic reaction that proceeds without changes in volume. As temperature increases, the equilibrium of the reaction shifts toward the formation of NO. Due to economical aspects, the oxidation of ammonia (Ostwald process) is more often used as a main source of NO for nitric acid production. This aspect is more reasonable analyzing the whole technological plant which can be utilized for production of nitrogen-based compounds such as ammonia, nitric acid, fertilizers (urea, tiourea, ammonium nitrate) [1–7].

The gas mixture containing hydrogen and nitrogen in the molar ratio of $H_2:N_2 \approx 3$ should be thoroughly purified from the poison admixtures (dust, water vapor) of catalyst before entering the ammonia synthesis plant. Also, inert gases such as argon and methane should occur in the amount as low as possible in the synthesis process. The sum of the oxygen content in the gases (SO_2, CO or CO_2) should not exceed 0.002–0.003 %. Nitrogen is obtained by the rectification and condensation of air (by physical method) and by binding of oxygen in the synthesis gas production process (by chemical method). Hydrogen can come from the following sources: gasification of solid fuels (hard coal, lignite, coke), gasification of liquid fuels (kerosene, heavy oils, refinery waste), conversion (CO, CH_4) or distribution of gaseous fuels (coke oven gas, natural gas, post-refinery gases), water electrolysis. The main reaction concerning ammonia synthesis is as follows:

$$N_2 + 3H_2 \leftrightarrow 2NH_3$$

This reaction is exothermic with additional reduction in volume of gases. That is why ammonia synthesis is a process that is carried out at high pressures. Therefore, the reactors construction, used for the production of NH_3, should include their operation under high pressure (10–30 MPa) and at elevated temperature (400–500 °C) and that they are exposed to the corrosive effects of chemical agents. In order to obtain a high reaction rate with the aid of an active catalyst, the temperature distribution in the reactor should be adjusted accordingly [1–7]. Two basic methods of temperature control are used in reactors:

–   by receiving heat using heat exchangers located in the catalyst bed,
–   using a stream of cold synthesis gas, which is mixed with the stream of reactants in the reactor, usually between individual catalyst beds.

The ammonia synthesis is also an example of typical heterogeneous catalysis process [1–7]. Catalysts consist mainly of iron, which contain promoters (e. g. Al_2O_3, K_2O and CaO). They increase the activity and durability of the catalysts. Such catalysts can operate in the temperature range of 380–550 °C. Iron catalysts are sensitive to certain gas pollutants, which reduce its activity (poisoning). Reversible poisoning is caused by oxygen and oxygen compounds (e. g. O_2, H_2O, CO, CO_2), while irreversible – by phosphorus, sulfur and halogen compounds. The synthesis mechanism is explained by the chemisorption of substrates on the catalyst surface. As a result of chemisorption, iron and nitrogen complexes are formed [1–7]. The course of the reaction can be represented by the following scheme:

the hydrogen is broken down into atomic form, and the hydrogen attachment begins:

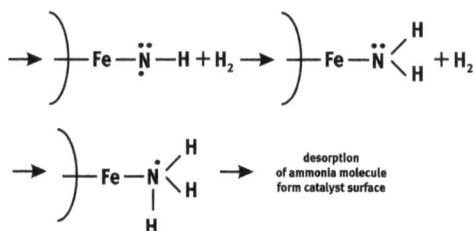

$$4NH_3 + 3O_2 \rightarrow 2N_2 + 6H_2O$$

In the synthesis of ammonia, due to the effect of pressure, three types of installations are distinguished but the medium pressure converters are most commonly used [1–7].
– low pressure – up to 200 atm;
– medium pressure – 300–500 atm;
– high pressure – up to 1000 atm.

Historically, the first converter used for ammonia synthesis was designed by Haber and Bosch but due to low material stability (carbon based steel) to the reaction conditions it was changed into converter made from chromium-nickel alloys. An example of converter used for NH_3 synthesis is presented in Figure 1.16.

After synthesis of ammonia the next step includes its oxidation. In 1839, Kuhlmann proposed the oxidation of NH_3 by air in the presence of platinum. Technical realization of this technology was done in 1907 by Ostwald. The main reaction of oxidation is presented below:

$$4NH_3 + 3O_2 \rightarrow 2N_2 + 6H_2O$$

Nitrogen oxide is only an intermediate. At equilibrium, as shown by thermodynamic calculations, the reaction products are nitrogen and water. In order to have a chance to fabricate the nitric acid it is desirable to stop the reaction when the nitrogen oxide concentration reaches its maximum value [1–7].

$$4NH_3 + 5O_2 \rightarrow 4NO + 6H_2O$$

Additionally, other reactions may occur in the reaction system. They can be as follows:

$$4NH_3 + 4O_2 \rightarrow 2N_2O + 6H_2O$$
$$4NH_3 + 6NO \rightarrow 5N_2 + 6H_2O$$
$$2NO \rightarrow N_2 + O_2$$
$$2NH_3 \rightarrow N_2 + 3H_2$$

In order to receive the high efficiency of NO production the selective catalyst should be used. In this particular reaction the rhodium-platinum alloy is most often used as

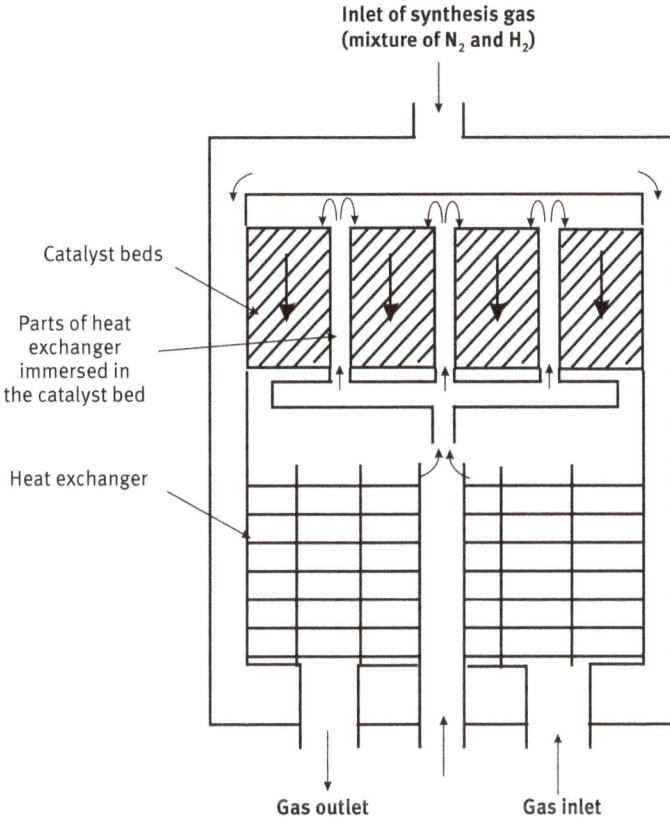

Figure 1.16: Converter for ammonia synthesis [1–7].

catalyst. Besides the selective solid catalyst other process parameters, such as temperature, pressure, contact time of reagents with catalyst bed and oxygen:ammonia molar ratio, need to be controlled. As presented in Figure 1.17, shifting of the reaction equilibrium toward NO will be favorable recovering the heat and working under normal pressure. Maximum yield of NO production is achieved at 850 °C. At lower temperatures, the oxidation rate decreases because N_2O is formed, on the other hand at temperature higher than 850 °C it also rapidly decreases as a result of NO decomposition. Pressure range should be from normal up to 10 atm – using a net packet (16–20 pieces) maximum yield of NO performance is almost the same as at normal pressure (96–98 %). Oxidation at higher pressures requires higher temperature (900 °C, at p = 8 atm) and longer contact time [1–7].

The contact time is the residence time of the NH_3 mixture with air in the catalyst zone (in seconds). There is a strictly defined contact time at which the highest NO efficiency is obtained. Optimal contact time is 1–$2 \cdot 10^{-4}$ s (the fastest reaction from

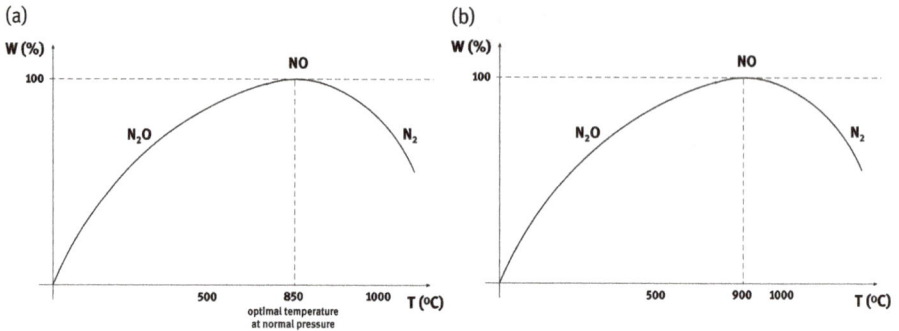

Figure 1.17: Influence of (a) temperature and (b) pressure on the efficiency of NO fabrication via ammonia oxidation.

known gas reactions in heterogeneous systems). Further increasing of the flow rate causes a decrease in NO efficiency – the consequence of NH_3 leaching through the rhodium-platinum bed is conversion to free nitrogen due to its reaction with its oxidation products [1–7]:

$$4NH_3 + 6NO \rightarrow 5N_2 + 6H_2O$$

Finally, the molar ratio of the reagents is also of great importance. At O_2:NH_3 molar ratio of 0.75 – NO does not appear but when the molar ratio equals to 1.25 – the efficiency of nitrogen monoxide is almost 100 %. The installation used for ammonia oxidation is presented in Figure 1.18.

NH_3 is introduced through the bottom with a series of nozzles and mixed with air. The gas mixture passes through a porolite filters (pipes of porous ceramic material) where residual of mechanical impurities retained. Above the filter there is a layer (approx. 300 mm) of porcelain Raschig rings for equal distribution of the mixture throughout the entire cross-section of the apparatus. The next section consist of catalyst bed (rhodium-platinum alloy), where the main reaction takes place. Dedusted and purified air is pumped to a heat exchanger via a blower, where it is heated up with the heat of waste reaction gases. The heated air mixes with NH_3 at the bottom of the apparatus. The post-reaction mixture with a temperature of 700–800 °C enters the heat exchanger and from there into the boiler, where the heat of the gases is used to produce steam. The gases leave it at 190 °C. Combustion of a mixture containing 10 % of NH_3 with a yield of 95 % gives the following gas composition: 9.3 % NO; 6.4 % O_2; 14.6 % H_2O and 69.7 % N_2. The efficiency of ammonia oxidation is crucial analyzing the amount of NO which in the next stage should be oxidized:

$$2NO + O_2 \leftrightarrow 2NO_2$$

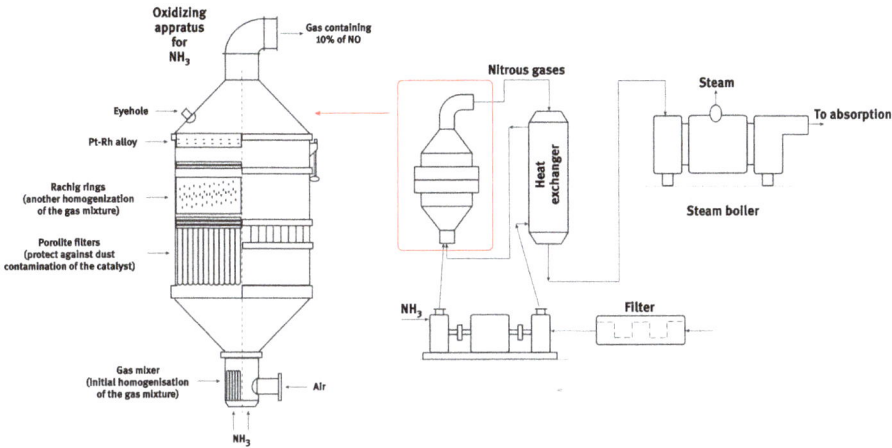

Figure 1.18: Installation used for oxidation of NH_3 [1–7].

This reaction is reversible, exothermic and it proceeds with a decrease in volume [1–7]. Shifting of the reaction equilibrium constant toward product will be promoted by decreasing the temperature and increasing pressure. Nitrogen oxides are furthermore absorbed at normal temperatures (ranging between 30 °C and 50 °C). There is no reversibility of this reaction in this temperature range. This reaction becomes reversible only at temperatures much higher. A characteristic feature of this particular reaction is a decrease in its speed as the temperature increases due to the fact that it actually consists of two stages:

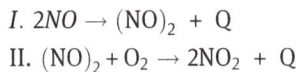

$$I.\ 2NO \rightarrow (NO)_2 + Q$$
$$II.\ (NO)_2 + O_2 \rightarrow 2NO_2 + Q$$

The reaction I – dimer $(NO)_2$ formation is very fast and exothermic. Reaction II is slower and is also exothermic. As the temperature increases, the equilibrium of reaction I shifts to the left. The amount of dimer $(NO)_2$ molecules that are expected to react according to a slow reaction decreases. So less and less NO_2 is produced. In total, the $NO \rightarrow NO_2$ oxidation reaction rate decreases with increasing temperature. The rate of NO oxidation is a matter of great importance in the production of nitric acid. Numerous attempts have been undertaken to increase the reaction rate using various types of catalysts (metal oxides, glass, activated carbon, coke, charcoal). Various types of filling of absorption towers were also used to accelerate NO oxidation, but without significant results. Therefore, the problem of faster NO oxidation on an industrial scale has not been solved yet.

However, the next stage includes final production of nitric acid. There are three types of installations for the production of HNO_3 [1–7]:

- non-pressure installations in which 50 % HNO_3 is produced (oxidation of nitrogen oxides and absorption is carried out at normal pressure);
- combined installations in which either oxidation is carried out at elevated pressure and absorption under normal pressure and vice versa;
- pressure installations – concentrated 98 % HNO_3 is fabricated.

The mixture containing NO and NO_2 (some amounts of N_2 and H_2O) is moved to absorption. In the second stage, dimerization of nitric oxide occurs which is favored by decreasing of the temperature (at 0 °C – 78 %, 30 °C – 28 %, 62 °C – 5 % NO_2 dimerizes). In the third stage, N_2O_3 is formed – a disproportionation reaction takes place.

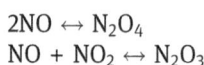

$$2NO \leftrightarrow N_2O_4$$
$$NO + NO_2 \leftrightarrow N_2O_3$$

Reaction related to the absorption furthermore take place. Nitric acid is formed based on the presence of NO_2, N_2O_4 and N_2O_3 which are well absorbed in water (NO is not absorbed).

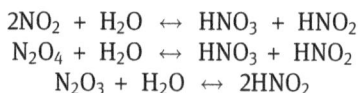

$$2NO_2 + H_2O \leftrightarrow HNO_3 + HNO_2$$
$$N_2O_4 + H_2O \leftrightarrow HNO_3 + HNO_2$$
$$N_2O_3 + H_2O \leftrightarrow 2HNO_2$$

Moreover, HNO_2 is formed which is unstable and undergoes a disproportionation of valence in a reaction environment.

$$3HNO_2 \leftrightarrow HNO_3 + 2NO + H_2O$$

So both absorptions can be written as:

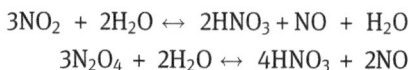

$$3NO_2 + 2H_2O \leftrightarrow 2HNO_3 + NO + H_2O$$
$$3N_2O_4 + 2H_2O \leftrightarrow 4HNO_3 + 2NO$$

In non-pressure installations diluted HNO_3 is produced. The technological plant used for this purpose is presented in Figure 1.19. Gases containing NO and NO_2 pass through gas coolers (1) where properly cooled are transferred to the so-called acid absorption installation (system of several absorbers with packing I–VI). All absorbers are sprayed with water. Gases pass sequentially through the absorber batteries. As a result of absorption, condensates of HNO_3 acid are obtained, and where there is a high NO_2 concentration the 50 % HNO_3 is obtained. Those less concentrated acids are used for spraying the towers which favors an increase of acid strength. Gases that leaving the last absorber contain a certain amount of NO and NO_2, which cannot be released into the atmosphere. That is why the alkaline absorption is supplemented. It

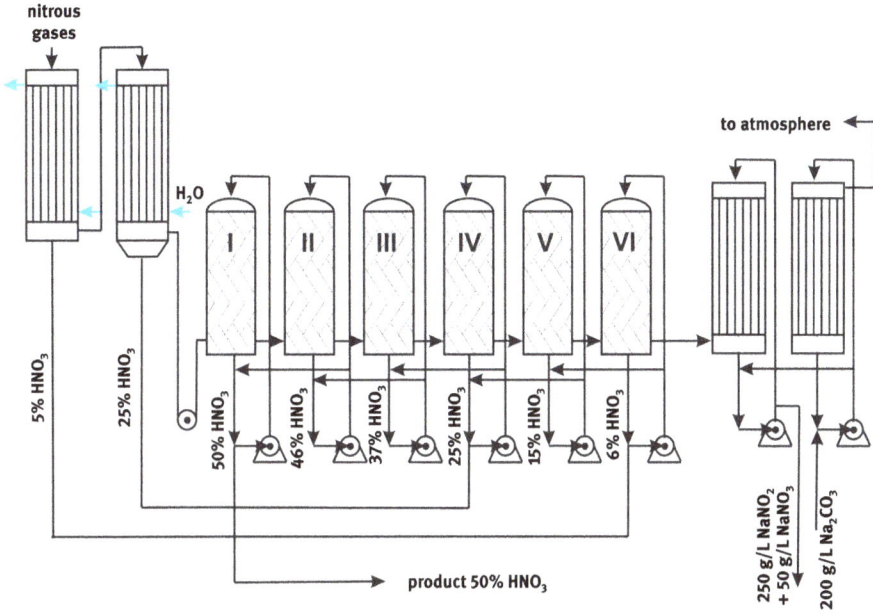

Figure 1.19: Non-pressure installation used for fabrication of diluted HNO_3 [1–7].

consists of two much smaller alkaline absorbers, to which gas containing a certain percentage of NO, NO_2, O_2, N_2 is fed. N_2 and O_2 escape into the atmosphere. The alkaline agent is Na_2CO_3 ($200\ g/dm^3$), which circulates between the towers. Reactions are as follows:

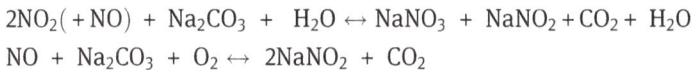

$$2NO_2(+NO) + Na_2CO_3 + H_2O \leftrightarrow NaNO_3 + NaNO_2 + CO_2 + H_2O$$
$$NO + Na_2CO_3 + O_2 \leftrightarrow 2NaNO_2 + CO_2$$

As a result of these processes, clean gases are released to the atmosphere. There is also toxic $NaNO_2$ – the solution is inverted. This process involves the reaction of $NaNO_2$ with 98 % HNO_3 according to the reaction:

$$NaNO_2 + 2 HNO_3 \leftrightarrow NaNO_3 + NO_2(+NO) + 8H_2O$$

NO and NO_2 are returned to the installation.

The production of concentrated nitric acid is based on the so-called "OXO" synthesis. Pure oxygen from air rectification is involved in the oxidation process. The pressure method is based on the dimerization reaction of NO_2 to N_2O_4 mentioned earlier. The dimer need to be converted from gas to liquid phase and subjected to the "OXO" synthesis process.

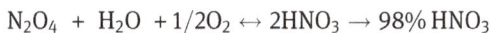

$$N_2O_4 + H_2O + 1/2O_2 \leftrightarrow 2HNO_3 \rightarrow 98\% \ HNO_3$$

This process is carried out in an autoclave – a pressure reactor working in a liquid phase. Ensuring optimal conditions requires: using the mixture with an appropriate excess of N_2O_4, temperature of 70–80 °C and O_2 pressure in the autoclave of 50–80 atm. The whole high pressure technological plant, for production of concentrated nitric acid is presented in Figure 1.20.

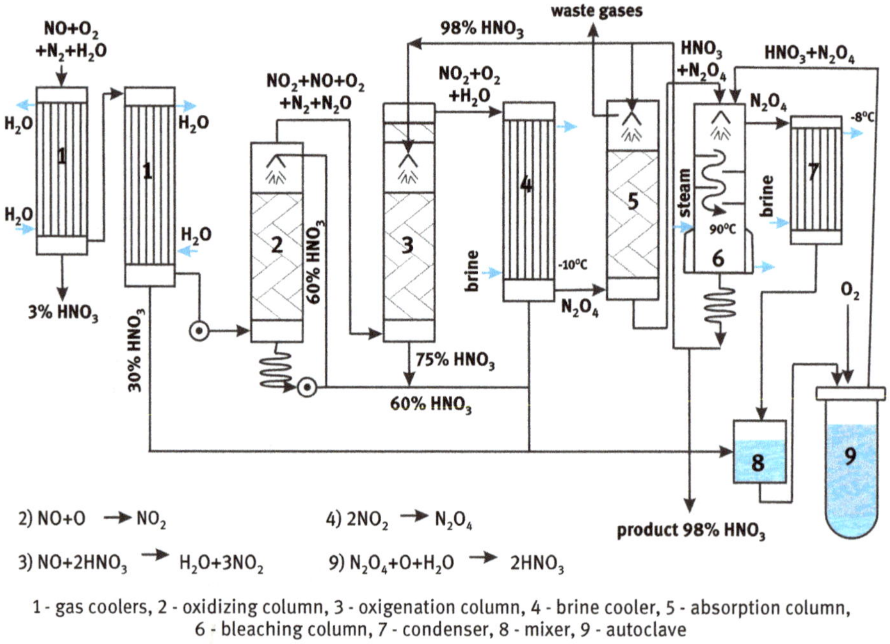

2) $NO+O \rightarrow NO_2$

3) $NO+2HNO_3 \rightarrow H_2O+3NO_2$

4) $2NO_2 \rightarrow N_2O_4$

9) $N_2O_4+O+H_2O \rightarrow 2HNO_3$

product 98% HNO_3

1 - gas coolers, 2 - oxidizing column, 3 - oxigenation column, 4 - brine cooler, 5 - absorption column, 6 - bleaching column, 7 - condenser, 8 - mixer, 9 - autoclave

Figure 1.20: Pressure installation used for fabrication of concentrated HNO_3 [1–7].

The nitrous gases are cooled before the oxidation of NO to NO_2, because absorption is an exothermic process. Diluted nitric acid condensates are formed here. 75% HNO_3 goes to the mixer before the autoclave. The cooled gases pass to the towers 2 – oxidation column and next are pumped to the free lower part of the absorber. In the oxidation column nitrogen oxides are oxidized to higher oxides in the presence of oxygen from air. The condensate resulting from the absorption leaves the column, is cooled and pumped to the oxidation column. The heat in the oxidation column is received as a result of continuous circulation of the condensate solution in the system. Thanks to the drip cooler, the gases are transferred to the oxygenation column, which is fed by the product of the installation – 98 % HNO_3. A disproportionation reaction, i. e. complete oxidation, takes place here:

$$NO + N_2O + 2HNO_3 + 2O_2 + H_2O \rightarrow 3NO_2 + 2HNO_3 + H_2O$$

The resulting condensate containing 75 % HNO_3 goes to the mixer. NO_2 exits the oxygenation column and is intensively cooled. Dimerization occurs by decreasing the temperature (NaCl brine). The dimer goes to the absorption column, where absorption is made in 98 % HNO_3. There is no chemical absorption, only physical one in 98 % nitric acid. The product of this part of installation – a mixture of HNO_3 and N_2O_4 – is moved to the installation heart – bleaching column. N_2O_4 is formed in the bleaching column, which furthermore goes to the condenser to transform into liquid N_2O_4 entering the mixer. The whole mixture is transferred to the autoclave for the synthesis process fed with pure O_2 (followed by the "OXO" process). About 98 % HNO_3 is formed and the unreacted N_2O_4 parts remain and enter the bleaching column once again. The bleaching column is a typical degassing device because the separation of N_2O_4 gases from liquid concentrated HNO_3 takes place there. It consists of two parts: upper – typical mechanical degassing where the gas is separated from the liquid during thermal separation (N_2O_4 is completely separated from concentrated HNO_3); the lower part is heated to about 90 °C. As a result, 98 % HNO_3 is obtained, which feeds the absorption and oxygenation column in order to receive the desired product of this technological plant [1–7].

The most recent research concerning this particular technology is in most cases focused on the emission of NO_x and energy aspects, including heat recovery. The same as in case of sulfuric acid production, the research are conducted on the main reaction that affects the efficiency of technological production of HNO_3. Perez-Ramirez and Vigeland [23] have focused their studies onto ammonia oxidation process using perovskite membranes. Authors worked at replacing noble metals by oxide catalysts for NH_3 oxidation. Oxides may offer the advantage of lower investment, simpler manufacture and reduced N_2O emission [24, 25]. The problem with industrial implementation of oxide catalysts concerns relatively low NO selectivity, rapid deactivation under relevant reaction conditions, and the lower optimal operating temperature as compared to noble metal catalysts. Therefore, Authors have made an attempt to use for the first time the lanthanum ferrite-based perovskite membranes for ammonia oxidation, with which NO selectivity's up to 98 % and no N_2O formation were attained [23]. The obtained results demonstrate the potential of lanthanum ferrite membranes for the high-temperature oxidation of ammonia with quite similar efficiency as commonly used Pt-Rh alloys. In addition, formation of undesirable N_2O is totally suppressed. Authors have justified that the implementation of an ammonia oxidation process based on oxygen-conducting membranes would constitute a major step change in nitric acid production and have a strong impact on the fertilizer industry. Moreover, energy savings for compression of the NO_x gas before the absorption step, which requires high pressure, further contributes to an improved nitric acid production process. The second important reaction in the production of nitric acid is the oxidation of NO towards NO_2.

Grande et al. [26] performed research on the kinetics of the catalytic oxidation of NO to NO_2 using a Pt/alumina catalyst, under conditions relevant to industrial nitric acid production. The idea was to replace the homogeneous oxidation process, which requires cooling of the process gas and a long residence time, with a more intensive heterogeneous oxidation process, allowing the heat of reaction to be recovered. Because of the well-known properties of NO conversion to NO_2 with low Pt or Ru concentration [27–29], Authors have selected those metals as an active site for the catalyst. Collected experimental data proved that a catalytic conversion of NO to NO_2 at higher concentrations is possible and efficient. A compromise between catalyst performance and pressure drop needs to be achieved for the catalytic process to accomplish the desired extra energy recovery [26–29]. The same reaction was a subject of the study performed by Salman et al. [30]. Authors have synthesized a series of perovskites – $LaCo_{1-x}Mn_xO_3$ and $LaCo_{1-y}Ni_yO_3$ by a sol-gel method. NO oxidation activity was tested using a feed containing 10 % of NO and 6 % of O_2. The assumption was to partly simulate nitric acid plant conditions. During the study it was confirmed that among the undoped perovskites, $LaCoO_3$ had the highest activity, followed by $LaNiO_3$ and $LaMnO_3$. Additional substitution of $LaCoO_3$ with 25 % of Ni or Mn was found to be the optimum that led to an enhanced NO oxidation activity. Authors have directly justified that perovskites are promising catalysts for NO oxidation at industrial conditions [30]. Finally, it should be emphasized that researches that are conducted on novel type of active catalyst seem to be reasonable. In many cases the low cost, ease of production and significant catalytic activity make such materials attractive candidates as alternatives to commonly known noble metal catalysts [31–34].

1.4 Phosphoric acid

The commercial manufacture of diluted phosphoric acid via so called "wet process" began in France and England in 1850. Stronger acid was produced via continuous process in 1915, based on the technology designed by Dorr Company [1–5, 8]. Additionally there were processes that produced concentrated thermal phosphoric acid from raw phosphorous. Purified phosphoric acid solvent extraction process was ran at industrial scale in 1970s. Basically, there are two main technologies that produce H_3PO_4, one is based on thermal treatment of phosphorous-containing minerals and second concerns acid leaching of the minerals. The difference is that first one allows to obtain concentrated acid and the second one – more diluted. Both of them require raw materials which contain phosphorous [1–5, 8]. Due to the economic aspect the more suitable is the production of phosphoric acid via extraction process, especially that sulfuric acid is needed for this purpose. Thus it is justified that both technological lines, production of phosphoric acid and sulfuric acid, may exist side by side.

The most important phosphorous containing minerals are apatite's (especially fluorapatite $Ca_{10}(PO_4)_6F_2$) and phosphate rocks. The general formula is: $Ca_5(PO_4)_3X$ where X – Cl, F, OH (most often F). The problem is a very durable apatite structure – practically insoluble [1–5, 8]. Apatite structure decomposition methods are presented in Figure 1.21.

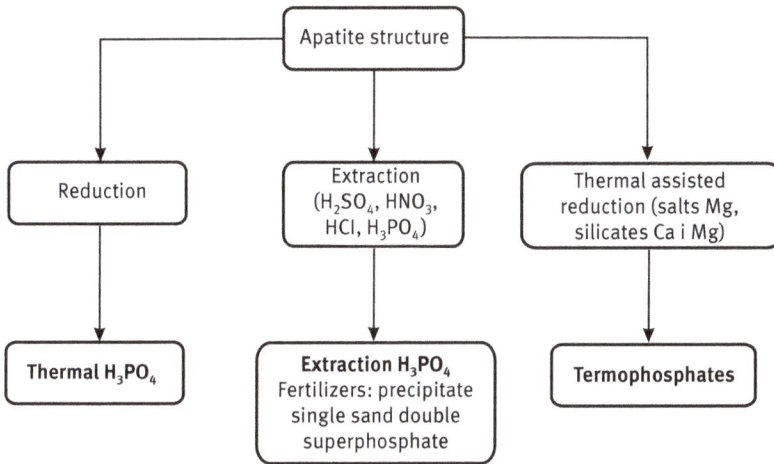

Figure 1.21: Apatite structure decomposition methods [1–5, 8].

In order to obtain the thermal phosphoric acid, thermal reduction of apatites, which results in the formation of phosphorous, need to be performed. White phosphorus is obtained by reducing mineral phosphates with carbon in the presence of SiO_2 (process temperature ranging 1400–1600 °C). Stage I includes reduction of tricalcium phosphate to calcium phosphide and after that stage II occurs, at which production of calcium oxide and phosphorus takes place:

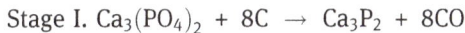

$$\text{Stage I. } Ca_3(PO_4)_2 + 8C \rightarrow Ca_3P_2 + 8CO$$
$$\text{Stage II. } 3Ca_3(PO_4)_2 + 5Ca_3P_2 \rightarrow 24CaO + 4P_4$$

A serious issue of this process is CaO, which strongly alkalizes the system – hence there is a need for the use of quartzite sand (SiO_2) binding CaO into hardly decomposable, neutral calcium metasilicates, which is a waste slag in the production of phosphorus.

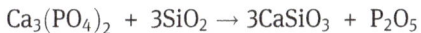

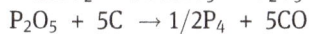

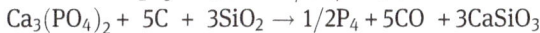

$$Ca_3(PO_4)_2 + 3SiO_2 \rightarrow 3CaSiO_3 + P_2O_5$$
$$P_2O_5 + 5C \rightarrow 1/2P_4 + 5CO$$
$$Ca_3(PO_4)_2 + 5C + 3SiO_2 \rightarrow 1/2P_4 + 5CO + 3CaSiO_3$$

Thermal phosphoric acid is prepared via the oxidation of phosphorus to P_2O_5 and the absorption of vapors of this compound in an aqueous solution of phosphoric acid [1–5, 8]. The process can be carried out in two ways:

– single-stage process – involves the oxidation of phosphorus vapors in a refractory chamber immediately after the outlet from the electric furnace, followed by the absorption of P_2O_5 vapors in a tower filled with Rachig rings sprinkled with a phosphoric acid solution;

– two-stage process – in this process the separation of P_4 + CO into pure P_4 is firstly performed, furthermore the P_4 is oxidized to P_2O_5 in the second stage – combustion of phosphorus takes place in a chamber with graphite lining and absorption of P_2O_5 is carried out at 150–200 °C in phosphoric acid containing solution, depending on the needs, 30–68 % P_2O_5. The reaction mixture containing P_4 and CO is separated based on the condensation phenomenon in water (CO does not absorb in water, so it will leave the system, while P_4 vapors strongly condense and change into liquid phosphorus). For this purpose, Ströder scrubbers are used (Figure 1.22).

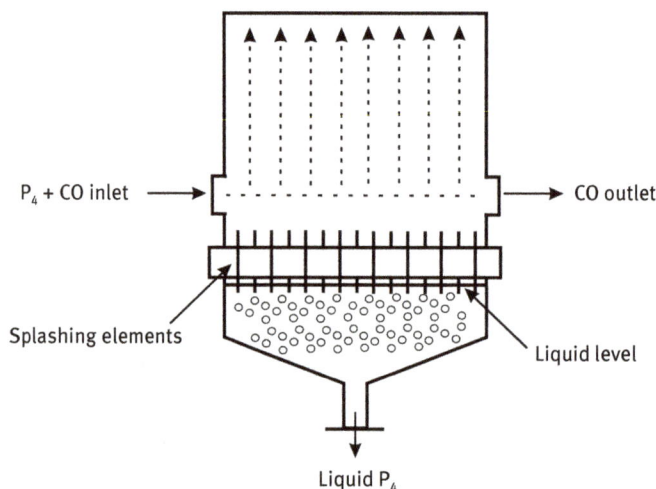

Figure 1.22: Ströder scrubber.

Absorption of P_2O_5 occurs in phosphoric acid due to the possibility of formation of mists of phosphoric acid after P_2O_5 contact with water steam. An alternative method is to conduct the process of obtaining thermal phosphoric acid by oxidation of liquid phosphorus and absorption of P_2O_5 in one tower. Disadvantages of the method includes its relative high cost – disproportionately expensive as compared to the extraction method as well as that it allows to obtain only concentrated phosphoric acid (diluted acids are also often used in various technological processes) [1–5, 8]. The scheme of technological plant used for production of thermal phosphoric acid is presented in Figure 1.23.

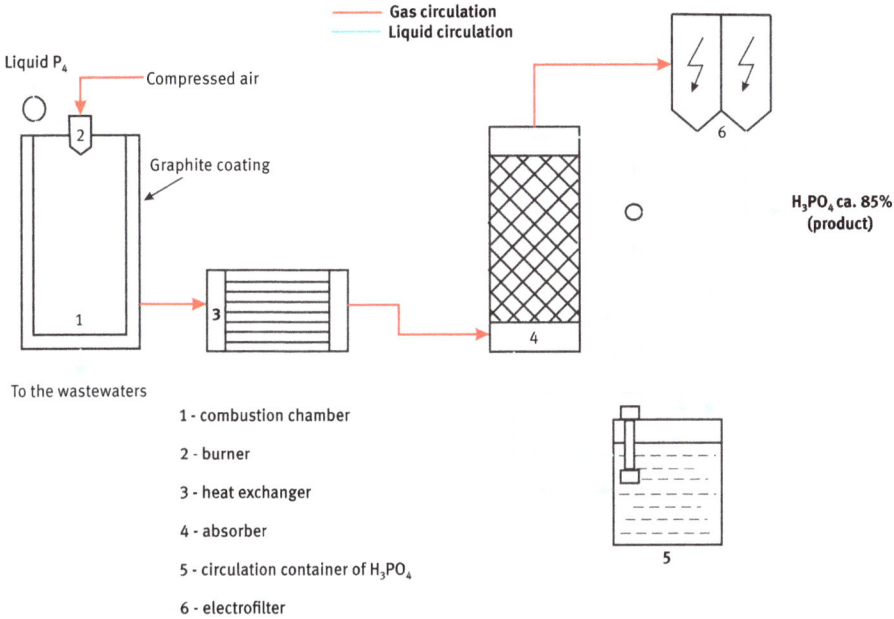

Figure 1.23: Two-stage process of production of thermal phosphoric acid [1–5, 8].

On the other hand, the extraction phosphoric acid is synthesized via extraction of H_3PO_4 from apatites or phosphorites with diluted H_2SO_4 solutions. Evolving hydrogen fluoride (resulting from the mineral structure) reacts with the silica contained in the mineral which results in the emission of nuisance fluorosilicic acid.

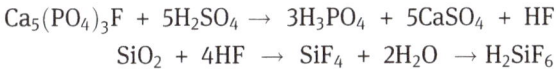

$$Ca_5(PO_4)_3F + 5H_2SO_4 \rightarrow 3H_3PO_4 + 5CaSO_4 + HF$$
$$SiO_2 + 4HF \rightarrow SiF_4 + 2H_2O \rightarrow H_2SiF_6$$

In this method we deal with two very dangerous waste by-products: fluorosilicic acid and phosphogypsum [1–5, 8]. The phosphogypsum form depends on the process parameters concerning synthesis of phosphoric acid:

- $CaSO_4 \bullet 2H_2O$ gypsum (gypsum process) – the mineral is treated with diluted H_3PO_4 at < 107 °C and then 78 % H_2SO_4 is introduced. In this process the H_3PO_4 with a concentration of 50 % (Dorr installation) is obtained;
- $CaSO_4 \bullet 1/2H_2O$ hemihydrate (hemihydrate process) – formed from solutions with a higher concentration of H_2SO_4 and at temperature slightly higher than 107 °C. In this process H_3PO_4 with a concentration of 60 % is produced;
- $CaSO_4$ anhydrite (anhydrite process) – requires the most drastic conditions (use of the most concentrated H_2SO_4 and temperature >120 °C), but there is a possibility for obtaining the phosphoric acid at a concentration of approx. 68 %. Due

to the strong corrosion of the apparatus and the need for reactor heating – a process is poorly widespread in industry.

In order to obtain concentrated phosphoric acid, it is important to obtain phospho-gypsum in the form of easy to drain and wash crystals. The installation used for extraction acid production, proposed by Dorr Company is presented in Figure 1.24.

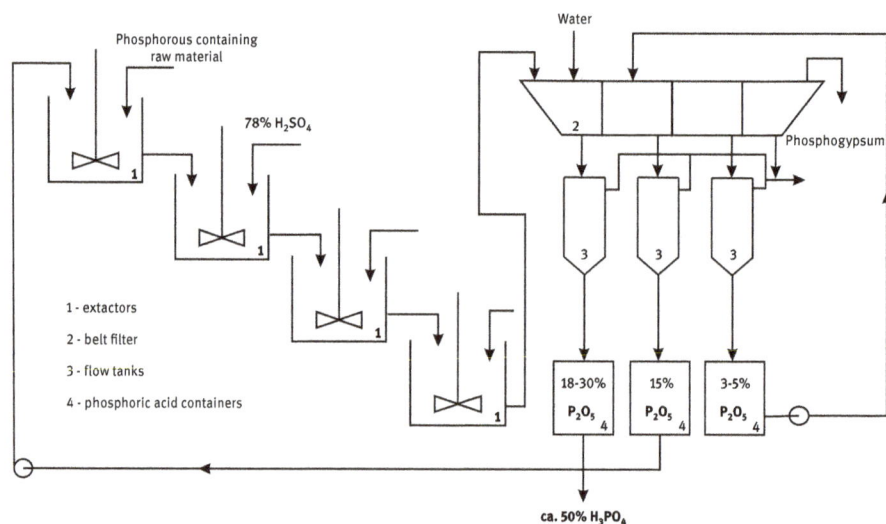

Figure 1.24: Production of extraction phosphoric acid in the Dorr company installation [1–5, 8].

Mentioned above technology is very unique because it is also a part (first stage) of the technological line used for production of phosphorous containing fertilizers such as e. g. superphosphate. Additionally, the energy and economic aspects are more justified as compared to the production of thermal H_3PO_4, by the availability of raw materials and energy expenditure related to its implementation on an industrial scale. There are only four scientific paper published to date, that concerns the production of phosphoric acid [35–38]. The last one is from 1997. There are not any approaches made so far that would enhance the main production of this particular inorganic acid. Probably it is related with quite good availability of raw materials as well as with the process simplicity and versatility. Some computer modeling aspects of acid production were mentioned and discussed but only based on the theoretical assumptions [39] Additional aspect that need to be mentioned and which may justified such small number of scientific papers, is that phosphorous acid is in most cases direct product used for phosphorous-containing fertilizers production.

1.5 Summary

The presented chapter deals with the technology concerning industrial production of major inorganic chemicals such as sulfuric, nitric and phosphorous acids. The background of the production, including main reactions and process conditions, was highlighted. The industrial production of mentioned acids is realized based on the technologies known from years. Insignificant changes were only made, that concerned novel type of catalyst, heat control or recovery, construction of the reactors. Insight into scientific literature concerning this particular reagent proved this statement. The literature concerning sulfuric acid production in most cases is related to the optimalization of main reactions conditions and whole plant efficiency. They concern also the improvement of oxidation of sulfur dioxide towards SO_3, so important analyzing the whole technological plant used for this acid production. Many works have been undertaken in order to analyze the catalyst behavior in this process as well to estimate optimum conditions of the oxidation reaction. The most important is also the environmental aspect concerning SO_2 emission. The most recent research concerning nitric acid technology are in most cases focused on the emission of NO_x, energy aspects including heat recovery as well as novel type of catalysts used for ammonia oxidation. The last aspect seems to be reasonable. In many cases the low cost, ease of production and significant catalytic activity make novel type of materials attractive candidates as alternatives to commonly known noble metal catalysts. The smallest amount of scientific papers concern industrial production of phosphoric acid. There are not any approaches made so far that would enhance the main production of this particular inorganic acid. Probably it is related with quite good availability of raw materials as well as with the process simplicity and versatility, and that phosphorous acid is in most cases direct product used for phosphorous-containing fertilizers production. One cannot forget about the importance of inorganic acids in many industries. Therefore, research on improving their production processes should be constantly carried out taking into account not only production efficiency but also the environmental aspect resulting from, among others, emissions of gaseous substances - only in this way it will be possible to implement industrial processes as sustainable technologies.

Acknowledgements: This research was supported by the Polish Ministry of Science and Higher Education (Grant No. 0912/SBAD/2006).

References

[1]   Buchel KH, Moretto H-H, Woditsch P. Industrial inorganic chemistry. Second completely revised version. New York: Wiley-VCh, 2003.
[2]   Benvenuto MA. Industrial inorganic chemistry. Germany: De Gruyter, 2015.

[3]   Hocking MB. Handbook of chemical technology and pollution control, 3rd ed. Amsterdam: Elsevier, 2005.
[4]   Ullmann F. Encyclopedia of industrial chemistry. Germany: Wiley-VCH Verlag GmbH & Co, 2002.
[5]   Moulijn JA, Makkee M, Van Diepen AE. Chemical process technology. New York:John Wiley & Sons, 2013.
[6]   Martin MM. Industrial chemical process analysis and design. Amsterdam: Elsevier, 2016.
[7]   Thiemann M, Scheibler E, Wiegand KW. Nitric acid, nitrous acid, and nitrogen oxides. Germany: Wiley-VCH Verlag GmbH & Co., 2011.
[8]   Gilmour RB. Phosphoric acids and phosphates. New York: John Wiley & Sons, 2019.
[9]   http://www.essentialchemicalindustry.org/.
[10]   http://www.webofkonowledge.com.
[11]   Suyadal Y, Oguz H. Oxidation of SO_2 in a trickle bed reactor packed with activated carbon at low liquid flow rates. Chem Eng Technol. 2000;23:619–22.
[12]   Nodehi A, Mousavian MA. Simulation and optimization of an adiabatic multi-bed catalytic reactor for the oxidation of SO_2. Chem Eng Technol. 2007;30:84–90.
[13]   Jorgensen TL, Livbjerg H, Glarborg P. Homogeneous and heterogeneously catalyzed oxidation of SO_2. Chem Eng Sci. 2007;62:4496–9.
[14]   Fabrizio S, Michele DA, Amedeo I. Modeling flue gas desulfurization by spray-dry absorption. Sep Purif Technol. 2004;34:143–53.
[15]   Vernikovskaya NV, Zagoruiko AN, Noskov AS. SO_2 oxidation method. Mathematical modeling taking into account dynamic properties of the catalyst. Chem Eng Sci. 1999;54:4475–82.
[16]   Bunimovich GA, Vernikovskaya NV, Strots VO, Balzhinimaev BS, Matros YS. SO_2 oxidation in a reverse-flow reactor: influence of a vanadium catalyst dynamic properties. Chem Eng Sci. 1995;50:565–80.
[17]   Gosiewski K. Unsteady states in metallurgical sulphuric acid plants after SO_2 concentration drop in the feed gas. Can J Chem Eng. 1996;74:621–625.
[18]   Hong R, Li X, Li H, Yuan W. Modeling and simulation of SO_2 oxidation in a fixed-bed reactor with periodic flow reversal. Catal Today. 1997;38:47–58.
[19]   Günther R, Schöneberger JC, Arellano-Garciam H, Thielert H, Wozny G. Design and modeling of a new periodical-steady state process for the oxidation of sulfur dioxide in the context of an emission free sulfuric acid plant, I.A. Karimi and Rajagopalan Srinivasan (Editors). Proceedings of the 11th International Symposium on Process Systems Engineering. 15–19 July 2012, Singapore, Elsevier B.V.
[20]   Kiss AA, Bildea CS, Grievink J. Dynamic modeling and process optimization of an industrial sulfuric acid plant. Chem Eng J. 2010;158:241–9.
[21]   Tejeda-Iglesias M, Szuba J, Koniuch R, Ricardez-Sandoval L. Optimization and modeling of an industrial-scale sulfuric acid plant under uncertainty. Ind Eng Chem Res. 2018;57:8253–8266.
[22]   Zhou J, Ding B, Tang C, Xie J, Wang B, Zhang H, et al. Utilization of semi-dry sintering flue gas desulfurized ash for SO_2 generation during sulfuric acid production using boiling furnace. Chem Eng J. 2017;327:914–23.
[23]   Perez-Ramirez J, Vigeland B. Perovskite membranes in ammonia oxidation: towards process intensification in nitric acid manufacture. Angew Chem Int Ed. 2005;44:1112–15.
[24]   Pérez-Ramírez J, Kapteijn F, Schöffel K, Moulijn JA. Formation and control of N_2O in nitric acid production: Where do we stand today? Appl Catal B: Environ. 2003;44:117–51.
[25]   Sadykov VA, Isupova LA, Zolotarskii IA, Bobrova LN, Noskov AS, Parmon VN, et al. Oxide catalysts for ammonia oxidation in nitric acid production: properties and perspectives. Appl Catal A: General. 2000;204:59–87.

[26] Grande CA, Andreassen KA, Cavka JH, Waller D, Lorentsen OA, Øien H, et al. Process intensification in nitric acid plants by catalytic oxidation of nitric oxide. Ind Eng Chem Res. 2018;57:10180–10186.

[27] Olsson L, Fridell E. The influence of Pt oxide formation and Pt dispersion on the reactions NO_2 ↔ NO + 1/2O_2 over Pt/Al_2O_3 and Pt/BaO/Al_2O_3. J Catal. 2002;210:340–353.

[28] Li L, Qu L, Cheng J, Li J, Hao Z. Oxidation of nitric oxide to nitrogen dioxide over Ru Catalysts. Appl Catal B: Environ. 2009;88:224–231.

[29] Hong Z, Wang Z, Li X. Catalytic oxidation of nitric oxide (NO) over different catalysts: an overview. Catal Sci Technol. 2017;7:3340–3352.

[30] Salman AUR, Hyrve SM, Regli SK, Zubair M, Enger BC, Lødeng R, et al. Catalytic oxidation of NO over $LaCo_{1-x}B_xO_3$ (B = Mn,Ni) perovskites for nitric acid production. Catalysts. 2019;9:429.

[31] Salman AUR, Enger BC, Auvray X, Lødeng R, Menon M, Waller D, et al. Catalytic oxidation of NO to NO_2 for nitric acid production over a Pt/Al_2O_3 catalyst. Appl Catal A: General. 2018;564:142–6.

[32] Hong Z, Wang Z, Li X. Catalytic oxidation of nitric oxide (NO) over different catalysts: an overview. Catal Sci Technol. 2017;7:3440–52.

[33] Després J, Elsener M, Koebel M, Kröcher O, Schnyder B, Wokaun A. Catalytic oxidation of nitrogen monoxide over Pt/SiO_2. Appl Catal B Environ. 2004;50:73–82.

[34] Zhou C, Feng Z, Zhang Y, Hu L, Chen R, Shan B, et al. Enhanced catalytic activity for NO oxidation over Ba doped $LaCoO_3$ catalyst. RSC Adv. 2015;5:28054–9.

[35] Gilbert RL, Moreno EC. Dissolution of phosphate rock by mixtures of sulfuric and phosphoric acids. Ind Eng Chem Process Des Dev. 1965;4:368–71.

[36] Morgunov EM, Masalovi NS, Alekseev TI. Determination of optimum conditions for decomposition of phosphite-containing slurry in phosphorous acid production. Ind Lab. 1968;34:1014.

[37] Gioia F, Mura G, Viola A. Analysis, simulation, and optimization of the hemihydrate process for the production of phosphoric acid from calcareous phosphorites. Ind Eng Chem Process Des Dev. 1977;16:390–9.

[38] Butenko YV, Koltsova EM, Vasilieva LV. Studying and modeling production of phosphoric acid from sodium phosphite solution. Russian J Appl Chem. 1997;70:1847–50.

[39] Abu-Eishah SI, Abu-Jabal NM. Parametric study on the production of phosphoric acid by the dihydrate process. Chem Eng J. 2001;81:231–50.

F. J. Alguacil and J. I. Robla

2 Innovative processes in the production of inorganic bases and derived salts of current interest

Abstract: Ammonia and sodium hydroxide are two important inorganic bases which served as the basis or precursors of other compounds with multiple uses. Some of their derived salts, i. e. ammonium nitrate, are of the paramount importance for daily life. Others salts, such as lithium carbonate, are gaining a primary role in the development of smart technologies, i. e. E-cars. This chapter described developments in the production of these useful compounds: ammonia, sodium and potassium hydroxides, related salts, i. e. ammonium nitrate, sodium and potassium carbonates, and finally, lithium carbonate.

Keywords: ammonia, chloralkali, lithium, potash, processing, salts

2.1 Introduction

The production of inorganic bases (mainly ammonia and caustic soda) is nowadays an important industrial business. There are a miriade of practical uses of these bases and their derivatives.

However, the basis for the production of these two bases is more or less the same those years ago. Ammonia was first produced around 1909 by the Haber–Bosch process; in this process N_2 and H_2 reacts, under given conditions and in the presence of an iron-catalysts, forming NH_3. Nowadays the process used more refined conditions but the same principle.

In the case of caustic soda (and potash), this chemical is really a sub-product in the production of chlorine gas, that was at the beginning and it is now; however, the technology had also been improved from the first processes and the use of mercury cells to the now more efficient membrane cells, and in a very lesser extent, diaphragm cells.

Salts derived form the above bases are normally produced by reacting the corresponding base with the corresponding chemical under known processes, and it is only in the last years that lithium carbonate enters the industrial scene due to the importance that e-cars have today, and being this salt an important component of the lithium-ion batteries associated with this new type of mobility.

This chapter reviews some of the most definitive advances in the production of these bases and their derived salts, including the production of lithium carbonate.

This article has previously been published in the journal Physical Sciences Reviews. Please cite as: Alguacil, F. J., Robla, J. I. Innovative processes in the production of inorganic bases and derived salts of current interest *Physical Sciences Reviews* [Online] 2020, 5. DOI: 10.1515/psr-2019-0031.

https://doi.org/10.1515/9783110656367-002

2.2 Ammonia [1, 1–10]

Ammonia has many uses and in different industries, as summarized in Table 2.1.

Table 2.1: Uses of ammonia.

Uses	Examples
Fertilizer	Ammonium nitrate, urea, ammonium salts
Chemical synthesis	Nitric acid, sodium bicarbonate, hydrazine
Explosives	Ammonium nitrate
Fibers and plastics	Nylon and other polyamides
Refrigeration	
Pharmaceuticals	Manufacturer of drugs (sulfonamide), anti-malarials, B vitamin
Pulp and paper	
Mining and metallurgy	Zinc, cobalt and nickel processing
Cleaning	

Approximately 85% goes to fertilizers for the food industry.

The largest producing countries of ammonia are China, India, Russia, USA, Trinidad-Tobago and Indonesia, whereas the global demand for regions is Asia-Pacific, Europe, North America, rest of world and Latin America.

As it is said in the Introduction, NH_3 is produced by the Haber or Haber–Bosch process. In this process, N_2 and H_2 react in presence of a catalyst (mainly based on Fe) forming NH_3. H_2 is produced by reaction of natural gas and vapor at high temperatures, whereas N_2 addition is by the air. NH_3 production at the cheapest cost is of the foremost importance for the industry. Hydrocarbons are the usual compounds to provide energy and H_2, though there are alternatives such as charcoal or electric power. Light hydrocarbons reforming is the most efficient route, summarizing more than 75% of world ammonia capacity based on gas as natural source. Thus, in actual vapor reforming plants, 40–50% of the overall energy consumption, related to NH_3 production, is in excess of the thermodynamic minimum. Losses due to compression are the responsible for more than half of this excess, whereas the useful minimum consumption is near 130% of the theoretical minimum.

Sizes of big single-train NH_3 plants are 1000–1500 t/day, though larger capacities are provided in new plants, i. e. 3000 t/day. Process efficiency in energy is reached by integrating it with the providing energy source. The ammonia plant can be alone or be integrated, as it can be seen latter, with other plants on the site, i. e.: urea or methanol plants.

Nowadays, large NH_3 plants have a great reliability with their equipment and related tools, whereas technical on-stream factors greater than 90% are the norm. Actually in Europe, ammonia is mainly produced by two types of processes:
i.   vapor reforming of natural gas or other light hydrocarbons,
ii. partial oxidation of heavy fuel oil.

Coal gasification and water electrolysis are not in use in the European NH_3 manufacture. Syn-ammonia synthesis process is non-dependent of the type of syn-gas gas manufacturing process: however, syn-gas quality has a marked influence on the circuit design and operating work.

2.2.1 Conventional steam reforming

The corresponding flow-sheet of the conventional vapor reforming operation is shown in Figure 2.1. More than 80% of world NH_3 production relies on vapor reforming.

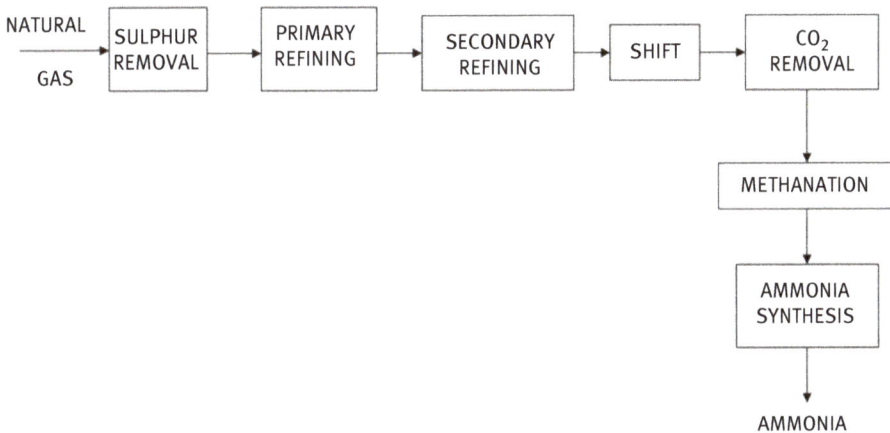

Figure 2.1: Flowsheet for ammonia production using natural gas as feed.

2.2.1.1 Conversion
Using CH_4 as feed, the reaction is

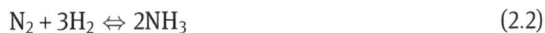

$$0.88CH_4 + 1.26Air + 1.24H_2O \rightarrow 0.88CO_2 + N_2 + 3H_2 \qquad (2.1)$$

$$N_2 + 3H_2 \Leftrightarrow 2NH_3 \qquad (2.2)$$

Syn-gas production and purification work under 25–35 bar pressure, whereas the NH_3 synthesis used pressures between 100 and 250 bar.

2.2.1.2 Sulfur removal

The catalysts used in the process are sensitive to S and related compounds. Feeds products usually contains up to $5\,mg\ S_xN/m^3$ as S-bearing compounds. Inlet gas is preheated until 350–400 °C, often in the primary reformer convection section, prior to its treatment in a de-sulfurization device. In this step, the sulfur compounds are hydrogenated to form H_2S, very often by means of a Co-Mo catalyst, and then adsorbed on pelletized ZnO:

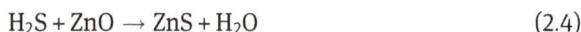

$$R - SH + H_2 \rightarrow H_2S + RH \tag{2.3}$$

$$H_2S + ZnO \rightarrow ZnS + H_2O \tag{2.4}$$

resulting in a gas, containing a concentration not exceeding 0.1 mg/L sulfur, which feeds the system, whereas ZnS remained in the adsorption bed. The necessary H_2, eq. (2.3), is recycled from the syn-site.

2.2.1.3 Prim-reforming site

A mixture of the sulfur-cleaned gas and process vapor, generated in a turbine, is heated until 500–600 °C in the convection site and prior to enter to the prim-reformer. In recent or revamped plants, the above mixture and before entering the prim-reformer, goes to an adiabatic pre-reformer and reheated operation is included in the convection section. In a number of plants, part of the process vapor comes by feed-gas saturation. The amount of process vapor is calculated by the S/C ratio (process vapor to carbon molar relationship), which needed to be lower than 3.0. Several parameters influenced this relationship, i. e. feed quality, purge gas recovery, prim-reformer capacity, shift operation, and the overall vapor balance.

The primary reformer is of a large number of high−Ni-Cr alloy tubes filled with Ni-catalyst. Being an endothermic reaction, addition of heat is needed to elevate the temperature to 780–830 °C at the reformer outlet. The gas exiting the primary reformer has a composition as follows:

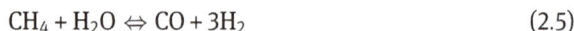

$$CH_4 + H_2O \Leftrightarrow CO + 3H_2 \tag{2.5}$$

with $\Delta H° = 206$ kJ/mol, and

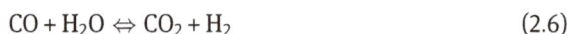

$$CO + H_2O \Leftrightarrow CO_2 + H_2 \tag{2.6}$$

this reaction being exothermic, with $\Delta H°$ of −41 kJ/mol. Natural gas and other fuels supplied the heat for the primary reforming process. The burned gas, after supplying the high level heat of the reforming process, exiting the radiant box with temperatures exceeding 900 °C. Approximately, a half of the heat value of the fuel is directly used into process. The waste heat of the flue-gas feeds the reformer convection unit. The energy requirement providing by fuel, in the ordinary reforming process, is 40–50% of the process feed-gas energy. Principal sites for emissions produced in the plant are generated by the flue-gas exiting the convection section at 100–200 °C. These consisted of CO_2, NO_x, and minor quantities of SO_2 and CO.

2.2.1.4 Sec-reforming

Due to the nature of the operational conditions and the related Chemistry, a mere 30–40% of the hydrocarbons feeding the operation is treated in the primary reformer. The above implies that temperature must be increased to increase conversion. This operation is performed in the sec-reformer by internal combustion of part of the gas with the process air, at the same time N_2 is provided for the syn-gas production. Often, the grade of primary reforming is adjusted in the form that the air supplied to the sec-reformer meets both the heat balance and the stoichiometric synthesis gas need. The process air is compressed to the reforming pressure and heated further in the primary reformer convection section to near 600 °C. The process gas is mixed with air and passed through a Ni-bearing catalyst. At the exit, temperature is almost of 1000 °C, and near 99% of the hydrocarbon feed (exiting the primary reformer) is converted, with a residual CH_4 content of 0.2–0.3% (based on dry gas) in the gas exiting the sec-reformer. The exiting gas is then conveniently cooled at 350–400 °C.

2.2.1.5 Shift conversion

The gas produced in the secondary reformer contained 12–15% CO (based on dry gas), with the CO converted in the shift section (eq. (2.6)). In the High Temperature Shift (HTS) conversion, the gas goes to a bed of iron-chromium oxides catalyst at around 400 °C, here, the CO content decreased to near 3% (based on dry gas), limited by the shift equilibrium of the operational temperature. Conversion can be improved by the use of Cu-bearing catalyst. Exiting gas, from the HTS, is cooled and goes across the Low Temperature Shift (LTS) converter. This converter contained a Cu-Zn oxides-based catalyst and works at near 200–220 °C. The CO content in the con-gas is in the 0.2–0.4% range (based on dry gas). The process efficiency increases as the residual CO content decreases.

2.2.1.6 Removing CO_2

The gas exiting the low temperature shift converter contains hydrogen, nitrogen and CO_2 together with vapor. This gas is cooled, and most of the excess vapor is condensed prior it enters the CO_2 removal system. This gas often contains 1500–2000 mg/L ammonia and 800–1200 mg/L of methanol. Amines, formic acid and acetic acid may be found in the condensate. All these compounds needed to be stripped, from the condensate, and/or recycled. The heat produced during cooling/condensation has various uses, i. e. the regeneration of the CO_2 scrubbing solution, with CO_2 removal by a chemical or a physical absorption process. The chemicals used in chemical absorption are, principally, aqueous amine solutions such as, mono ethanolamine (MEA), thought is not labeled as the best available technology, activated methyl diethanolamine (aMDEA) or hot potassium carbonate solutions. Physical solvents are glycol dimethylethers (Selexol), propylene carbonate and others. In new ammonia plants the following CO_2 removal processes can be considered as usefulness technologies:

i. aMDEA standard 2-steps operation,
ii. Benfield process (HiPure, LoHeat),
iii. Selexol or related physical absorption processes.

Variations of the above concepts must be also considered, whereas Swing Adsorption (PSA) conception is the best available technology in a series of new plants; consideration of this option also assumed that the PSA unit is not the single function of the CO_2 removal. Overall range of heat consumption, in new chemical absorption process, is 30–60 MJ k/mol CO_2. The physical absorption operations are designed for zero heat consumption, but if compared with the chemical processes, the mechanical energy requirements need to be considered.

2.2.1.7 Methanation

The amounts of CO and CO_2, which are found in the syn-gas, are toxics to the syn-ammonia catalyst and need to be eliminated, as methane, in the methanator:

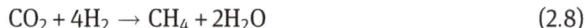

$$CO + 3H_2 \rightarrow CH_4 + H_2O \tag{2.7}$$

$$CO_2 + 4H_2 \rightarrow CH_4 + 2H_2O \tag{2.8}$$

The synthesis showed in eqs. (2.7) and (2.8) occurred at near 300 °C in reactors containing a Ni-based catalyst. The reactions product is an inert gas, however, H_2O needed to be removed prior to enter to the converter. The operation is performed in two sequential steps, (i) cooling and condensation downstream of the methanator, and (ii) condensation and absorption in the produced NH_3 a convenient drying unit.

2.2.1.8 Syn-gas compression and NH_3 synthesis

Recent NH_3 operations use centrifugal compressors to perform the syn-gas compression, this is carried out by vapor steam turbines, using the vapor produced in the NH_3 plant. The condensation of NH_3 is performed in a compressor which used a vapor turbine. NH_3 is synthesized on Fe-catalysts using pressures in the range 100–250 bar range, and temperatures of 350–550 °C, in a reaction given by eq. (2.2) with $\Delta H°$ (298 K) = −92.4 kJ/mol. At each single operation about 20–30% reacts, being this attributable to the non-favorable equilibrium conditions. The separation of the as-formed NH_3 from the gas is done by cooling/condensation, loop pressure is maintained by substitution of the gas with fresh gas. Much heat exchange is needed due to the exothermic reaction, together with the large temperature range found in the loop. The apparition of recent Ru-catalyst is promising since they have a greater activity per unit of volume, with the possibility of increase conversion and lower operating pressures. As recently as 2018, it was published about the properties of a new ruthenium catalyst which is formed by calcium amide with a small amount of added barium, $Ba\text{-}Ca(NH_2)_2$, together with Ru nanoparticles immobilized onto it. Results presented in Table 2.2 indicated that with the use of this new catalyst, and working at temperatures below 300 °C, the efficiency of ammonia synthesis is about 100 times greater that of the obtained with conventional Ru catalysts.

Table 2.2: Comparison of ammonia synthesis activity using three types of catalysts.

Catalysts	Ammonia synthesis rate, mmol/g h
Conventional Ru (Cs-Ru/MgO)	0.072
Industrial Fe	2.2
$Ru\text{-}Ba\text{-}Ca(NH_2)_2$	7.5

Temperature: 260 °C. Pressure: 9 atm.

Also, it can be observed that its performance is more than three times greater than the corresponding of industrial Fe catalyst.

Reforming with methanation, as the final purification step, produces a syn-gas, which contained CH_4 and Ar and in un-dissolved amounts in the condensed NH_3. An important amount of these inert compounds is eliminated by a purge stream from the loop, resulting in the lowering of inert to 10–15%. Water is used to eliminate NH_3 from the gas prior to its use as fuel or prior to be used as H_2. Condensation operation on NH_3 is not achieved completely if the operation utilized water or air, and it is dependent on loop pressure and the temperature of the cooling agent. NH_3 in gaseous form is the coolant agent in an important number of NH_3 producing plants. NH_3 vapor is liquated in a coolant compressor.

2.2.1.9 Vapor and power system

Vapor reforming NH_3 plants produced a great level of excess heat, being used for vapor production in the reforming, shift conversion, and syn-sections, also in the convection section of the prim-reformer. Most of this excess heat is utilized for high pressure vapor production, being useable in turbines to drive compressors and pumps, and as process vapor exiting the turbine system. Recent vapor reforming NH_3 plants may work as self-sufficient from an energy point of view, but often, a small vapor export and electricity import are the norm.

Some processes reduced prim-reforming using part of the second-reformer facilities, being this attributable to the *marginal* low efficiency of the prim-reformer.

Features diverging from the ordinary concept are summarized as follows:
– Decreased firing in the prim-reformer.

- Increased process air flow to the second-reform operation.
- Methanation followed of cryogenic purification.
- Decreasing the level of inerts of the make-up syn-gas.

2.2.2 Partial oxidation of heavy oils

If heavy compounds than nafta are feeding, partial oxidation with oxygen is carried out in the syn-gas manufacture. A scheme of a partial oxidation process is shown in Figure 2.2.

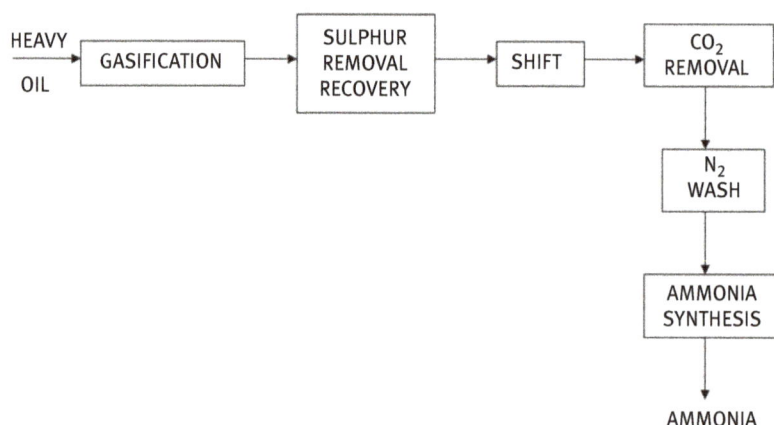

Figure 2.2: Flowsheet for ammonia manufacture using heavy oils as feed.

2.2.2.1 The process

The partial oxidation operation is used to gasify heavy feeds, i. e. residual oils and coal. High viscosity hydrocarbons and plastic wastes are used as fractions of such feed. The partial oxidation process can be considered as an alternative for future uses of these wastes. O_2 is provided by a partial oxidation step carried out in an in the regulation of temperature. The reaction is as follows:

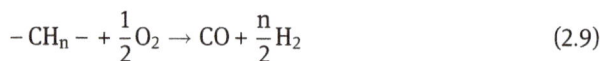

$$- CH_n - + \frac{1}{2}O_2 \rightarrow CO + \frac{n}{2}H_2 \qquad (2.9)$$

By-products are CO_2, CH_4 and soot. Chemicals containing S are converted to H_2S, whereas mineral compounds, present also in the feed, evolved into ashes. Waste heat recovery is followed by the elimination of solids from the process gas, being the soot is recycled-back to the feed. Ashes are eliminated with the condensate, and/or together with a part of the soot. In some European plants, the soot is separated from soot water in a filtration step, which avoided ash build-up in the air separation unit, whereas N_2 addition is carrying out in the liquid N_2 nitrogen wash, with the objectives of the removal of impurities present in the syn-gas, and to obtain the correct H_2/N_2 relationship in it. This oxidation is a not-catalytic driven and occurs at pressures exceeding 50 bar, and temperatures in the 1400 °C range. Some vapor is needed downstream gasification cycle. Heavy metals, such as V, Ni and Fe are recovered. H_2S presented in the gas is separated, in a selective absorption step, and reprocessed, in a Claus unit, to zero valent sulfur. The shift conversion used two high temperature shift catalyst beds with intermediate cooling. Vapor used in the shift conversion is supplied by a cooler-saturator, and by vapor injection. CO_2 is eliminated by the use of an absorption compound which resembled that used in the S elimination stage. Traces of the absorption compound and CO_2 are eliminated from the process gas, prior to the last purification operation, using liquid N_2 washing. In this step, most of the impurities are removed and N_2 is added to reach the H_2 to N_2 stoichiometric ratios. NH_3 synthesis resembled the operation in vapor reforming plants, however, in the case of NH_3, the operation is easier to perform and more efficient, due to the high purity of the syn-gas coming from the liquid N_2 wash units and the syn-loop without purges.

2.2.2.2 Vapor and power system
A vapor-feeding compressor needed auxiliary boilers. The flue-gas generated in these power plants is the principal source of emissions containing SO_2, NO_x and CO_2. Special compressors, i. e. electric driven, decreased the site emissions.

2.2.3 Modern conventional flow-sheets

In general terms, the principal process stages are the same for every plants, being the major differences between them related to the design of various equipment and refinements of the various process steps. Hydrogen recovery is used in most modern plants.

2.2.3.1 Haldor Topsoe ammonia production
The features of the technology offered included:
i.   side fired reformer,
ii. radial flow converters,
iii. S-series of converters.

Some of the offered processes are:

–   IMAP Ammonia+™

This process (Figure 2.3) employs an "in-line" ammonia-methanol concept. The production of methanol is a once-through process carried out at high pressure to

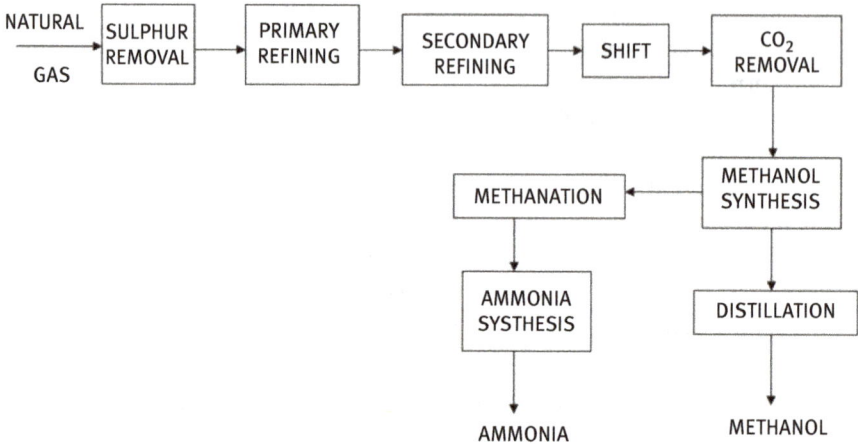

Figure 2.3: Haldor Topsoe IMAP Ammonia+™ process.

reach maximum conversion and efficiency. This process can be complemented by a flexible skid-mounted UFC85 urea formaldehyde unit to create a fully integrated solution at fertilizer complexes. With this process design, ammonia is the main product and methanol is the by-product (up to 35%).

The design can be used for grassroots plants and revamps of existing plants, especially where there is local demand for small amounts of methanol, or when methanol prices are comparatively favorable, if no methanol is needed, this part of the process can be bypassed. According with the company, the process benefits are: shared reforming section, in-line design and once-through process, high conversion and efficiency, very flexible solution.

– SynCOR Ammonia™

This is a new process for grassroots plants (Figure 2.4), offering, among other benefits, greater economies of scale. The process appeared to be a technological must in NH_3 manufacture.

As we aware, every NH_3 plants rely on ordinary two-step reforming, which are based on tubular vapor methane reforming and an air-burned secondary reformer, whereas SynCOR Ammonia™ used the unique Topsoe SynCOR™ solution. This is an idea based on oxygen-burned autothermal reforming, which eliminates the use of the two reformers. This differentiated well the process against conventional ammonia plants (Table 2.3).

In addition, the conventional plant relative energy consumption of 100 is reduced to 97, and the relative make-up water of 100 is reduced to 40–50.

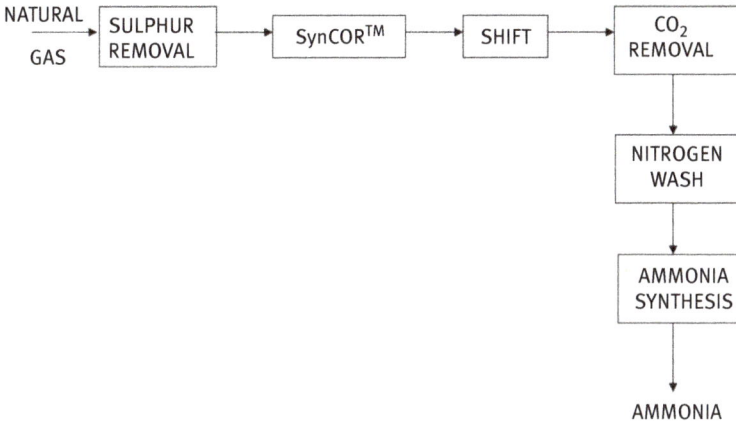

Figure 2.4: Haldor Topsoe SynCOR Ammonia™ process.

Table 2.3: Variations between conventional ammonia plants and SynCOR Ammonia™.

Step	Conventional plant	SynCOR Ammonia™
Desulfuration	standard	standard
S/C ratio	3.0	0.6
Reforming	tubular steam reformer and air blown secondary reformer	Pre-reformer and oxygen blown SynCOR™
Shift	high temperature and low temperature shifts	two high temperature shifts in series
CO_2 removal	standard	standard
Syn-gas cleaning	methanation	N_2 wash with purge and N_2 addition
Syn-ammonia	syn-ammonia circuit with purge	inert free synthesis circuit with no purge
Purge gas treatment	NH_3 wash followed by H_2 recovery	no treatment required

– Haldor Topsoe low energy ammonia process

This process, having practically the same scheme of Figure 2.1, uses steam and air reforming and is applicable for both small and large scale capacities up to 3500 t/day of ammonia.

– Green ammonia: the Haldor Topsoe concept of solid oxide electrolyzer cell

The next generation of ammonia plants developed by Haldor Topsoe uses a solid oxide electrolysis cell (SOEC) to fabricated syn-gas, which I used to feed the proprietary company existing technology of ammonia synthesis sites. Thus, NH_3 was fabricated using air, water and renewable electricity. A small industrial scale green-NH_3 plant is expected to build by 2025, and commercially available by 2030. The new project was awarded funds from the Danish Energy Agency, allowing Haldor Topsoe to demonstrate the system with its academic partners (University of Aarhus, Denmark).

This new concept is an advanced integration of electrolyzer and Haber–Bosch unit, reducing the capital operational expenditure and the operational expenses in relation to other electrolyzer technologies but similar to conventional NH_3 plants using natural gas. In its operation, the necessity of an air separation unit is avoided, fulfilling one of the economical concepts mentioned before. Large scale green ammonia plants will be benefit for the above, whereas small scale plants are expected to improve their economics if the reduction of the inversion is combined with the electrolyzers scaling-up. Relative to the expenses, often influenced by energy costs, the new plant concept will produce green NH_3 an energy consumption of near 26GJ which compared very well with the value of 36 GJ which is the overall standard today. The technology is claimed to be carbon-free.

Another difference between the new technology and the conventional ones using alkali electrolytic cells (AEC) is the Chemistry involved in the process. Figure 2.5 and Figure 2.6 showed schemes of the steps which feed the ammonia synthesis step:

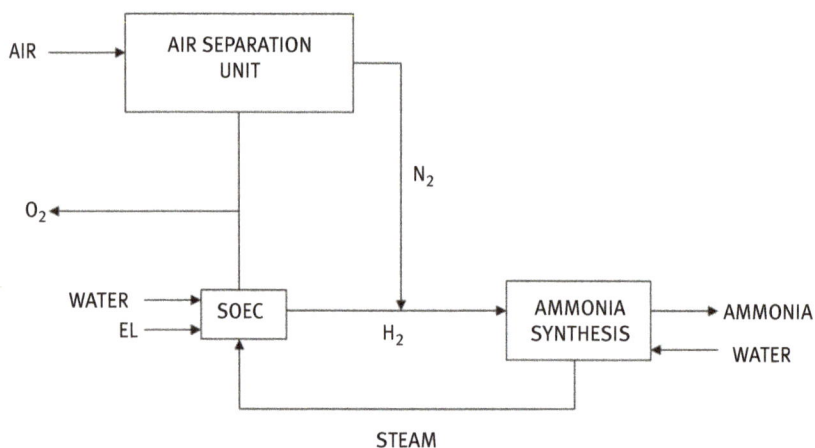

Figure 2.5: SOEC and air separation unit feeding the ammonia synthesis step.

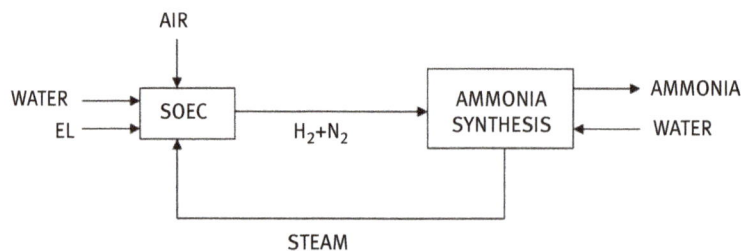

Figure 2.6: SOEC unit feeding the ammonia synthesis step.

Whereas the SOEC step transfers O^{2-} through its membrane, the AEC step transfer OH^- ions, this difference makes the new technology to be an important cost saving change in the design of a green NH_3 plant, which evidently is to avoid of the air separation unit. In the Haldor Topsoe integrated technology, the SOEC produces the requirements of H_2 and N_2 needed in the HB unit. In the SOEC step, N_2 is generated from the air input by burning H_2 produced in the electrolyzer, here, O_2 is removed from air generating vapor. Accordingly, the electrolyzer produced H_2 in excess to the requirements of the syn-NH_3 sites, and the process becomes more efficient in energy terms.

Other difference, between SOEC and AEC, is that SOEC consumes power in the form of electricity or heat. The vapor generated by burning H_2 may be partially recycled to the system, which provided water input and heat energy to the electrolyzer. The new process operates at a lower voltage. In conclusion, SOEC presented high energy efficiency and the process eliminated the air separation unit.

With the built of 60 ammonia plants since 2000 year, some recent projects participating of Haldor Topsoe technology are shown in Table 2.4.

Table 2.4: Recent ammonia plants using Haldor Topsoe technology.

Plant	Country	NH_3 production, t/day	Year
Turkmenhimiya	Turkmenistan	2000	2018
United Chemical Co. Shchekinoazot	Russia	370	2018
Tatarstan	Russia	1964	2016

2.2.3.2 ThyssenKrupp industrial solutions AG ammonia production

Recent plants built by the corporation are summarized in Table 2.5.

Table 2.5: Recent ammonia plants built by ThyssenKrupp industrial solutions AG.

Plant	Country	NH_3 production, t/day	Year
SAFCO	Saudi Arabia	3300 + 11%	2017
Ma´aden	Saudi Arabia	3300	2016
CF Industries	USA	2200	2016
SASOL	South Africa	930	2016
CF Industries	USA	3300	2015
CF Industries	Canada	1300 + 11	2015
QAFCO 4	Qatar	2300	2013

– Uhde ammonia process

With basically the same scheme than that shown in Figure 2.1, this process has a number of significant features such as:

i. a top-burned prim-reformer with a cold exit manifold,
ii. sec-reformer containing a circumferential vortex burner, and
iii. a magnetite-based three-bed radial heat exchange synthesis containing one or two synthesis converters.

Depending of plant location and special requirements, the total plant consumption is in the order of 27.6–30.1 GJ.

Besides the above, there are other 11 plants built since 2000 year.

– Uhde dual-pressure ammonia process

This process was developed in collaboration with Johnson Matthey-Catalysts. The most important innovation of this process is the addition of medium pressure once through syn-NH_3 connected in series with a conventional high pressure syn-NH_3 circuit. Some of the highlights of the process can be summarized as:
i. more energetically efficient (4%) than the conventional Uhde process,
ii. process characteristics, allowing the reduction of piping sizes in the high pressure loop,
iii. considering plants in the order of 3000 t/day of ammonia production, the syngas compressor is of the same size that those used in lower capacity, i. e. 2200 t/day, plants,
iv. one third of H_2, recovered from the purge gas, is converted to NH_3 in the once through step,
v. the operation utilizes magnetite-based catalysts in all the stages.

2.2.3.3 Others flowsheet

– Braun

The Braun process used a primary reformer with exiting temperature of 700 °C, and excess air addition to the secondary reformer. Methane slip is in the 2–2.5 mol% range. Excess N_2 and CH_4 are cryogenically removed down the methanator, whereas H_2O is removed from the ammonia synthesis step due to its poisoning characteristics against catalysts. The S/C ratio can reduce to 2.7–3.0.

– ICI AMV (Ammonia V) process

The main features are summarized as: using excess air. Excess nitrogen goes to the circuit. This operates at pressures of 80–110 bar. Usage of ICI LP catalyst (KATALCO™ 74–1). Saturation of feed gas. S/C ratios of 2.8.

- ICI LCA (Leading Concept Ammonia) process

The process is defined for plants of near 500 t/day, in which the excess air is used to close the heat balance and a PSA system to eliminate CO_2 and excess N_2.

- Linde Ammonia Concept (LAC)^{TM} process

In this offering, no secondary reformer is present in the process, and N_2 is cryo-genically produced and added after a PSA unit. The process uses an isothermal shift and the methanation step is not considered. Inert free-ammonia syngas generation was produced with the correct H_2 to N_2 ratio, adjustable and CO_2/NH_3 product relationships. Design feedstock flexibility from light hydrocarbon feed-stocks (LAC.L) to heavy hydrocarbon feedstocks (LAC.H). One example of this type of processing is the ammonia plant of Sadara Complex (Saudi Arabia), with a capacity of 650 t/d and starting at 2016. A scheme of the Linde Concept is shown in Figure 2.7.

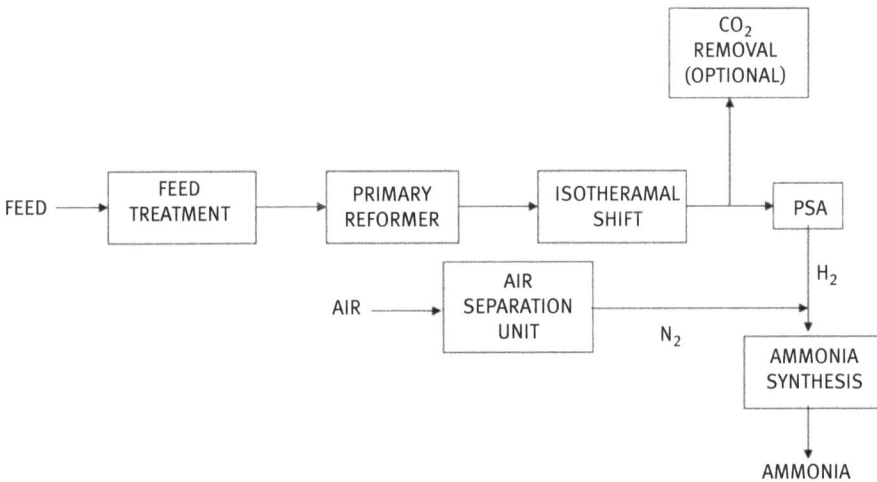

Figure 2.7: The Linde ammonia concept.

- Casale

Under this label, two processes for ammonia synthesis are offered. First, is the STANDARD process, designed for plant capacities of 2500–3000 t/day ammonia production. This process, which used natural gas as feed, is based on the chemical reforming route (with hydrogen recovery).

The second process, named MEGAMMONIA®, developed in conjunction with Lurgi, was designed for plant capacities higher than 3000 t/day ammonia. The sequential of the main steps of the process is: air separation unit, desulfurization, autothermal reforming, high temperature shift, CO_2 removal, nitrogen wash, compression and ammonia synthesis. Using natural gas as feed the process used an autothermal reforming to generate the syngas at a higher pressure than the standards. A physical process for the CO_2 removal (Rectisol®) is used, the step which works at a higher pressure is more efficient than standard chemical processes, and a shift an ammonia synthesis converters, which in combination with a nitrogen wash unit, allowed to keep the size of these converters (and of the entire synthesis loop) at standard sizes. Either the autothermal reforming or the CO-shift unit used Casale technology. The CO-shift unit consisted of two axial-radial beds of conventional high temperature catalyst in series. The CO_2 is removed by absorption in cold methanol (Rectisol® process). Syn-ammonia unit is of Casale axial-radial design, having the ammonia converter three beds configuration with two interchanges. The overall consumption may be around 27.6 GJ.

– KBR purifier and KAAP™

The technology uses a purifier (cryogenic syngas Purifier™) which is located after the methane dryer stage; however, the process can be operated without this purifier. KAAP™ is located after the compression unit, the recycle purge is added to the methane dryer step. KAAP™ uses a ruthenium catalyst, which apparently is more active than traditional iron catalysts, thus, the technology allows lower synthesis loop pressure and the use of a single-barrel syngas compressor.

The KRES™ (KBR Reforming Exchanger System) has been developed to replace the traditional primary reformer. This system operates with air instead of oxygen, because downstream purifier needed excess nitrogen, Within the KRES™ system there is no necessity of an air separation unit.

Recent KBR ammonia plants are summarized in Table 2.6.

Table 2.6: Recent KBR ammonia plants.

Plant	Country	NH_3 production, t/day	Year
BASF group, Yara Int.	USA	2200	2018
Dyno Nobel	USA	2300	2016
PT Pupuk Sriwidjaja	Indonesia	2200	2016
Pupuk Kaltim	Indonesia	2700	2015
Petrobras	Brazil	2200	2015
Moron	Venezuela	1800	2014
Turkmenhimiya	Turkmenistan	1200	2014
China Coal	China	1630	2014

Recently (published date 24 January 2018), the company had been awarded with a contract, by Indorama, for an ammonia plant to be built in Nigeria.

2.2.4 The future

Broadly speaking, in NH_3 manufacture not a unique process can be considered as best available technology. However, there are some points to take into account, thus, ammonia synthesis needed processes which considered:
- Reduction of power consumption.
- Improvement of stack losses.
- Cut-off energy consumptions for CO_2 removal.
- Low S/C ratios.
- More efficient and cheaper catalysts.
- H_2 recovery is a necessity.
- Improvement in the environmental emissions.
- Electrochemical membrane reactors; photochemical routes to feedstocks; ionic liquid reaction media

The treatment of natural gas with vapor and air is the easiness operation of NH_3 syngas production. The comparative consumptions of natural gas reforming, heavy oil and coal gasification are showed in Table 2.7.

Table 2.7: Comparative consumptions of ammonia synthesis gas production using various feeds and standard procedures.

	Natural gas	Heavy oil	Coal
Energy consumption	1.0	1.3	1.7
Investment cost	1.0	1.4	2.4
Production cost	1.0	1.2	1.7

In the basis of actual resources of fossil raw materials, it is expected that natural gas still continues being the feed for ammonia in future years. Thought future is unpredictable, at the half of the present century, and based on world reserves and consumption rate, coal will take place as feed material instead of natural gas. Heavy oil may be considered due to its feed flexibility and under determined environmental concerns i. e. unavailability of natural gas, and if the partial oxidation operation solves the waste problem of heavy residues and plastics recycle.

It is expected that the actual 175,000,000 t/year of ammonia production will increase in 1–2% during years.

2.3 Sodium hydroxide [11–14]

Sodium hydroxide (caustic soda) is the by-product yielded from the Cl_2 generation. Chemically speaking, this hydroxide is alkaline and of pure ionic character (formed by

Na$^+$ and OH$^-$ ions), thus, it is easily soluble in water. NaOH is a very useful product as precursor of sodium chemicals, being the key component in many important chemical processes. Some chemicals are: sodium formate, cyanide, chloroacetate, stannate, picramate, metasilicate, orthosilicate, phenolate, propionate, sulfite, molybdate, formaldehyde sulfoxylate, fluoroacetate, oleate, napthenate, lauryl sulfate, stearate and monosodium glutamate. Its uses in processes including: aluminum ore, cobalt hydroxide, xylenesulfonic acid, chemical wood pulp, polycarbonate, hydrazine, vanillin, epoxy resin, water treatment, disinfection, effluent treatment, galvanization, rubber reclamation, regeneration of ion exchange resins and fruit and vegetable products.

The key ingredients needed to make chlorine and sodium hydroxide are: salt, water and electricity. Industrially it is produced by different processes namely:

2.3.1 Membrane cell process (the most widely used)

Figure 2.8 showed a scheme of a membrane cell for the production of Cl$_2$ and NaOH. In the device, the anode and the cathode are separated by an ion-exchange membrane. The membrane role is to allow Na$^+$ ions (and water) to go across the membrane and reached the cathode. In the anode

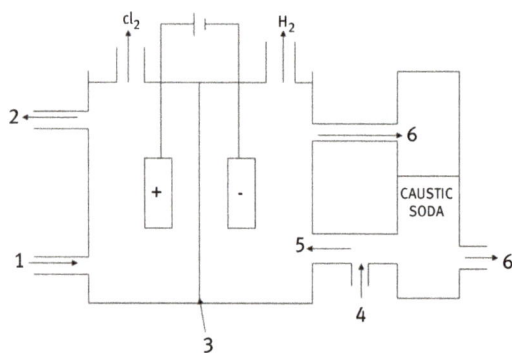

$$2Cl^- \rightarrow Cl_2 + 2e^- \qquad (2.10)$$

Figure 2.8: Membrane cell. 1 – Saturated brine, 2 – Depleted brine, 3 – Non-permeable ion exchange membrane, 4 – Pure water, 5–30% NaOH, 6 – 33% NaOH.

whereas on the cathode, H$_2$ splitted, collecting the bubbles conveniently:

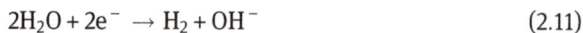

$$2H_2O + 2e^- \rightarrow H_2 + OH^- \qquad (2.11)$$

The remained NaOH phase leaves the cell at near 30% concentration prior to a second concentration step, until 50%.

In the anode, Cl_2 exits and the "spent" brine is re-saturated with more "rich" salt prior to its purification using ion-exchange technology. Gaseous Cl_2 can be contaminated with O_2, and in this case must be purified by liquefaction and evaporation. In this process, energy consumption is the lowest of the three processes (membrane, diaphragm and mercury), and the amount of vapor needed to concentrate the soda solution is also low (less than one ton per ton of soda).

2.3.2 Diaphragm cell process

In this operation, a permeable diaphragm separates the anode and the cathode (Figure 2.9). The entering brine material goes to the anode compartment flowing across the diaphragm toward the cathode compartment. The chemical reactions involved in the anode and the cathode are exactly the same as those described in eqs. (2.10) and (2.11), respectively.

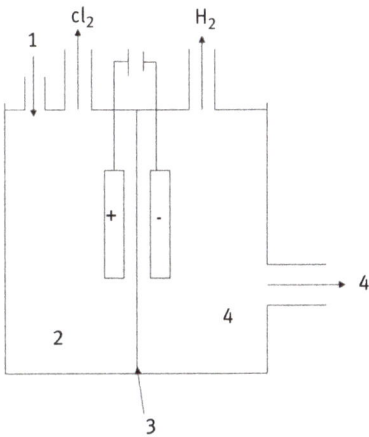

Figure 2.9: 1 – Saturated brine, 2 – Diluted brine, 3 – Permeable diaphragm, 4 – Diluted caustic soda and diluted brine.

Diluted caustic brine exits the device. If necessary, soda can be concentrated until 50%. Normally, this is done by evaporation at the cost of three ton of vapor per ton of soda. Salt by-product is reusable to saturate dilute brine. If Cl_2 contains some O_2, the purification step occurred in the same manner that in the membrane cell process.

2.3.3 Mercury cell process

This operational mode for the production of Cl_2-NaOH is phased out and it is not in operation in Europe.

By 2017, near 80% of the industrial capacity of chlorine and sodium hydroxide manufacturing process is claimed to membrane process, 12% to diaphragm, near 6% to mercury and less than 5% to others technologies.

A new process for the production of NaOH from seawater desalination brine had been recently proposed [11]. The new process included the new steps of nanofiltration, electrodialysis, evaporation, chemical softening and ion exchange prior to enter the membrane electrolyzer (Figure 2.10).

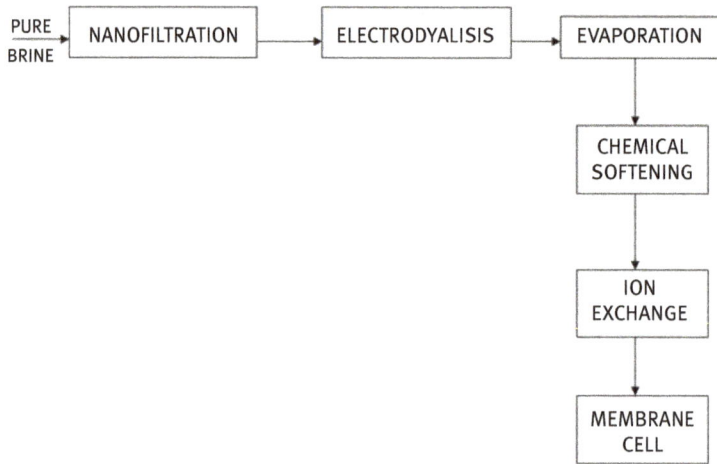

Figure 2.10: Brine to membrane cell for NaOH production.

In the nanofiltration step, the brine was purified from sulfate and a fraction of the accompanying calcium and magnesium. The resulting permeate is concentrated *via* the electrodialysis and evaporation steps. In the exiting solution, remaining calcium and magnesium are completely removed by the chemical softening, using sodium carbonate or sodium hydroxide, and ion exchange, using a cationic resin, steps. On the partial pure and concentrated brine HCl is added prior entering the membrane cell to yield sodium hydroxide, chlorine and hydrogen. Accordingly to the authors, the process presented advantages as follows:
i.   decreasing of effluent generated in the reverse osmosis step,
ii. reusing the electrodialysis eluate,
iii. avoid the concentration-dilution cycle,
iv. it has a potential revenue from side products, and
v.   it reduces the transportation costs.

Global caustic soda capacity by world region distributed as: Asia-Pacific (55%), Europe (20%), North America (20%), Middle East, Africa and Latin America (5%). Asia-Pacific provides more than half of global caustic soda production, with the market expecting to grow at 3% per year. This market, together with these of Latin America and Middle East are expected to increase with decrease of the Europe and North America markets.

Future: Concentrated caustic soda could be used to efficiently store heat in homes for later release when it gets cold at night.

2.4 Potassium hydroxide

The manufacture of potassium hydroxide used potassium chloride instead of sodium chloride in the very same electrolysis process. Thus, the reactions involved in the process are the very same as described in eqs. (2.10) and (2.11) for the anode and the cathode, respectively.

KOH is used as electrolyte in batteries, the manufacturing of soaps, chemicals and fertilizer production, petroleum refining and in de-icing and cleaning solutions. Its uses also included the pH balance.

2.5 Salts [15–24]

2.5.1 Ammonium nitrate (AN) and calcium ammonium nitrate (CAN)

AN is amply utilized as a fertilizer containing N_2, basically due to the high percentage of nitrogen (33.7%) presented in its molecule. This salt is only manufactured by reaction of NH_3 gas and HNO_3 solutions. The resulting AN solution has various uses, depending of its application:

i.   storage as a solution and then used in downstream plants or sold as such,
ii. transformation into solid AN by prilling or granulation, and
iii. forming a mixture with a solid filler.

Calcium carbonate is often used as filler. This salt can be found as ground limestone, dolomite or as a secondary product of nitrophosphate processing. When mixed with AN a product named calcium ammonium nitrate (CAN) is formed, this product can be prilled or granulated. Granular products with AN and NH_3 or $CaSO_4$ are found.

One of the parameters to consider in the manufacture of AN is the strength of the HNO_3 feed, this strength varies from 50% to 70%. Often AN is synthesized using the HNO_3 fraction coming from the own site, though purchasing it can be another option.

The end product can leave the production site both, as loose bulk or in various pack sizes.

Product must conforms to current European Union specifications, including (i) decreasing the sensitivity of AN to heat or detonation with non-addition of external compounds, (ii) the oil retention must pass a specified test, (iii) the combustible material must be less than 0.2% for product containing more than 31.5% N and less than 0.4% for product between 28% and 31.5% N, (iv) the pH value of a 10% solution must be greater than 4.5, (v) less than 5% of product must be smaller than 1 mm and less than 3% smaller than 0.5 mm, (vi) chlorine content tolerances less than 0.02% by weight, (vii) heavy metals must not be added and traces incidental to the process should not exceed the limit fixed by the regulations.

AN declared as EU Fertilizer may only be supplied to the final user in packages, however, the legislation of the AN receiving country must be consulted for the precise details of particular requirements. The production process (Figure 2.11) consisted of three operations:

i.   neutralization,
ii. evaporation,

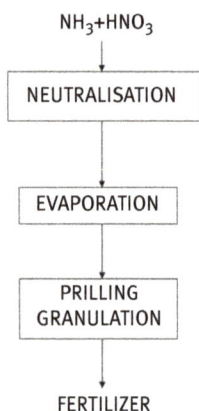

NH_3+HNO_3

↓

| NEUTRALISATION |

↓

| EVAPORATION |

↓

| PRILLING GRANULATION |

↓

FERTILIZER   Figure 2.11: Steps for AN production.

iii. solidification (prilling and granulation).

In the case of AN manufacturing no single route can be identified as best available technology; several reasons for the above include: (i) the selected solidification process depended of certain commercial considerations, (ii) the goal of the operational units is the prevention or emissions, but it may be under the terms for the vendor to render the emission harmless by end-of-pipe treatment.

2.5.1.1 Neutralization or production of AN solution

The reaction of NH_3, in gaseous form, and HNO_3 produced AN and steam, this reaction is of exothermic nature.

Gaseous NH_3 may contain minimum quantities of inert impurities such as H_2, N_2 and CH_4. These impurities needed to be vented from the neutralizer system at levels depending of the given process.

HNO$_3$ can be pre-heated using equipment with corrosion resistance, especially if the concentration of nitric acid is near the lower limit of 50%. Pre-heating operation is carried out by the use of steam or hot condensate from the AN processing. The amount of pre-heat can be estimated from the concentration of HNO$_3$ and the required concentration of the resultant AN solution by means of the enthalpy balance. The formation of AN can be carried out in one or two stages. In this second option, the first neutralizer operates at a low pH value, that is, in acidic conditions, whereas in the second stage the conditions are of a neutral pH value. Whereas the equipment operated at different pressures and temperatures, in most neutralizers, these operational variables, including concentration, are in closed relationship of the boiling point of AN solutions.

Neutralizers, as used in Europe, have different configurations, i. e. circulating systems, pipe reactors, etc. There are several environmental factors which influence the choice of neutralizer: (i) a single-stage neutralizer is simpler and cheaper (ii) a two-stage neutralizer produced most of the boil-off steam in the first stage and most of the NH$_3$ emission from the second stage, decreasing the total emission of ammonia, (iii) neutralization, at an elevated pressure, produced steam at a higher temperature (and AN at a higher concentration)., thus, favoring the use of such steam in downstream processes, i. e. evaporation and drying, (iv) the control of the neutralizer is a critical parameter. The pH and the temperature must both be strictly controlled to limit the losses from the equipment, consequently, all neutralizers unit must included pH and temperature controls to avoid any environmental issue, (v) impurities must be controlled, within the operating temperature of the neutralizer, impurity control is of the foremost importance also to avoid control safety and associated environmental issues. This is because some plants do not recycle AN screenings to the neutralizer. Recycling is especially risky if the there are contamination by an organic anticaking additive. In general terms, an AN acidic solution is less stable than an alkaline one.

In neutralizers unit, best available technologies may consider: (i) in the event of water addition to the unit, this water should be used to the safe recycling of AN solution, (ii) impurities should be rigorously excluded, though some recycling may be practicable, (iii) the steam formed within the neutralizer contained both ammonia and AN in concentrations of a few thousand mg/L of each. The careful design of the neutralizer equipment can ten-fold reduced them.

The steam, by-product of AN fabrication, exiting the neutralizer needed to be purified, or it can be condensed and then purified. The steam is used, in the evaporator (see below), or it may be used to preheat and evaporate NH$_3$ or be used to preheat HNO$_3$. The next operations have been used commercially for the purification of the steam: (i) drop separation methodologies (knitted wire mesh demister pads, wave plate separators, fiber pad separators), and (ii) scrubbing units (packed columns, venturi scrubbers, irrigated sieve plates).

The formation of AN in the neutralizers is a removal challenge since the particles are very fine. Often, a combination of drop separators and scrubbers are used for the above. In the scrubbers, it is required the addition of acid, often nitric acid, to neutralize any free ammonia and optimization of its removal.

The condensate must be cleaned using various technologies, including ion exchange resins; however, safety concerns with the use of this technology may be considered. The recycle of organic resins to the AN process may be avoid, and the resin may not be nitrated. The choice of the cleaning methodology is dependent on the necessity of nitrate elimination, and this is dependent on the receiving water. The final condensate exiting the neutralizer, can be differently discharged: (i) to drain, (ii) to a HNO_3 plant to be used as absorption water, if nitric acid fulfill all the requirements, (iii) to other uses on the site, i. e. in the manufacture of solution fertilizers, (iv) to boiler water feed, i. e. after further purification, (v) to an aquatic environment for control/analysis purposes or for subsequent evaporation by the heat of natural sources or for disposal to land, though neither of these is practicable in many countries in Europe because of the climate or the amount of land required.

Biological treatment is an alternative for the removal of N_2 from fertilizer site effluents, however, this is not the used norm on a commercial basis in Europe, except in the case of an existing public utility or on a large integrated chemical site.

The final product, from the neutralizer, is AN solutions, with concentrations depending of the feed materials and the operational uses. It is stored without further processing, but, if it is to be used in the manufacture of solid AN, CAN or NPK fertilizer, it is normally concentrated by evaporation. NPK is a three component fertilizers providing nitrogen, phosphorous and potassium.

2.5.1.2 Evaporation

This step (Figure 2.11) is necessary to eliminate water, normally found, in the AN solution. The acceptable water content is dependent on each given process, but it is below 1% for a prilling product, however, amounts up to 8% is needed on some granulation processes. Evaporation is always performed using vapor usually coming from the neutralizers or from a vapor raising facility located within the plant location.

Undesirable AN decomposition is avoided by the use of saturated vapor heated at a given temperature. This step can be carried out at atmospheric pressure or under vacuum; in this second case, allowed the reuse of neutralizer vapor at the cost of more capital investments. In the step, some NH_3 is lost from the AN dissolution, and this needed to be replaced in advance to the next step. The vapor, boiling off, is contaminated with NH_3, this chemical needed to be removed, whereas some AN, in the form of drops, are also found.

Evaporators used in practice included circulatory systems, shell and tube heat exchangers and falling film types (which presented some features such as: decreasing of the working volume and a shorter residence times). These equipment produced contaminated vapor, which needed to be decontaminated previously to its discharge to the environment. Methodologies used in the purification included: (i) drop separators, (ii) scrubbers and (iii) condensation of the vapor.

The evaporator produced AN solutions of the adequate concentration, and at a temperature which eliminated the risk of crystallization. Sometimes, the solution is cooled to decrease effluents emissions from downstream equipment.

2.5.1.3 Prilling and granulation

Prilling is the formation of granules by means of the solidification of drops of fertilizer materials. Granulation refers to methodologies which used agglomeration, accretion or crushing, to produce a granular fertilizer.

The prilling operation technique is used in a number of plants producing AN or even CAN. Granulation of AN is carried out in a plants used only to perform this task, or in CAN producing plants. As in the case of AN, there are plants which only granulated CAN, whereas this product can be produced in plants producing NPK fertilizers.

The feed of AN to a prilling plant needed to be almost anhydrous. Drops of the product are formed, falling to a prill tower. Air is allowed to flow upstream, and the drops cool and become solid. Drops are generated by (i) a rotating perforated bucket, or (ii) a static system of fixed orifices. If CAN is manufactured, ground $CaCO_3$ is added previously to the generation of the drops.

Several effluents resulted from the loss of NH_3, AN and $CaCO_3$ (the latter in CAN production) to the air stream. Decreasing the melting temperature reduced emissions, whereas NH_3 is often eliminated neutralizing it in a wet scrubber.

Tiny AN particles are finding within the air, they are eliminated by the use of dedicated equipment. However, AN fume is lost from the prills surfaces, and being in the sub-micron sizes, it is non-easy eliminated. Its presence is determined by a persistent blue haze, which it is seen at larger distances from the plant. This problem is avoided by the use of irrigated candle filters (with candles incorporating fine glass fiber), but at millionaire (Euros) cost. In CAN prilling towers, candles are not used for the elimination of the effluent, because, insoluble $CaCO_3$ fouled the surface of the filter in an undesirable time. Similar situation is found if insoluble materials are added to the AN. Other scrubbing equipment did not improve the elimination of the effluent.

Granulation is a more difficult operation than prilling, and a variety of equipment is used in the industry, including: rotating pans and drums, fluidized beds and other more specialized equipment. From an environmental point of view and having effluent of the same nature, that in prilling processing, the advantage of granulation is that the amount of air to be treated is smaller, and the related equipment is cheaper and easier to install, being also lower the energy consumption. If the AN feeding the granulator has elevated moisture contents, the emission contained only coarser particles, against the presence of fumes, and consequently, the scrubbing operation is performed using cheaper equipment in comparison to candle filters. Granular products are made in a wider range of particle size than prills.

A number of granulation processes used AN containing up to 8% water, being this water eliminated within the process, and at a lower temperature.

In AN/CAN sites, granulators include: rotary pans and drums, spherodisers, pugmills and fluidized beds. If CAN is manufacture, the filler are added to the process before the granulator, and AN is added in the granulator spraying a hot concentrated solution, and no further drying of the granules are required. This product is screened, and the fines and crushed oversize are back to the granulator.

CAN and CAN/NPK granulators included drums and pugmills. The filler is mixed with AN solution prior to granulation or even in the granulator. Granules from this process are drying in a fluidized bed or rotary drier; being an autothermal process, additional heat is not necessary in this drying step. Gases coming from the granulator and from the drier are cleaned by a combination of dry cyclones or bag filters and wet scrubbers. Candle, venturi and cyclon devices are frequently used for the latter. Candle filters are suitable if the emission contains a large proportion of sub-micron particles, but they can not be used in CAN plants.

NH_3 and AN emissions into air and generated in the prilling and granulation sections of AN and CAN plants can be decreased by a range of suitable equipment. The resultant emissions are dependent upon two principal factors: (i) the efficiency of the equipment used in the operation, and (ii) the volume to be treated.

For the fine particles produced in the prill tower, candle filters are normally required, these decrease particulate emissions to 15 mg/m^3 of air. For coarser materials, as in granulation plants, dry equipment, i. e. bag filters or dry cyclones are better options.

Modern plants used lower air volumes; however, the reduction in air volume on an existing plant tends to be expensive.

Prill and granulator towers very often produce a product which requires further cooling, this step is done using rotary or fluid bed coolers, being the air cleaned by high efficiency cyclones, bag filters or wet scrubbers.

AN and CAN tend to caking during storage and need to be conditioned to prevent it. Anticaking agents are internal to the finished particle or applied, as a coating, to the outside; in addition, these additives are also useful to prevent dust formation and moisture pick-up during storage. Their characteristics vary according to the necessities of individual plants.

Losses to the atmosphere are a general problem but particular to any plants, some examples are: loss of AN to drain in a number of locations, losses from pump seals, leaks from flanges and valves, etc. The proper plant design and maintenance must be a necessity to avoid these different losses and to help to decrease waste products generated for the plant.

Packs of solid AN are stored in warehouses approved by specific authorities of each country, for AN duty. Since both, bulk AN and CAN, are hygroscopic, they needed to be protected from moisture.

AN solutions are stored prior to use in downstream plants or prior to sale. This storage is done at a temperature above the crystallization point of the solution. NH_3 gas is often added in quantities which maintained the solution at the correct pH value, because AN solutions tend to lost NH_3 during storage.

2.5.2 Sodium carbonate

Sodium carbonate or soda ash is a key product in many industries: glasses, soap, paper making, bleaching products, etc. Though sodium carbonate can be found naturally, it is often fabricated by the Solvay process (Figure 2.12), which uses purified brine and ammonia as starting reagents.

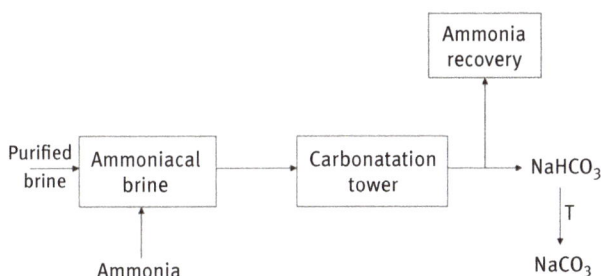

Figure 2.12: Production of sodium carbonate from purified brine.

The reactions occurring in the tower are

$$NH_3 + H_2O + CO_2 \rightarrow NaHCO_3 \qquad (2.12)$$

$$2NaHCO_3 \rightarrow Na_2CO_3 + H_2O + CO_2 \qquad (2.13)$$

A final and key step is the recovery of ammonia:

$$Ca(OH)_2 + 2NH_4Cl \rightarrow CaCl_2 + 2H_2O + 2NH_3 \qquad (2.14)$$

2.5.3 Potassium carbonate

This salt is used as a pH buffer, fire suppressant, glass, welding and as a source of potassium intake of animal farms. It is prepared by electrolysis of KCl. On as-produced KOH, the carbonated salt is formed:

$$2KOH + CO_2 \rightarrow K_2CO_3 + H_2O \qquad (2.15)$$

Asia-Pacific region accounted for more than a half of global potassium hydroxide precursor demand, followed by North America and Europe.

2.5.4 Lithium carbonate

Actually, 35% of the lithium production goes towards battery production, other industries which used lithium compounds are ceramics, glass, greases, etc. Lithium consumption for batteries is being increasing due to the use of rechargeable Li batteries, electric tools and especially in e-cars and hybrid cars. One of the most important derivatives of lithium is lithium carbonate.

Besides several uses, including the treatment of bipolar disorder (lithium carbonate is included in the WHO list of essential medicines), this salt is a key component in the manufacture of lithium-ion batteries, since it serves as precursor of lithium cobalt oxide ($LiCoO_2$) which formed the cathode in a number of these batteries. The preparation of the oxide consisted in the reaction of the stoichiometric concentrations of lithium carbonate and cobalt (II,III) oxide (Co_3O_4), or even zero valent cobalt, at temperatures in the 600–800 °C range, and further annealing of the product at 900 °C under oxygen atmosphere.

Lithium carbonate is basically produced by three general routes: the treatment of brines, the treatment of certain ores/clays and the recycle of batteries or electrolytes.

In the case of brines, the top producers' countries are Australia, Chile and China. Lithium carbonate is produced by a general scheme of steps as it is shown in Figure 2.13. The recovery time is about 18–24 months, since solar evaporation is a slow process.

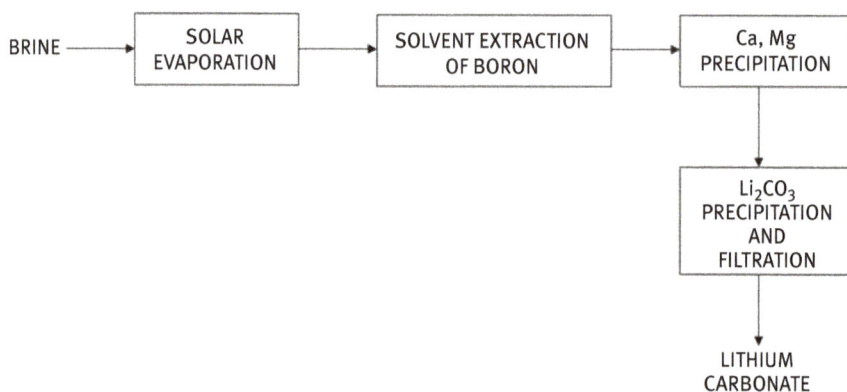

Figure 2.13: Production of lithium carbonate from brines.

Other possibilities about the lithium recovery from seawater brines are by membranes or solvent extraction technologies.

In the case of geothermal brines, CMI (USA), had developed the CIM Lithium Project, in which lithium is extracted from the brines, purified, concentrated and converted along the next steps (Figure 2.14).

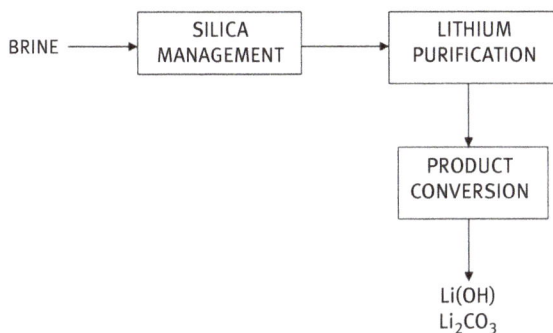

Figure 2.14: Basic steps for lithium processing from geothermal brines.

Various types of adsorbents/ion exchangers are proposed to purified Li from these geothermal brines:

i.   spinel type lithium manganese oxide (Li-Mn-O), which is a cation exchanger,
ii. spinel type lithium titanium oxide (Li-Ti-O), also a cation exchanger, and
iii. lithium aluminum layered double hydroxide chloride (LDH) with $LiCl·2Al(OH)_3$ stoichiometry, though stoichiometries as $[LiAl_2(OH)_6]^+ X^- ·nH_2O$ where $X = Cl^-$ or Br^- also applied, and adsorbent.

The production of battery-grade lithium compounds directly from brines was investigated in a laboratory-scale process by the use of ion exchange and chemical precipitation. Thus, brine, without impurities, can be used in an industrial membrane electrolysis process. Various sequences and conditions were investigated to purify lithium-concentrated brines, principal impurities found in the natural brines were Ca^{2+}, Mg^{2+} and sulfate, being all of these remove under the operational conditions used. In the chemical precipitation process, lithium-concentrated brine reacted with the precipitating agents at different steps. Then, the pulp obtained, and after sedimentation, it was filtered, allowing the removal or decreasing the presence of impurities in the lithium brine. Removal efficiencies for Ca^{2+}, Mg^{2+} and SO_4^{2-} were of 98.93%, 99.93%, and 97.14%, respectively. Further, ion-exchange technology decreased the concentration of Ca^{2+} and Mg^{2+} to values less than 0.001 g/L.

The treatment of brines and seawater can be also accomplished by precipitation, solvent extraction, membrane technologies, ion exchangers and adsorbents. Green solvents, as ionic liquids seemed to be, are suitable to recover Li from these natural compounds.

In the case of ores/clays, two mixed pyro-hydrometallurgical process had been proposed to recover Li from a Li-rich ore. One used roasting and leaching, whereas the other used chlorination and leaching steps, in both cases the end product is lithium carbonate. In the case of spodumene ($LiO_2·Al_2O_3·4SiO_2$), either pressure leaching or bioleaching had been proposed as a method to dissolve this lithium material.

Li-ion based batteries are another resource for lithium production. The recycling of these batteries can be also done by pyrometallurgical, hydrometallurgical (including bioleaching step) and hybrid pyro-hydrometallurgical processing.

No less than 13 pyrometallurgical processes had been proposed by different companies for the treatment of batteries, and particularly 3 of them are specifically developed to lithium batteries. The pyrometallurgical processing includes the mechanical and thermal steps, whereas the thermal steps can be approached calcination, roasting, reduction or chlorine processing.

It was described, a process for the recycling of Li-ion batteries and the production of lithium carbonate. The steps are: nitration, selective roasting, water leaching and Li_2CO_3 preparation. In the process, metal-bearing materials in Li-ion battery waste are transformed into their nitrates salts, roasted to transform them to their non-soluble oxides, and solubilization of $LiNO_3$ in water. The best conditions for the above steps are: nitration (70 °C, 5 h, acid-to-scrap ratio of 30 mmol/g), selective roasting (250 °C, 1 h) and 4-stage cross-current water leaching (25 °C, liquid-to-solid ratio of 2:1). Li solubilizes up to 93%, whereas extraction of Co, Ni and Cu is less than 0.1%. On the as-obtained Li solution (34.1 g/L Li), a carbonation step at 95 °C for 30 min, is carried out, resulting in the formation of the Li-carbonate salt, having a richness up to 99.95%, which is above the usual minimum requirements for battery grade Li_2CO_3.

In general terms, the hydrometallurgical recycling of Li-ion batteries comprises various steps including:

i.   pretreatment of the spent battery: dismantling the cells, thermal treatment and mechanochemical process. It can be pointed out that some of the battery components are flammable and toxic,
ii. leaching of metals from the cathode material, and
iii. separation of lithium and accompanying metals via a suitable process and depending of the metal concentrations in the solution obtained in (ii), i. e. solvent extraction, liquid membranes or impregnated exchange resins and adsorbents, all the above with or without the use of ionic liquids, chemical precipitation, electrochemically ion exchangers and adsorbents.

A process to prepare Li_2CO_3 from a Li-rich electrolyte generated in the aluminum industry was described elsewhere. The described process involved the next steps: H2SO4 leaching, elimination of impurities and precipitation of Li_2CO_3. Near 98% Li leaching efficiency was obtained in 1 h using 6% sulfuric acid at 80 °C, and with a liquid-to-solid 2:1 relationship. The leaching solution, containing lithium sulfate, was treated with NaOH, until pH 11, and further by 3 g/L EDTA (ethylenediaminetetraacetic acid). Lithium carbonate (99.5% purity) was yielded by precipitation with 290 g/L sodium carbonate at 95 °C and 50 min reaction time.

In one European patented process, which simplified flow-sheet is showed in Figure 2.15, lithium carbonate is yielded from the treatment of lithium-bearing solutions.

Li solution

↓

| CHEMICAL PURIFICATION | → Ca²⁺, MG²⁺ |

becomes:

CHEMICAL PURIFICATION → Ca^{2+}, MG^{2+}

↓

BIPOLAR ELECTRODYALISIS → HCl, H_2SO_4

↓ Li concentration

ELECTRODYALISIS

↓

CARBONATATION

↓

HOT WATER WASHING → Na^+, K^+

↓

PURE LITHIUM CARBONATE

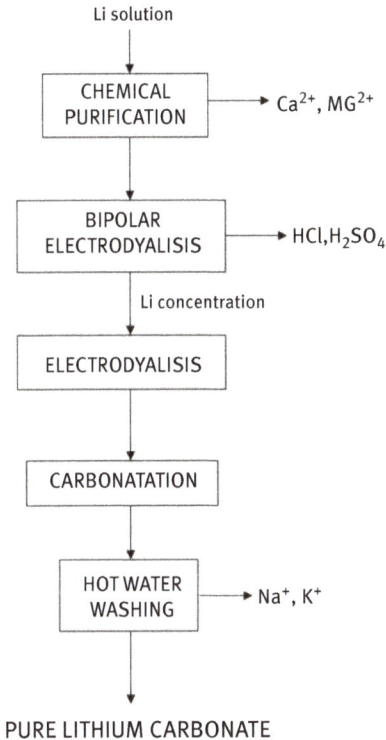

Figure 2.15: Process to yield lithium carbonate from lithium-bearing solutions.

Accordingly with this flow-sheet, the lithium-bearing solution is purified from accompanying divalent metals, and from the resulting solution, first lithium hydroxide and secondly lithium carbonate is yielded as end product. The purification of the lithium solution, from divalent cations, is carried out by bipolar electrodyalisis, which also converted lithium to lithium hydroxide. This product is converted to the carbonate salt by a further carbonation step.

References

[1] Brightling J. Ammonia and the fertiliser industry: The development of ammonia at Billingham. Johnson Matthey Technol Rev. 2018;62:32–47. DOI: 10.1595/205651318X696341.
[2] EFMA. Production of ammonia, BAT for pollution and control in the European fertilizer industry. Book No 1. 2000.
[3] Kitano M, Inoue Y, Sasase M, Kishida K, Kobayashi Y, Nishiyama K, et al. Self-organized Ruthenium-Barium Core-Shell Nanoparticles on a Mesoporous Calcium Amide Matrix for Efficient Low-Temperature Ammonia Synthesis. Angew Chem Int Ed. 2018;57:2648–52. DOI: 10.1002/anie.201712398.
[4] Tikovsky R, EVP Linde Engineering. Technion Conference. Haifa. November 2017. www.linde-engineering.com accessed: www.chemeng.technion.ac.il.
[5] ThyssenKrupp Industrial Solutions AG. Ammonia Technology. Downloaded July 2019.

[6]   www.casale.ch.
[7]   www.catalystgrp.com.
[8]   www.kbr.com.
[9]   www.mcgroup.co.uk.
[10] www.topsoe.com.
[11] Du F, Warsinger DM, Urmi TI, Thiel GP, Kumar A, Lienhard JH, V. Sodium hydroxide production from seawater desalination brine: process design and Energy efficiency. Environ Sci Technol. 2018;52:5949–58. DOI: 10.1021/acs.est.8b01195.
[12] Whitfield R, Brown F, Wood L. The economics benefits of sodium hydroxide chemistry in the production of organic chemicals in the United Stats and Canada. HIS Markit. Prepared by the American Chemistry Council. October 2017.
[13] www.eurochlor.org.
[14] www.ihsmarkit.com.
[15] EFMA. Production of ammonium nitrate and calcium ammonium nitrate. BAT for pollution and control in the European fertilizer industry. Book No.6. 2000.
[16] Grágeda M, González A, Grágeda M, Ushak S. Purification of brines by chemical precipitation and ion-exchange processes for obtaining battery-grade lithium compounds. Int J Energy Res. 2018;42:2386–99. DOI: 10.1002/er.4008.
[17] Harrison S, Simbol Final DOE GTO DE-EE0002790 Report. 2014. Accessed via P.Paranthaman. Lithium extraction from geothermal brine solution. OAK RIDGE Nat. Lab. USA. November. 2018. accessed: www.cmi-webinar.
[18] Li L, Deshmane WG, Paranthaman MP, Bhave R, Moyer BA, Harrison S. Lithium recovery from aqueous resources and batteries: A brief review. Johnson Matthey Technol Rev. 2018;62:161–76. DOI: 10.1595/205651317X696676.
[19] Minera Exar s.a., Perez WA, Ruiz HACB. Method for the production of battery grade lithium carbonate from natural and industrial brines. US Patent 20130101484. 25th March 2013.
[20] Paranthaman P, Lithium extraction from geothermal brine solution. OAK RIDGE Nat. Lab. USA. November. 2018. accessed: www.cmi-webinar.
[21] Park SK, Park KS, Lee SG, Jung WC, Kim KY, Lee HW. Method for producing lithium hydroxide and lithium carbonate. European Patent Application EP3326974A1. Date of publication: 30. 05.2018.Bulletin 2018/22.
[22] Swain B. Recovery and recycling of lithium: A review. Sep Purif Technol. 2017;172:388–403. DOI: 10.1016/j.seppur.2016.08.031.
[23] Wang W, Chen W, Liu H. Hydrometallurgical preparation of lithium carbonate from lithium-rich electrolyte. Hydrometallurgy. 2019a;185:88–92. DOI: 10.1016/j.hydromet.2019.02.013.
[24] Wang W, Chen W, Li Y, Wang K. Study on preparation of lithium carbonate from lithium-rich electrolyte. Miner, Met Mater Ser. 2019b;923–7. doi:10.1007/978-3-030-05864-7_112.

Magdalena Regel-Rosocka

3 Technology of simple hydrocarbon intermediates

Abstract: Plastics, thanks to their versatility, and high resource efficiency have become important materials in such branches as packaging, transportation, medicine, building and construction, etc. Although the role of recycling of various plastics in their manufacturing has been recently increasing (as a consequence of strong enhancement for circular economy, particularly in Europe), still production of monomers as substrates for plastics manufacturing is growing. It is predicted that, despite ongoing efforts to reduce, reuse, recycle and even ban plastic materials, improving living standards, population growth, and the lack of ready alternatives support the market for plastics for the next 20 years. The most important monomers produced on industrial scale belong to simple unsaturated hydrocarbons (alkanes/olefins), called also hydrocarbon intermediates because they are substrates for further processes of organic synthesis or polymerization. A great variety of current technological approaches to olefins production was shown, i.e. steam cracking (SC), methanol-to-olefins (MTO), dehydrogenation (PDH, Catofin, Oleflex, STAR, FCDh), methathesis (OCT). The continuous improvement and development of on-purpose processes is a response to dynamic changes on feedstock market of petrochemical raw materials, consumer needs, and environmental regulations. It was emphasized that boom in dehydrogenation processes (particularly, PDH) results from shift to ethylene production in steam crackers, a gap between supply of propylene, butylenes and higher olefins compared to the continuously growing demand for their derivatives.

Keywords: olefins, ethylene, propylene, butenes, steam cracking, propade dehydrogenation (PDH), Catofin, Oleflex, fluid catalytic cracking, methanol-to-olefins (MTO), steam active reforming, fluidized catalytic dehydrogenation, metathesis

3.1 Introduction

Plastics, thanks to their versatility, and high resource efficiency have become important materials in such branches as packaging, transportation, medicine, building and construction, etc. Production of only low- and high-density polyethylene (PE-LD, PE-HD) reached 14.5 mln tons, compared to 335 mln tons of global plastics production (excluding polyethylene terephthalate (PET) fibers, polyamide (PA) fibers, polypropylene (PP) fibers, polyacryls-fibers) in 2016 [1]. Although the role of recycling of

This article has previously been published in the journal Physical Sciences Reviews. Please cite as: Regel-Rosocka, M. Technology of simple hydrocarbon intermediates *Physical Sciences Reviews* [Online] 2020, 5. DOI: 10.1515/psr-2019-0033.

https://doi.org/10.1515/9783110656367-003

various plastics in their manufacturing has been recently increasing (as a consequence of strong enhancement for circular economy, particularly in Europe), still production of monomers as substrates for plastics manufacturing is growing. It is predicted that, despite ongoing efforts to reduce, reuse, recycle and even ban plastic materials, improving living standards, population growth, and the lack of ready alternatives support the market for plastics for the next 20 years [2]. The most important monomers produced on industrial scale belong to simple unsaturated hydrocarbons (alkanes/olefins), called also hydrocarbon intermediates because they are substrates for further processes of organic synthesis or polymerization. Alkanes are characterized as organic compounds with at least one double bond between carbon atoms.

Variety and a great number of their applications influence enormously huge volume of their production in the organic industry, with ethylene as the largest petrochemical by volume in 2018 (185 mln tons per annum) [2, 3]. Examples of the crucial syntheses involving ethylene (ethene) and propylene (propene) are shown in Figure 3.1.

Steam cracking (SC) is the basic process focused on production of such light olefins as ethylene (mainly), propylene, butylenes and butadiene (as by-products) (Figure 3.2). Production of low levels of propylene and higher olefins in steam cracking results from drop of prices of natural gas, mainly from shale gas resources, and shift to ethylene production from this light feed. Therefore, other processes to fabricate high levels of propylene and higher olefins are developed. Among them propane or butane dehydrogenation (PDH, Oleflex, Catofin, FCDh) are very important for propylene or butadiene (Catadiene) synthesis, fluid catalytic cracking (FCC) is an important propylene and butylene source (Figure 3.2) [6, 7] or methanol-to-olefins (MTO) process as a developing method for production of ethylene or propylene [8, 9].

The main feedstocks to produce olefins are methane (from natural or shale gas) and oil fractions. It seems that, in spite of declining reserves of crude oil and increasing environmental restrictions, petrochemical production of olefins still will be the leading one. Some alternative feedstocks (e. g. coal, biomass, waste streams) are considered of course, and industrial processes for them are developed. However, implementation of these raw materials is very limited due to lack of economic viability.

3.2 Steam cracking

SC, called also pyrolysis, is the leading technology for olefin production, mainly because of well-established technology, high capacities, whose economics can hardly be challenged [10]. High availability of relatively cheap natural gas, e. g. ethane, resulting from the recent boom in shale gas production in USA, has influenced type of the feedstock converted in SC. A shift from oil-based naphtha to shale-based ethane for steam cracking in the US has caused increase in ethane crackers and focus on ethylene production, while the naphtha crackers have been either dismantled or

a)

| | INTERMEDIATES | PRODUCTS |

Ethylene
$H_2C=CH_2$

H_2O

Cl_2
$CH_2ClCH_2Cl \xrightarrow{-HCl} CH_2=CHCl$

O_2
H_2C-CH_2 (ethylene oxide) $\xrightarrow{H_2O} CH_2(OH)CH_2OH$ ethylene glycol

$CH_2=CHCN$ acrylonitrile

C_6H_6
$C_6H_5CH_2CH_3 \xrightarrow{-H_2} C_6H_5CH=CH_2$
ethylbenzene → styrene

O_2, (PdCl$_2$/CuCl$_2$/HCl)
CH_3CHO acetate aldehyde

O_2, CH_3COOH (Pd/Al$_2O_3$)
$CH_3COOCH=CH_2$ vinyl acetate

HCl

PRODUCTS:
- Ethylene polymers (PE-LD, PE-HD, copolymers)
- Poly(vinyl chloride) PVC
- Polyestrs, synthetic detergents, solvents
- Polyacrylonitrile
- Ethanol
- Polystyrene, various synthetic rubbers
- Acetic acid and anhydride, alcohols
- Poly(vinyl acetate), copolymers
- Chloroethane

b)

| | INTERMEDIATES | PRODUCTS |

$H_2C=CH_3$ (CH$_3$) Propylene

O_2, NH$_3$
$CH_2=CHCN$ acrylonitrile

C_6H_6
$C_6H_5C(CH_3)_2H \xrightarrow{O_2} C_6H_5C(CH_3)_2OOH$
cumene → cumene hydroperoxide

$CO+H_2$ (oxo synthesis)
$CH_3CH_2CH_2CHO \xrightarrow{H_2}$
$(CH_3)_2CHCHO \xrightarrow{H_2}$

Cl_2, H_2O
$CH_3CH(OH)CH_2Cl \rightarrow H_3C-HC-CH_2$ propylene oxide $\rightarrow PG \rightarrow$

ROOH Halcon process
TBA or styrene

PRODUCTS:
- Propylene oligomers (C_6H_{12}, C_9H_{18}, $C_{12}H_{24}$)
- Polypropylene, copolymers
- Polyacrylonitrile
- Phenol+acetone
- n-Butanol
- Isobutanol
- Polyesters→PUR
- Polystyrene, synthetic rubbers

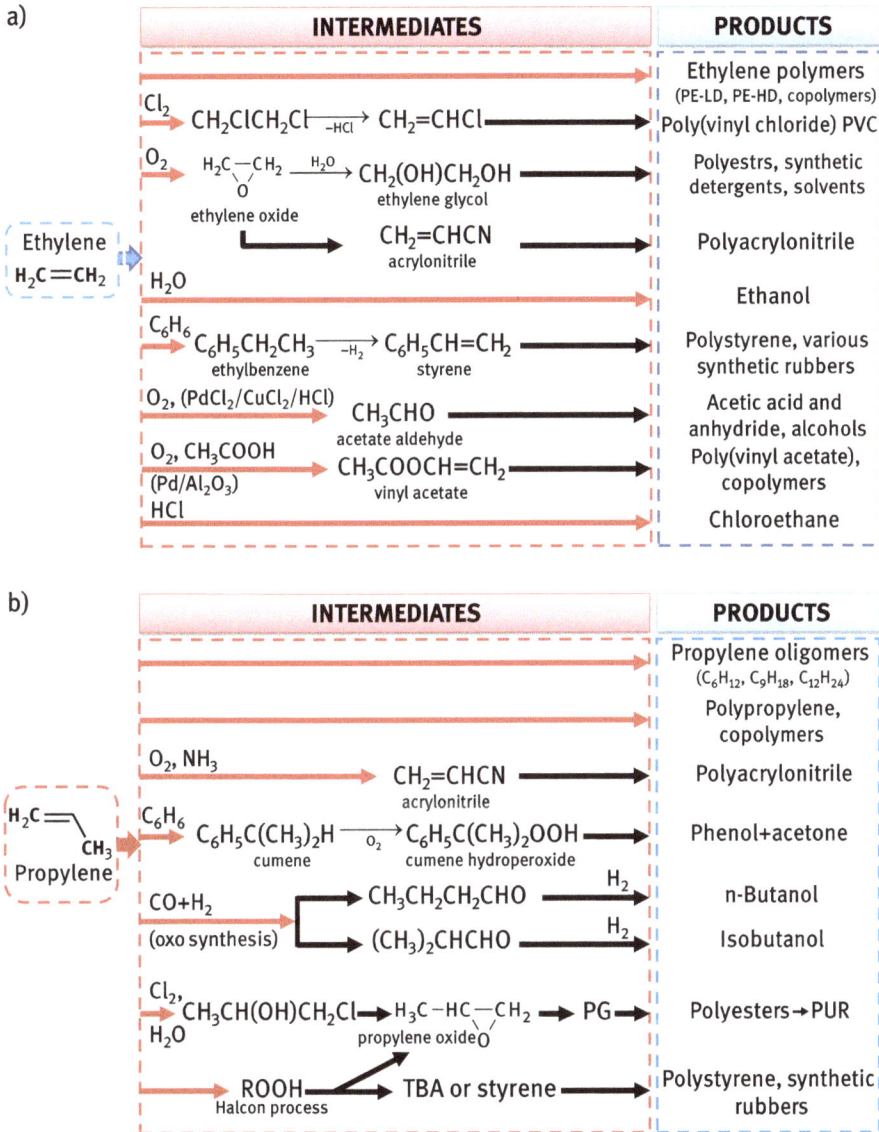

Figure 3.1: Examples of the crucial syntheses involving a) ethylene (ethene) and b) propylene (propene) (PUR – polyurethanes, TBA – t-butyl alcohol) [1, 4].

converted into another feed. As SC of ethane produces most ethylene in the market and a negligible amount of other olefins, the change has negatively affected supplies of propylene, and caused rise in price of this olefin. This market situation has accelerated development of other technologies for on-purpose propylene production, e. g. light paraffin dehydrogenation [11]. Of course, the choice of the feedstock

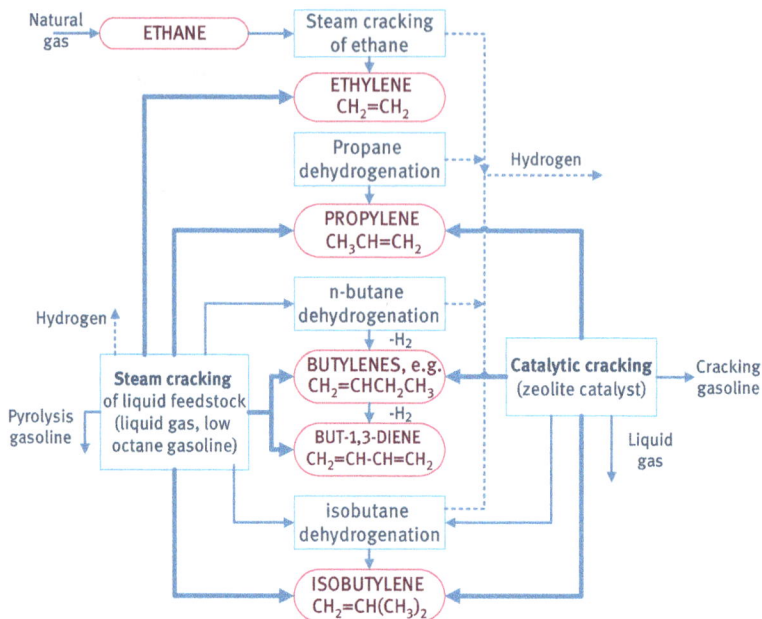

Figure 3.2: Classical industrial routes of light olefins production (according to [4, 5]).

for SC depends on the region where it operates, and the availability of the particular feed there, its current market price and the desired end products. Thus, in Europe still liquid oil fractions are the main feedstock for olefin production by SC.

The SC reactions are endothermic dehydrogenation of saturated hydrocarbons (alkanes) in high temperatures (750–900 °C – depending on the feedstock) at low pressure (0.17–0.25 MPa) [5, 12, 13] without catalyst, running according to the radical mechanism. The more ethane the feedstock contains, the higher cracking temperature is necessary to dehydrogenate hydrocarbons. Hence, light gaseous feeds such as ethane and propane will be cracked at temperatures near 850 °C or even 900 °C, while the heavier feedstock such as naphtha and gasoil – at the lower (near 750 °C). The crucial parameter, from the point of view of the process selectivity, is a very short residence time of the reaction mixture at high temperature in the tubular furnace. Too long residence of hydrocarbons at such high temperatures leads to side reactions proceeding to coke formation. In modern furnaces the residence times are between 0.08 and 0.25 s, and are controlled by the reactor flow rate and tube diameter. In order to reduce the residence time, the reactor tube diameters are small, i. e. 3–15 cm [5, 12].

To suppress coke formation inside the tubes, addition of superheated steam dilutes the feedstock, increases its temperature, and reduces partial pressure of hydrocarbons. The ratio of steam to the feed varies from 0.2 to 1.0 kg steam/kg of hydrocarbons.

The SC reactors are long (45–90 m long depending on the feedstock), and contain from 16 to 128 coils (tubes) per furnace. The tubular furnace is divided into two sections, radiant (higher temperature) and convective (lower temperature). There are different designs for the steam cracking furnaces offered by some engineering companies, e. g. Lummus Technology, Linde AG (Pyrocrack), KTI-Technip, Stone and Webster, and Kellogg [5, 12, 14]. The furnace construction is still updated to fulfill expectations of the petrochemical industry, i. e. improving flexibility and reduction in energy demand of the processes [14–20]. For example, Lummus Technology has developed a state-of-the-art SC furnace in response to the growing ethylene demand. Since mid-1960s (the first SRT heater), seven generations of SRT pyrolysis furnaces have been commercialized. Reduction in energy demand, i. e. decrease in fuel required to heat up the furnace, is achieved by preheating the air in burners using heat of the exhaust gases from the furnace. Lummus Technology has built more ethylene furnaces than any other ethylene process licensor, representing more than 40% of worldwide capacity.

The simplified scheme of steam cracking is shown in Figure 3.3.

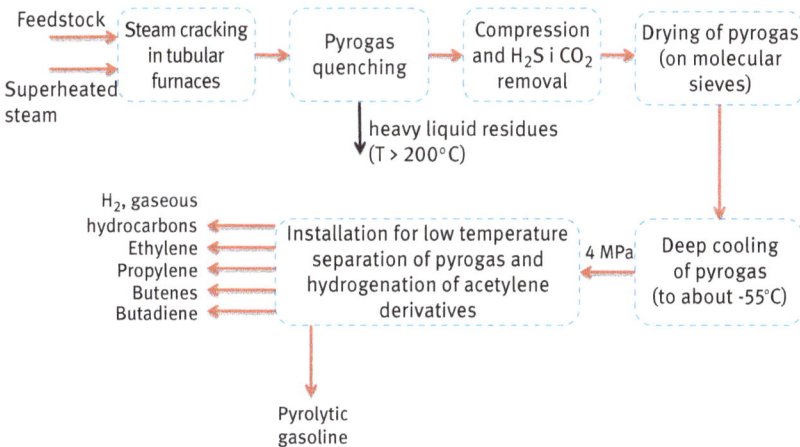

Figure 3.3: Block diagram of steam cracking process for olefins production.

The feedstock and superheated steam are introduced to the convective section of the tubular furnace, where they are gradually heated to 500–650 °C, and flow to the radiant section, where the temperature of the gas mixture rises rapidly to the desired cracking temperature around 750–900 °C. In the radiant section, the hydrocarbons are cracked to various compounds, such as olefins, aromatics, pyrolysis fuel oil (pyrolysis gasoline) and the heavier products, and just after leaving the tubular furnace the reacting gases must be rapidly and instantaneously cooled down (quenched) to freeze the undesired reactions (coke formation, olefins decomposition). The pyrolytic gas (pyrogas) is quenched rapidly by 150–200 °C in a special design heat exchanger, called transfer line exchanger (TLE or TLX) [5, 12]. The quenching is followed by further cooling with steam condensation and separation of heavy liquid residues (boiling temp. >200 °C) containing mainly BTX (benzene, toluene, xylenes) aromatics, if liquid feedstock was cracked.

Further, the pyrogas is compressed and purified from H_2S and CO_2 by adsorption, dried and cooled down to low temperature (about −55 °C), and directed to product fractionation. The fractionation is proceeded by low temperature rectification at high pressure, which leads to achieving the high-grade olefins. The low temperature pyrogas separation installations are much more complicated than typical refinery gas separation units. They are also costly because are equipped not only with low temperature rectification columns but also turbo compressors, special installations for cooling of ethane-ethylene and propane-propylene fractions, reactors for selective hydrogenation of acetylene (pyrolysis by-product). It shows that pyrogas separation is a complicated and multistep process, and it is as important as its generation in the tubular furnace to obtain polymer-grade (>99.8%) ethylene and propylene with required yield. Apart from separation of light olefins, also heavier fractions are obtained, e. g. C_4 fraction – mixture of saturated and unsaturated hydrocarbons having similar boiling temperatures, or pyrolytic gasoline mixture of aromatics – mainly benzene, toluene and xylenes (BTX). It should be mentioned that BTX fraction from the steam cracking of liquid feedstocks is one of the sources of aromatics for various industrial purposes. Thus, it is predicted that a shift to lighter feeds, such as shale gas, will strongly impact the availability of these aromatics [21]. C_4 fraction contains a mixture of saturated and unsaturated hydrocarbons having similar boiling temperatures, e. g. after SC of gasoline C_4 fraction contains 35–50% of butadiene, 20–30% of isobutene, 12–15% of 1-butane, 8–10% of 2-butane, 3–5% of n-butane, 1–2% of isobutene [5]. To separate these products, extractive distillation with such organic solvents as dimethylformamid, acetonitrile, N-methylpirolidone is applied, while isobutene is removed by reaction with 40–60% H_2SO_4.

With the passing of tubular furnace operation with hydrocarbons in very high temperatures, the coils (tubes) are coked by products of undesired condensation reactions. Thus, regular decoking is required to remove organic coke from the furnace coils, typically every 3–4 months, depending on the type of feed and the reaction severity [13]. Decoking is typically done by coke combustion in a steam/air stream at the isolated furnace. The decoking effluent is directed to decoking drums, where the coke fines are separated from the effluent gases. As the decoking is an undesirable operation due to loss of the main products (ethylene and propylene), high operating and maintenance costs, and shortened life of the coil (constant thermal changes of the coil), the coke formation should be minimalized by maintenance of optimal conditions during SC.

Changes in supplies and prices on petrochemical feedstock market enhance the producers to modify and adjust the installations to operate flexibly. On the one hand, a trend to deepen the petrochemical processing by SC of hydrocracked vacuum distillates is observed, on the other hand, steam crackers are adjusted to light feedstock pyrolysis. Feedstock change negatively affects economy of ethylene production in an unmodified unit. Thus, four groups of feeds are reported to be processed efficiently, each group in a by design unit, i. e. liquid gas (ethane-propane), propane-gasoline, gasoline-kerosene, kerosene-hydrogenated vacuum distillates.

Some new projects for ethane cracking can be given as examples of the industrial flexibility. In 2019, Qatar Petroleum has announced development of a new Petrochemicals Complex in Ras Laffan Industrial City, where an ethane cracker with a nameplate capacity of 1.9 million tons of ethylene per year will be constructed, making it the largest ethane cracker in the Middle East and one of the largest in the world [22].

Recently, there is an increasing interest in direct processing "crude oil-to-chemicals", which bypasses the traditional naphtha refining. Bypassing refinery saves 5–10 US$ per barrel processing costs and 5–14 days processing time. As a consequence of this trend, the first world-scale facility with steam cracker fed by crude oil, instead of oil fractions, has been commissioned by ExxonMobil and Saudi Aramco [22–24]. To feed crude oil to the cracking furnace, the unit needed to be modified to include a flash pot between the convective and radiant sections of the furnace. The crude oil is pre-heated in convective section of the furnace and then flashed (partially vaporized in a flash pot outside the furnace), essentially "topping" the lighter components from the crude. This extracted vapor (76%) is then fed back into the furnace's radiant coils and cracked in the usual way. The heavier liquid (24%) that collects at the bottom of the flash pot is either transferred to the adjacent refinery, or sold into the merchant market. The cash-cost of this Aramco crude-to-olefins process is estimated to be 200 US$-per-ton cheaper than for a naphtha cracker. It shows that continuous development in this well-established process is still possible and necessary to adapt to changing market situation.

3.3 Methanol-to-olefins (MTO) process

The gap between supply of C_3, C_4 and higher olefins compared to the growing demand of their derivatives created an interest in the, so called, on-purpose olefins production technologies [25].

One of the countries looking for feedstock independence and developing alternative to crude oil-based processes, is China, where coal is a raw material for production of olefins (nearly 40% of the country olefin capacity in 2018 [10]), with methanol as an intermediate (coal → synthesis gas → methanol → olefins). The process is called MTO (methanol-to-olefins), and a great interest in coal-to-olefins (CTO) technologies has been reported in China recently. Though, MTO process has been known and developed since 1970s of the twentieth century (by Exxon Mobil or

Dalian Institute of Chemical Physics), a first industrial MTO unit commissioned by UOP (Universal Oil Products) Honeywell Company started just in 2013 in Nanjing, China [7]. Up to now, eight advanced MTO units of annual capacity of nearly 3.2 mln t of ethylene and propylene have been developed as a result of abundant availability of coal as the low-cost advantageous feedstock for petrochemical production in this country. The share of coal-based petrochemical production in total production has increased from 3% in 2010 to approximately 16% in 2018. Methanol is considered as Chinese resource not only for production of olefins but also of synthetic gasoline.

However, active approach of China to reduce carbon dioxide emission and turn into clean technologies, makes the competitiveness of CTO projects declining. Additionally, rising prices of methanol negatively influenced profitability of MTO plants, causing temporary shut downs of some installations or suspension of realization of the planned projects, e. g. in Nigeria or Norway. Presently operating MTO plants worldwide are presented in Table 3.1. Changes in the number of ethylene plants based on methanol (MTO) in China are shown in Figure 3.4.

Table 3.1: Worldwide MTO plants in twenty-first century [26–33].

Company	Location	MTO capacity (kt/yr)	Status
Yangmei Hengtong	Shandong, China	300	Operating
Nanjing Wison	Jiangsu, China	295	Operating
Zhongyuan Ethylene	Henan, China	200	Shut down in 2018
Ningbo Fund	Zhejiang, China	600	Shut down in 2018
Zhejiang Xingxing	Zhejiang, China	600	Shut down in 2018
Jilin Connell Chemical Industry	Jilin, China	300	Planned to start in 2019
Statoil	Tjeldbergodden, Norway	no data	Pilot plant in 2001, not operating
Viva Polymers Ltd.	Lagos, Nigeria	1200	Planned to start in 2012, not realized

The feedstock for methanol production depends on the region and resources available, e. g. it is produced from natural/shale gas by steam reforming, from residues of crude oil or from coal and biomass by gasification in the oxygen atmosphere. However, it seems that syngas and methanol production from coal and biomass is too expensive because it yields in hydrogen-deficient gas and emits huge volumes of CO_2 to the atmosphere [10]. More information on methanol production is given in the chapter *Technology of large volume alcohols, carboxylic acids and esters*.

MTO process is a two-step catalytic conversion of methanol by DME (dimethyl ether), as intermediate – reaction 1A, to olefins (reaction 1B) according to the following reactions [35–37]:

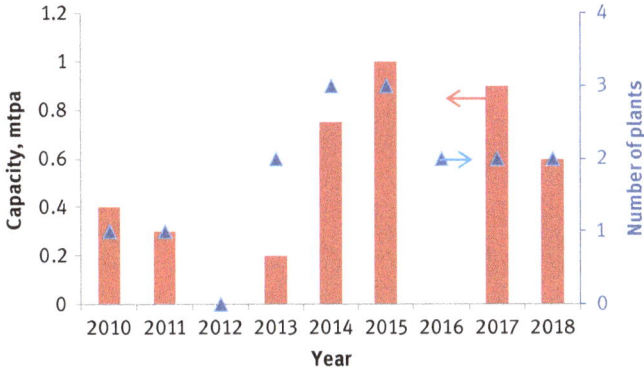

Figure 3.4: Ethylene plants based on MTO process in China in 2010–2018 (according to [33, 34]; data on capacity of some plants is not available).

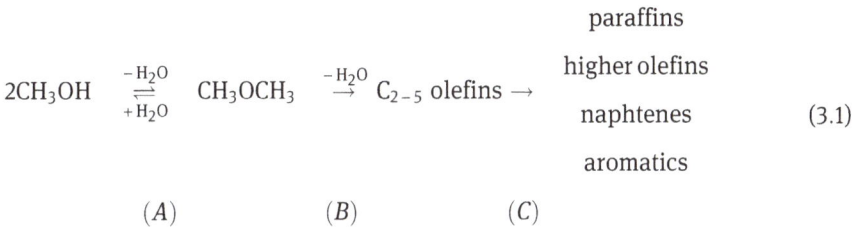

$$2CH_3OH \underset{+H_2O}{\overset{-H_2O}{\rightleftharpoons}} CH_3OCH_3 \overset{-H_2O}{\rightarrow} C_{2-5} \text{ olefins} \rightarrow \begin{array}{l} \text{paraffins} \\ \text{higher olefins} \\ \text{naphtenes} \\ \text{aromatics} \end{array} \qquad (3.1)$$

$$(A) \qquad\qquad (B) \qquad\qquad (C)$$

Main products of the MTO reaction are ethylene, propylene, butylenes. However, depending on the conditions and catalyst, also higher olefins, paraffins and naphtenes can be formed, and after dehydrogenation of naphtenes – aromatics can be generated (reaction 1C). Also water as a co-product, and such by-products of the reactions as coke (remaining on a catalyst), small amounts of H_2 and CO_x are present in the MTO product streams.

The ratio between ethylene and propylene production in the range from 1 to 1.8 depends on the catalyst and reaction parameters. It is noteworthy to emphasize that MTO leads to almost 100% methanol conversion to olefins with minimal by-products, which is important, especially when compared with various undesired products of Fischer-Tropsch synthesis of gasoline.

General flowsheet of MTO process is shown in Figure 3.5.

The dehydration of methanol is carried out at 330–500 °C and near atmospheric pressure (0.11–0.15 MPa). As conversion of methanol to olefins is highly exothermal (the heat of reaction is −196 kcal/kg methanol feed) the heat transfer must be efficient to keep the reaction temperature in the designed range. If not, coke formation increases and deactivates the catalyst [39, 42, 43].

Unless MTO reactions were enhanced by catalysts, they would be very slow and the process would be unprofitable. The most important part of the MTO unit is a fluidized bed reactor to which methanol is fed in a gaseous phase. Thanks to

Reaction section ⋮ **Recovery section**

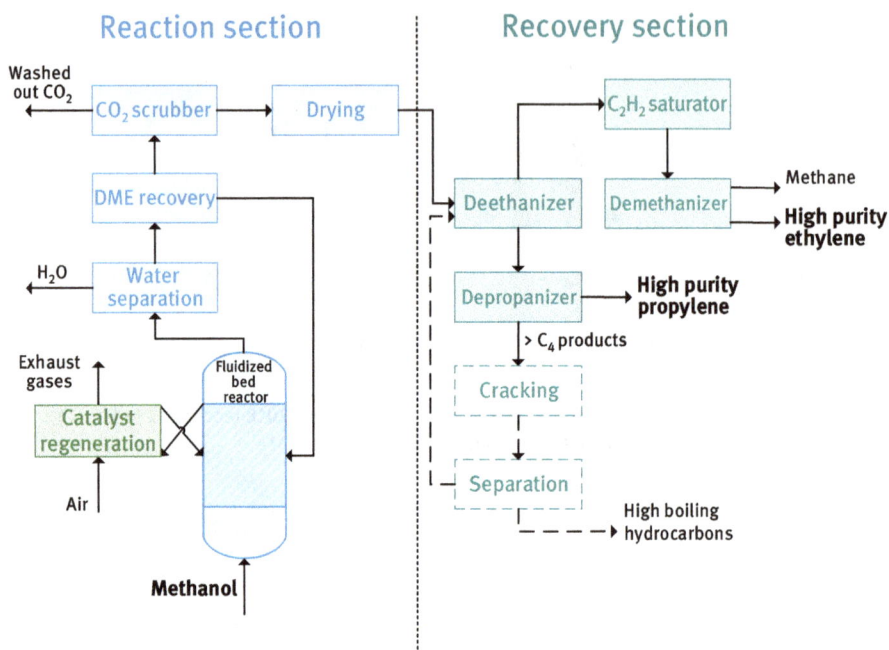

Figure 3.5: Scheme of MTO technology (according to [9, 35–41]), dashed sections of cracking and separation are optional.

fluidized bed, heat generated in the exothermic reaction is efficiently removed, and regeneration of the catalyst can be continuously carried out. Also, other solutions are proposed to control temperature of this exothermic process, i. e. use of additional tubular reactor with co-feeding of light olefins to improve reactor stability [44] or application of such operating conditions that restrict feed conversion to 5–25% [45].

Coke, a disadvantageous by-product, which is formed on a catalyst, must be burnt out in a regenerator, to regenerate and activate the catalyst [46]. Water steam generated during the catalyst regeneration is used for temperature regulation in the MTO unit. The MTO products leaving the reactor are cooled with air or water. The olefin-rich gas is purged with alkali, and then dried, transported to separation section which includes demethanizer, deethanizer, depropanizer, separators (splitters) of C_2 and C_3 fractions. Optionally, additional cracking section (marked with dashed line in Figure 3.5) for higher hydrocarbons (>C_4) can be applied to increase ethylene and propylene production [41, 47]. The integrated MTO and olefin cracking process, achieving high yield of light olefins, low yield of hydrocarbon by-products and high flexibility of the technology, was developed by Total Petrochemicals and UOP [10, 48].

The advantages of this technology of olefin production, indicating it as a technology for the future, are the following:

- significant increase in yield and feedstock efficiency,
- low operation cost,
- nearly double return on investment,
- half of the cash cost of production compared to traditional cracking methods.

However, the environmental concerns about CO_2 emission from coal treatment may slow down MTO development, particularly in China.

A special modification of MTO is conversion of methanol-to-propylene (MTP developed by Lurgi) [9, 36, 49]. In relation to MTO process, in MTP additional purification of crude methanol before dehydration is necessary, and therefore an additional distillation column is added to the installation. Also, another type of reactor is used, i.e. fixed bed reactor. Five or six such reactors are used to ensure efficient heat exchange and stable process conditions. Such a configuration of fixed bed reactors enables regeneration of the catalyst because one of the reactors (so called "swing reactor") can replace any of them for the period of coke removal by burning it out [9].

The heart of the MTO/MTP process is a catalyst. Selection of an appropriate catalyst influences composition of the products, which can range from light olefins (MTO) to heavy hydrocarbons – gasoline fraction (MTG, methanol-to-gasoline). Zeolite catalyst (molecular sieves), composed of oxides of phosphorus, silica and aluminum (silicoaluminophosphate), of defined pore size (MTO-100 by UOP, SAPO 34) and moderate acidity, limits size of the formed hydrocarbon molecules to C_4. This feature of the catalyst is called shape selectivity, and means that pore size of zeolites limits size of product molecules formed inside or restricts substrates that have access to the pores of the catalyst, which enhances selectivity of MTO process for light olefins [36, 50, 51]. The SAPO-34 catalysts are characterized by CHA (chabazite) framework type of a double six-ring structure in an ABC stacking arrangement. The pore windows have a diameter of about 4.0 Å, and cylindrical cages within the structure of approximately 10×6.7 Å [36, 52]. The selectivity for ethylene and propylene over SAPO-34 catalyst is reported to amount to 90%, and simultaneously formation of butylene and higher olefins, also branched hydrocarbons is largely restricted. Some modifications of the catalyst can improve ethylene selectivity by introducing such additional metals into the zeolite structure as silicon, magnesium, zinc, iron, cobalt, manganese and their mixtures, which changes acidity of the catalyst [53].

Generally, the SAPO molecular sieves are synthesized by hydrothermal crystallization from a reaction mixture containing silica, aluminum, phosphorus (or admixtures of other introduced metals) and a templating agent (e.g. amines and quaternary ammonium compounds, a preferred templating agent is tetraethylammonium hydroxide). After producing crystals of the catalyst (reaction preferably in 100–200 °C for 24–48 h), the product is recovered by centrifugation or filtration, and then it is heated at 200–700 °C to remove the templating agent from the pores, and activate the catalyst [53].

Due to smaller pores of the SAPO-34 molecular sieves, which catch aromatics formed inside, compared to ZSM-5 (with bigger pores), a catalyst for methanol conversion into gasoline, coke formation on SAPO-34 is higher, and aromatics removal is more difficult than from ZSM-5 [42]. Therefore, many researches have been carried out to improve performance of SAPO-34 catalyst by diminishing its deactivation due to coke formation. It was reported that the selectivity to ethylene increases with increasing coke content, while selectivity of higher olefins decreases in the following order: $C_6 > C_5 > C_4, C_3$ [54]. The key parameters affecting catalyst deactivation by coke formation are SAPO-34 crystal size (small size is better to achieve low coke selectivity), and operating temperature [54]. Therefore, some efforts were put to reduce crystal sizes (nanosize) or introduce secondary larger pores (>2 nm) into the zeolite crystals to form hierarchical structures [42]. Changes in the structure the microporous SAPO-34 zeolite crystals by mesopores or macropores enhance the mass transport efficiency of the reactants and products, and extend lifetime of the catalyst by slowing rate of coke formation. Moreover, the hierarchical structure of SAPO-34 decreases acidity of the catalyst, which is positive for its catalytic performance. However, up to now, industrial methods for preparation of zeolites with ordered meso/macroporous hierarchical structures with narrow pore size distributions are not well developed. Typical SAPO-34 catalysts are prepared by aminothermal [55–57], hydrothermal [58], ionothermal [59] or phase-transformation [60] synthetic methods.

Jasper and El-Halwagi [9] estimated environmental impact of MTO and MTP taking into account as a feedstock methanol or natural gas (with intermediate process of methanol manufacturing). They concluded that the MTP process emitted about 12% more than the MTO process when the total annual emission (8 and 9.2 t/t for MTO and MTP, respectively) was concerned. However, if the normalized emissions per ton of product (13.4 and 12.7 t/t for MTO and MTP, respectively) or ton of propylene (20.9 and 16.1 t/t for MTO and MTP, respectively) were considered, the MTP process produced less emissions both per ton of total product and per ton of propylene than MTO [9, 32]. Other data for olefins production from coal (CTO, MTO) confirmed high CO_2 emission, which was estimated for 6–10 t per ton of olefin, and was higher than CO_2 generated from SC (1 t/t) [10]. Such high emission of greenhouse gases from MTO units, particularly coal-based, may limit their development in the light of efforts toward reducing CO_2 levels.

3.4 Dehydrogenation of alkanes

Recently, as a result of increasing shale gas share in the feedstock market, natural gas prices, especially in USA, have dropped with increasing supply, and olefin cracker operators have shifted to the lighter feed. As a consequence, they have focused on ethylene production, which expands a gap between supply of propylene, butenes and higher olefins compared to the continuously growing demand for their derivatives.

Moreover, access to the cheap gas is limited, which is reflected by the changes in its prices, i. e. in the years 2005–2012 natural gas prices rose 35% in the EU and dropped 66% in USA, and the cost of petrochemical production was down 50% in USA and up 20% in Europe compared to the prices five years earlier [6, 10, 21].

SC, as it was mentioned in the previous section, and fluid catalytic cracking (FCC) of naphtha, light diesel, and other oil by-products are considered to be the most common production methods of light olefins. However, a great increase in share of other production methods in this market has been observed in the last decade (Figure 3.6). In 2007 less than 3% of propylene was produced by on-purpose technologies, while FCC and SC were responsible for the bulk of the propylene produced [37]. Honeywell UOP predicts that on-purpose propylene technologies will supply 30% of the propylene by the year 2025, which shows the current boom in propylene production [62].

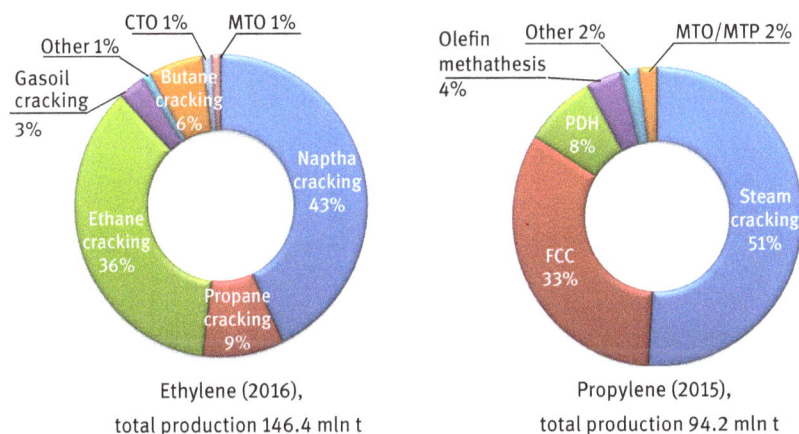

Ethylene (2016),
total production 146.4 mln t

Propylene (2015),
total production 94.2 mln t

Figure 3.6: Industrial sources of ethylene and propylene [23, 61].

Dehydrogenation of alkanes to olefins can generally proceed with either oxidative or non-oxidative reactions. Oxidative dehydrogenation is highly exothermic and it is difficult to control the process, thus, desired product selectivity and quality are low. Non-oxidative processes (i. e. direct dehydrogenation (SC) or catalytic dehydrogenation (FCC, MTG, Catofin, Oleflex, STAR, FCDh), on the other hand, are endothermic and need a continuous heat supply to initiate the endothermic reaction. Catalytic dehydrogenation of alkanes to alkenes provides highly selective method for production of specified olefins (mainly propylene). As a consequence of the demand (in 2014, 5 mln t of propylene was produced by dehydrogenation), more than 10 new plants has been announced to be constructed (Table 3.2). The economic viability of light hydrocarbon dehydrogenation is dependent on the available feedstock and the price of olefins produced [89]. Dehydrogenation units are often integrated with installations for MTBE, TBA or other C_4 products.

Table 3.2: Examples of worldwide operating or constructed dehydrogenation plants [10, 11, 63–88].

Company	Location	Capacity (kt/yr)	Status
PetroLogistics	Houston Ship Channel, USA	550	PDH, Catofin, operating from 2010
SK Gas Corporation and Advanced Petrochemical Company	Ulsan, South Korea	600	PDH, Catofin, operating
Advanced Petrochemical Company (APC)	Al-Jubail, Saudi Arabia	455	PDH, Catofin, operating
Saudi Polyolefin Co. (SPC)	Al-Jubail, Saudi Arabia	455	PDH, Catofin, operating
Hengli Petrochemical Refinery	Dalian, China	500 800	propene, Catofin, operating isobutene, Catofin, operating
Ningbo Haiyue New Material Co	Ningbo, China	600	PDH, Catofin, maintenance shutdown
Dongguan Juzhengyuan Technology	Guangdong, China	600	PDH, Catofin, under construction
Jinneng Science & Technology	Qingdao, China	900	PDH, Catofin, under construction
Dow Chemical	Freeport, USA	750	PDH, Oleflex, operating
National Petrochemical Industrial Company (NATPET)	Yanbu, Saudi Arabia	420	PDH, Oleflex, operating
Al-Waha Petrochemical Company	Al-Jubail, Saudi Arabia	450	PDH, Oleflex, operating
Tecnimont	Tobolsk, Russia	510	PDH, Oleflex, operating
Zhejiang Satellite Petrochemical	Jiaxing, China	450	PDH, Oleflex, operating
Fujian Meide Petrochemical	Fuzhou, China	660	PDH, Oleflex, operating
Shandong Luqing Petrochemical	Binzhou, China	116 104	propene, Oleflex, operating isobutene, Oleflex, operating
HMC Polymers	Rayong, Thailand	300	PDH, Oleflex, operating
BASF and Sonatrach	Tarragona, Spain	350 now, 420 from 2021	PDH, Oleflex, operating
Grupa Azoty Polimery Police	Police, Poland	429	PDH, Oleflex, under construction
Williams Energy Canada	Fort Saskatchewan, Canada	525	PDH, Oleflex, under construction
Borealis Group	Kalo, Belgium	750	PDH, Oleflex, under construction

Table 3.2 (continued)

Company	Location	Capacity (kt/yr)	Status
Sidi Kerir Petrochemicals Company (SIDPEC)	Amerya, Egypt	500	PDH, Oleflex, under construction
Coastal Chemical Inc.	Cheyenne, USA	100	isobutene, STAR process, operating
Polybutenos	Ensenada, Argentina	40	isobutene, STAR process, operating
Egyptian Propylene & Polypropylene Company (EPPC)	Port Said, Egypt	350	propene, STAR process, operating
MEPEC	Iran	450	propene, STAR process, under construction
Salman-e-Farsi Petrochemical Co. (SFPC)	Iran	425	propene, STAR process, under construction
Formosa Plastics	Texas, USA	545	propene, STAR process, under construction
PetroLogistics	Gulf Coast, USA	500	PDH, Dow FCDh, under construction

Two commercial units based on the STAR process technology (ThyssenKrupp Industrial Solutions, formerly Uhde) have been commissioned for the dehydrogenation of isobutane integrated with the production of MTBE [78, 88]. Catalyst systems, such as chromia catalyst in Lummus Catofin and platinum–tin catalysts in UOP Oleflex and STAR are used in commercially available PDH processes [90].

Although, Dow Chemical elaborated Fluidized Catalytic Dehydrogenation (FCDh) process for on-purpose propylene production, PDH in Dow's Chemical plant in Freeport is licensed by UOP Honeywell providing Oleflex technology (Table 3.2) [81]. However, recently PetroLogistics has announced its plan to construct the first PDH plant licensed by Dow Chemicals FCDh technology [82].

As bonds in alkanes are very strong, hydrogen removing from such a molecule by breaking of two carbon-hydrogen bonds with the simultaneous formation of a hydrogen and carbon-carbon double bond (reactions 2 and 3, propane and isobutane dehydrogenation, respectively) needs providing high energy.

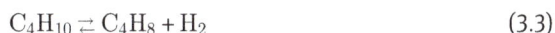

$$C_3H_8 \rightleftharpoons C_3H_6 + H_2 \qquad (3.2)$$

$$C_4H_{10} \rightleftharpoons C_4H_8 + H_2 \qquad (3.3)$$

The enthalpy required to dehydrogenate alkanes decreases as the chain becomes longer, i. e. $\Delta H^{0}_{298} = 137$ kJ/mol for ethane, 124.3 kJ/mol for propane, and 117.6 kJ/mol for isobutane dehydrogenation [11]. The process of olefin formation is highly endothermic running to higher volume of products than substrates, thus, according to Le Chatelier's principle, high temperatures and low paraffin partial pressure should favor it. Additionally, to decrease the activation energy of dehydrogenation reactions, metal-based catalysts are applied in industrial units. The main types of such catalysts are as follows [6, 11, 62, 89, 91]:

– platinum-based supported catalyst with promoters (and group VIII metals),
– chromium oxide-based supported catalyst with promoters,
– gallium supported oxide or included in zeolites,
– vanadium oxide-based catalysts,
– molybdenum oxide-based catalysts,
– carbon-based catalysts.

Only the first two types, i. e. platinum- and chromium oxide-based catalysts on alumina support are applied in the industry.

Industrial implementation of the dehydrogenation of paraffins is complicated by equilibrium constraints, side reactions, and coke formation. Therefore, one of the main technological challenges is design of commercial reactors and processes focusing on efficient heat delivery to the reactor [11, 89, 92]. On the one hand, high temperature is necessary to dehydrogenate alkanes, on the other hand, high temperature promotes side reactions and coke formation, which deactivates a catalyst. An additional drawback is that olefins are considerably more reactive than their paraffinic counterparts, which can further lead to unwanted side and secondary reactions. Three types of side reactions can occur, i. e. hydrogenolysis, cracking, and isomerization. In hydrogenolysis, the addition of hydrogen into a C – C bond within a paraffin results in the formation of two smaller alkanes. Thus, the dehydrogenation catalyst is suppressed with time-on-stream, making it necessary to regenerate the catalyst frequently to preserve sufficient activity. This is particularly important when the process operates at high alkane conversions, due to the fact that polymerized olefins are coke precursors. Hence, two strategies have been developed in industry to lower paraffin partial pressure to suppress the undesired side reactions. These are reducing the pressure below atmospheric (Catofin, Oleflex process) or diluting the paraffin feed with steam (STAR process) [11].

Generally, commercialized dehydrogenation is realized in two worldwide spread routes, i. e. Catofin (Lummus Technology) and Oleflex (Honeywell UOP). Also, STAR process licensed by ThyssenKrupp Industrial Solutions (formerly Uhde) is operating in 3 locations, and FCDh by Dow Chemicals is announced to be constructed for the first time in the world in USA. In these plants isobutylene, n-butenes, and, mainly, propylene are generated from isobutane, n-butane, or propane, respectively. The feedstock containing C_3–C_5 paraffins may come from various resources, e. g. natural gas, liquid petroleum gas from crude oil fractionation or MTO by-products.

According to McDermott/Lummus Technology (supplier of Catofin technology) there are worldwide [91]:

- 9 Catofin plants producing 5 mln t of propylene (e. g. Petrologistics, Texas, USA; Advanced Petrochemical Company, Jubail Industrial City – the largest petrochemical complex in the world, Saudi Arabia),
- 5 Catofin plants producing 3 mln t of isobutene (substrate for MTBE production),
- 2 plants for integrated production of propylene and isobutene.

Since the first commercialization of Oleflex technology in 1990, Honeywell UOP has commissioned 29 Oleflex units for on-purpose propylene and isobutylene production. The global production capacity of propylene from Oleflex technology now stands at nearly 7 mln tons per year from 19 operating units [83].

Butadiene is produced in the only commercially available dehydrogenation technology called Catadiene (Lummus Technology). Up to now 17 Catadiene plants have been commissioned throughout the world, however, only two of them still operate in Russia producing about 270 000 t per year of butadiene [93, 94]. The last Catadiene plant was built in the USSR in 1985 [95].

3.4.1 Catofin and Catadiene (Houdry) processes

Catofin process origins from one of the earliest petrochemical technologies, developed by Eugene Houdry in 1942 to remove hydrogen from paraffins with high conversion and selectivity, and it was called from the name of the inventor the Houdry process. Later he modified the process to dehydrogenate n-butane into butadiene, which is now known as the Catadiene process. Technologies for production of isobutylene by dehydrogenation of isobutane, and propylene by dehydrogenation of propane are known as Catofin. Now the exclusive licensing partner of this process is Lummus Technology, and it has been developing catalysts for dehydrogenation for the last 20 years, as well [6]. A simplified scheme of Catofin process for PDH is shown in Figure 3.7.

The feedstock, after preheating in the heat exchanger (1), is vaporized and raised to reaction temperature, i. e. 550–650 °C, in the charge heater (2), and then enters the adiabatic fixed bed reactor (3) (thickness of a catalyst layer is 1–3 m). Depending on the feedstock, the reactor works under low pressure ranging from 16 to 100 kPa, which is advantageous for good selectivity of dehydrogenation reaction. As high temperature as 650 °C gives high conversion into olefins, e. g. 70% for PDH. A characteristic technological solution for Catofin process is an internal heating of the reaction mixture inside the reactor. Vaporized hydrocarbon feed is

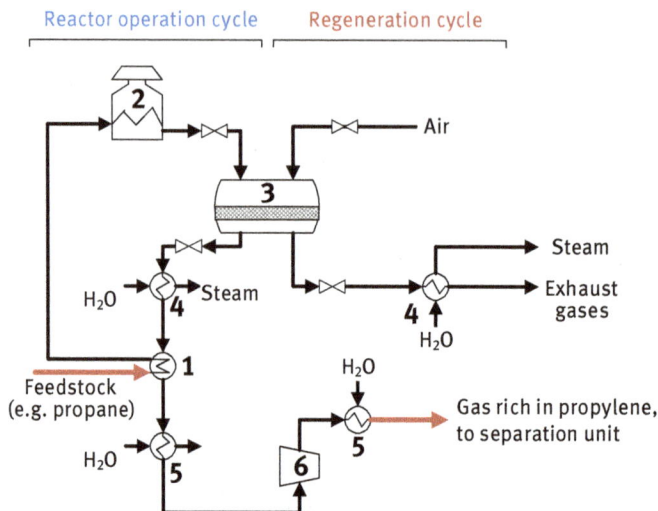

Figure 3.7: Simplified scheme of Catofin process: 1 – heat exchanger, 2 – charge heater, 3 – reactor (representing a series of parallel adiabatic reactors, e. g. five to eight), 4 – boiler, 5 – coolers, 6 – compressor (based on [4, 11, 91, 96, 97]).

contacted with the catalyst (usually of spherical or a horizontal cylinder shape [97]) having high thermal capacity, and uses the catalyst heat for endothermic reactions. As reactor is not externally heated, therefore the catalyst must be periodically warmed up. The Catofin installation consists of a series of 5 to 8 parallel reactors to ensure continuous work of the whole process. Each reactor works in two cycles (shown in Figure 3.8), i. e. reaction (dehydrogenation) for short cycles (7–15 min each) [6, 98, 99] and regeneration cycle for 7–15 min, too. This technology provides more than 45 and 50% conversion of propane and isobutane, respectively, and selectivity exceeding 88% to propylene and 89% to isobutene [91].

The regeneration cycle aims not only in heating up the catalyst but in combustion of coke, which deposits on the catalyst and deactivates it. Hot air (700 °C) flow

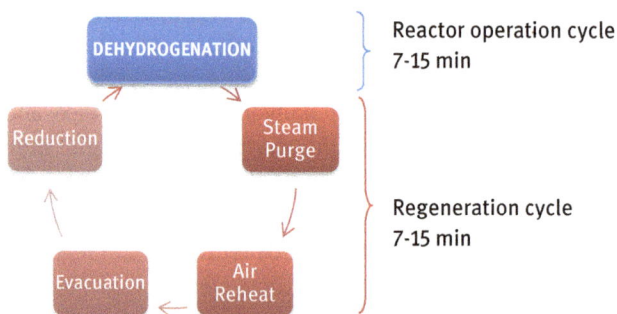

Figure 3.8: Cyclic operation of dehydrogenation Catofin process (according to [6, 89]).

and coke burning are the main sources of heat which is stored by the catalyst bed during this stage. After reheating and regenerating the catalyst, the oxidized catalyst must be reduced by passing a reduction gas (hydrogen or hydrocarbons) through the catalyst bed. The oxidation of the reducing gas is also an additional source of the heat. To prevent an explosion of the unit, it is important to protect vaporized hydrocarbons from the contact with hot air by providing automatic control of the cycles and complete tightness of the installation.

In the installation analogous to Catofin, Catadiene process can be carried out with high n-butane conversion (>38%) and selectivity to n-butylenes (>80%) and butadiene (>65%) [94].

The catalyst working in such demanding conditions (high temperature, frequent changes of temperature) must be robust, and additionally should feature high activity, high selectivity and long life. Catofin process uses, as the dehydrogenation catalyst, Cr_2O_3 on Al_2O_3 support, and it is in use for 2–4 years [6, 11]. The former catalysts, Catofin STD, have been improved by inhibition chromium migration into alumina lattice, and the latest generation of the catalysts, Catofin PS, performs even 50% longer operation [6].

To improve heat exchange, which is crucial and critical for the endothermic dehydrogenation process, an additional novel material was designed and applied to store additional heat in the catalyst bed [100, 101]. The role of the new material, called "heat generating material" (HGM), is to increase selectivity and yield of Catofin units by producing the heat while remaining inactive to the feed and products, and with respect to dehydrogenation or side reactions such as cracking or coking. HGM is typically a granular material of similar particle size to the catalyst, and is characterized by a high density and high heat capacity. This material has been proposed to be selected from oxides of various metals (M): copper, iron, chromium, molybdenum, vanadium, yttrium, scandium, tungsten, manganese, cobalt, nickel, silver (transition metals), bismuth (post-transition metal) and cerium (lanthanide metal), and combinations thereof [89, 101, 102]. However, recently, also semimetals such as boron, silicon, germanium, antimony and tellurium have been reported to be the inert heat generating components of "catalytic composite" that comprises a dehydrogenation catalyst, a semimetal and a support [103]. During the whole process, the HGM oxide undergoes reduction with heat generation (reaction 4) and oxidation (during regeneration cycle) producing the additional heat (reaction 5) [6, 103].

$$MO_x + H_2 \rightarrow M + H_2O \qquad \Delta H < 0 \qquad (3.4)$$

$$M + O_2 \rightarrow MO_x \qquad \Delta H < 0 \qquad (3.5)$$

It is emphasized that HGM not only enhances dehydrogenation yields and saves energy, but also reduces emissions. Thanks to addition of the inert material, the same paraffin conversion is achieved at significantly lower air inlet temperatures (during regeneration), and increases olefin selectivity by decreasing the feed preheating temperature, and eliminating exposure of the product olefins to the high temperatures, compared to process without HGM. Operating at lower temperature can reduce the deactivation rate of the catalyst and thus extends its lifetime [6, 89, 102, 103]. Additionally, application of the inert oxides has led to increase in Catofin yield by an increase in the number of reactors that are operating in the dehydrogenation mode at the same time. HGM has been applied in several Catofin units for isobutylene production since 2011, and, in 2015, the first PDH Catofin plant using HGM was successfully commissioned in Ningbo, China [102].

3.5 Oleflex process

Oleflex (Olefin flexibility) by Honeywell UOP, similarly to Catofin, is a technology to produce propylene, isobutylene and butenes or even C_5 olefins by catalytic dehydrogenation of appropriate alkanes from liquid gas or light gasoline fractions. The difference between the processes lies in composition of the catalyst, heating of reaction mixture and way of the catalyst regeneration. A scheme of Oleflex process is shown in Figure 3.9.

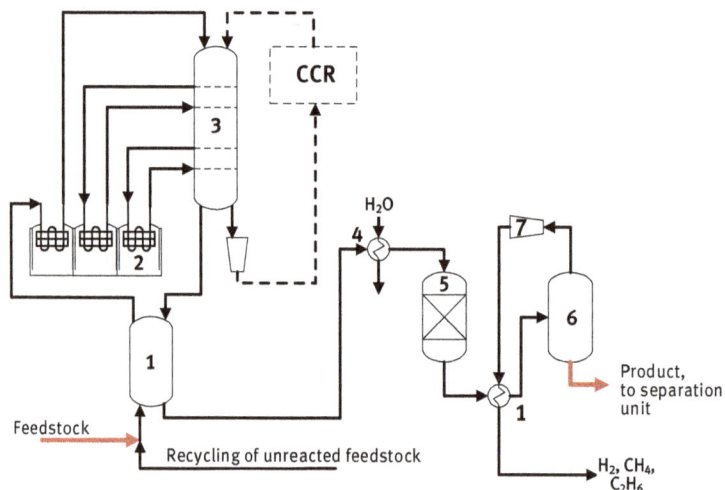

Figure 3.9: Simplified scheme of Oleflex process: 1 – heat exchangers, 2 – charge/interstage heaters, 3 – series of three dehydrogenation reactors, 4 – cooler, 5 – dryer, 6 – fractionation column, 7 – decompressor, CCR – continuous catalyst regeneration unit (based on [4, 11, 96, 97, 104]).

Oleflex uses a series of moving bed adiabatic reactors with interstage external (unlike Catofin) heating of the reaction mixture. Feedstock heated in (1) and (2) to reaction temperature, i. e. 525–705 °C [11], is provided to a series of reactors (three or four) (3) working at 0.1–0.3 MPa and moves in a conflow with particles of the catalyst through the system of the reactors with interstage heaters (2) in between. The interstage heaters (2) between reactors (3) are necessary to provide the thermal energy consumed by the endothermic dehydrogenation reactions. Products of the reactions are cooled in (1) and (4), dried and separated in column (6) from light gases, i. e. hydrogen, methane and ethane. Afterwards, the purified olefins are moved to a separation unit (not shown in Figure 3.9), where polymer-grade propylene is obtained after removal of other hydrocarbons in deethanizer and propane-propylene splitter. However, it is pointed out as a drawback of the gas condensation that some amounts of C_3 hydrocarbons are lost with large amounts of hydrogen formed and discharged as offgas [105].

In Oleflex process the catalyst continuously circulates between the reactors and the regeneration unit (CCR, continuous catalyst regeneration). The circulation is so slow that catalyst layer in each reactor is considered almost stationary, while vapor of hydrocarbons flows radially. In this mode, a small amount of the catalyst from the reactor is removed and moved to the CCR unit where coke burning and platinum redistribution assisted by chloride addition take place. Then, the regenerated catalyst is returned to the top of the first reactor. The entire cycle lasts 5–10 days [78]. The process is continuous with no need to stop the reactor for regeneration cycle and providing effective activation of the catalyst [97]. This continuous circulation of the catalyst is distinctive for the Oleflex process.

Such a technological solution of the dehydrogenation process with the active and stable catalyst provide high selectivity of the reactions, i. e. 89–91% for propylene and 91–93% for butylene [4]. The platinum-based catalyst used in Oleflex was developed on the basis of catalysts applied for UOP Pacol process for the production of linear olefins for the manufacture of biodegradable detergents [97].

The key role of dehydrogenation catalysts is to accelerate the main reaction and to control side reactions. Despite the fact that unmodified platinum-based catalysts are highly active in dehydrogenation reactions, their selectivity to olefins and catalyst stability are not satisfactory. Therefore, a promoter – mainly tin, which is able to modify the catalytic properties of platinum, is added to obtain an optimal industrially applied catalyst on alumina support. Alumina support has acidic sites that accelerate isomerization, cracking, oligomerization, and polymerization of olefins, and facilitates coke formation. Tin presence minimizes reactions of hydrogenolysis and isomerization, suppresses the formation of by-products, and neutralizes the acidity of the support. The modifier also improves the stability against coking by heavy carbonaceous materials, thus increasing lifetime of the catalyst [11, 97].

A series of the Oleflex catalysts patented by Honeywell UOP is marked as DeH catalysts, from which now DeH-16 is recommended for use [106–108]. Various improvements in DeH catalysts performance has been done for almost 30 years since the first PDH Oleflex plant NPC was commercialized in Thailand [108]. The plant started with DeH-6, which appeared to be susceptible to the presence of CO in the hydrogen recycle stream used for reducing the metal oxides at the catalyst reduction step. To avoid poisoning of the metal function of the catalyst, purified PSA hydrogen was supplied to the reduction zone. However, DeH-6 was not stable enough, and it was replaced by improved DeH-8 (0.75% Pt content), and next, with DeH-10 of reduced platinum (0.6%) and modifier content. Further investigation led to application of DeH-12 and DeH-14 of lower Pt content (0.45%), characterized by slower selectivity decline than the previous ones, longer catalyst lifetime, and improved economics of the operation. The newest generation of the catalysts, DeH-16 (1.6 mm diameter spheres and 0.3 wt% Pt content), is designed to minimize the formation of high molecular weight hydrocarbons, side products of dehydrogenation, and to extend the catalyst lifetime (now it reaches 3 years). Also, as a consequence of 30% lower content of Pt in comparison to the previously produced catalysts, an improvement in operability and profitability of the Oleflex process has been reported [106].

It should be emphasized that Oleflex is the leading on-purpose polymer grade propylene production technology in the world, and provides the lowest cash cost of production and highest return on investment when compared to competing PDH processes. Among the advantages of the process, according to Honeywell UOP, are [7, 62]:

- low feedstock consumption and low energy usage, which reduce operating cost,
- the only continuous process in industry and operation with a highly active and stable catalyst at positive pressure utilizing only four reactors, which reduces capital cost,
- recent design enhancements and the ability to change catalyst on-the-fly without stopping propylene production, which assure high reliability,
- the highest return to investment compared with competing technologies.

3.5.1 Other dehydrogenation processes

Steam active reforming – STAR process (ThyssenKrupp Industrial Solutions, formerly Uhde) is used for dehydrogenation of light hydrocarbons such as propane or butane to propylene and butylenes, respectively. Similar method was patented by BASF [105]. Up to now, the STAR technology is used for the commercial-scale production of isobutylene in two plants and of propylene only in one plant in Egypt (Table 3.2) [11, 78, 88, 90, 97]. The reactor setup in the STAR process is completely different from Oleflex and Catofin technologies, and is shown in Figure 3.10.

All the industrial processes for PDH have limited conversion of substrates, in the range of 32–55%, due to temperature and pressure limitations negatively affecting propylene selectivity [90]. The key issue of the STAR process is lowering partial pressure of paraffins by adding steam to achieve as effective conversion as could be

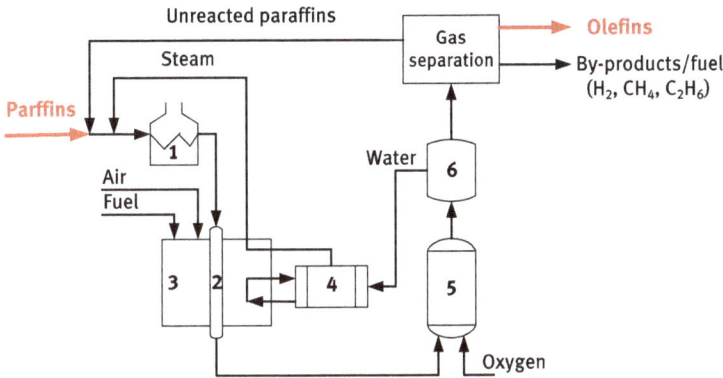

Figure 3.10: Scheme of Steam Active Reforming (STAR) process: 1 – heater, 2 – tubular reactor, 3 – oven, 4 – steam generator, 5 – oxyreactor, 6 – condenser (according to [11, 88, 96, 97, 105]).

obtained under vacuum, but still operating at atmospheric pressure, i. e. 0.6–0.9 MPa [11, 96]. Potential benefits of superheated steam presence in the reactor are as follows [11, 96, 97]:

- steam dilution of the reaction mixture positively affects dehydrogenation equilibrium, shifting it towards olefins,
- steam is a heat carrier for the endothermic dehydrogenation reactions,
- steam interacts with coke at the surface of the catalyst, and partly regenerates it keeping the catalyst active.

The reaction is carried out at 500–600 °C in the presence of Pt (0.01–5 wt%)-Sn (0.1–5 wt%) catalyst stabilized by alkaline support of $Zn-NaAlO_2$ with Ca/Mg-$NaAlO_2$ as a binder to resist a contact with superheated steam [11, 96]. The STAR catalyst, due to its basic nature, maintains its activity for dehydrogenation in a contact with high temperatures, steam, and oxygen, and does not convert the hydrocarbons to carbon oxides and hydrogen [88]. Although coke deposition on the catalyst surface is slow, regeneration of the catalyst takes place every 7 h of the dehydrogenation and lasts 1 h [11, 96]. Hence, STAR process is also cyclic, similarly to Catofin. ThyssenKrupp emphasizes that thanks to steam presence in the STAR process, "only 14.7% additional reactor capacity is needed to account for regeneration requirements, which is the lowest of all commercial technologies" [88].

The STAR catalyst is placed as a fixed bed in a multi-tubular reactor (2) heated externally by an oven (3). The reactor design is similar to a typical steam reformer that is operated until the catalyst deactivates by coke deposition. When one reactor is switched for catalyst regeneration, another one ("swing") multi-tubular reactor continues the dehydrogenation cycle [97]. The gas mixture from the tubular reactor (2) is moved to the second reactor (adiabatic), so-called oxyreactor (5), a refractory vessel filled with STAR catalyst, where hydrogen formed in the first reactor is partly combusted by the oxygen-steam mixture, and further paraffin conversion takes place [88]. Oxygen (mainly as 90% oxygen enriched air) is added in a molar ratio of 0.08 to 0.16 with respect to the hydrocarbon supply [96]. Thus, the equilibrium of dehydrogenation is shifted towards higher yields of olefins due to consumption of hydrogen, one of the dehydrogenation products, in the following reaction:

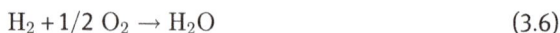

$$H_2 + 1/2\ O_2 \rightarrow H_2O \tag{3.6}$$

Additionally, this exothermic reaction provides the heat required for conversion of propane to propylene. This method for equilibrium shifting, to increase product yield, is used also in other dehydrogenation processes, e. g. SMART (Styrene Monomer by Advanced Reheat Technology) process for styrene production [97, 109].

After cooling down the gas from oxyreactor (5), the steam is removed by condensation (6), olefins are separated from hydrogen and by-products, which can be used as a fuel to heat the tubular reactor oven (3).

One more dehydrogenation process which is expected to be commercialized soon by PetroLogistics in the US (see Table 3.2) is Fluidized Catalytic Dehydrogenation (FCDh) developed by Dow Chemical Company [11, 110–112]. The idea of FCDh process is presented in Figure 3.11. In various sources also FBD (Fluidized Bed Dehydrogenation) process, similar to FCDh, is mentioned, however, in FBD chromium oxide alumina-supported catalyst is reported to be applied instead of Pt-based one [11, 96]. This section focuses only on FCDh, as it about to be commercialized.

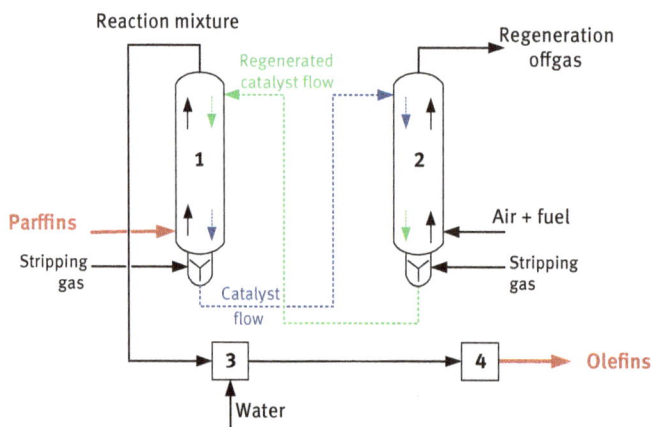

Figure 3.11: Scheme of Fluidized Catalytic Dehydrogenation (FCDh) process: 1 – fluidized bed reactor, 2 – regenerator, 3 – jet scrubber, 4 – gas separation (according to [11, 110–112]).

Generally, fluidized catalytic cracking (FCC) aims in gasoline production from oil fractions containing high-boiling hydrocarbons (more than 12 carbons in a chain). However, recently, there has been growing interest in FCC application for on-purpose production of light olefins, mainly propylene. As an example of modification of FCC can be given a project of PetroLogistics and Dow Chemical Co., which provides technology of 500 000 t/year PDH unit with a novel reactor design based on FCC to carry out reaction of fluidized catalytic dehydrogenation (FCDh) [63, 110–112].

The FCDh process is carried out in a circulating fluidized bed system with a Ga-Pt alumina-supported catalyst. The fluidized bed catalyst in microspheroidal form must have suitable fluid-dynamic characteristics and mechanical resistance to attrition [96, 110]. The Pt level is lower than in Oleflex catalyst, and thus this catalyst can be more attractive because of its lower price. In addition, the Ga-Pt catalyst tolerates reasonable levels of feed impurities, such as sulfur, which eliminates the need for costly additional gas purification.

The idea of FCDh process is shown in Figure 3.11. The FCDh reactor does not require external feed heating, as it uses fluidized catalyst (average diameter of particles less than 0.1 mm) to rapidly heat up the propane to the reaction temperature (550–600 °C) just in the reactor [11, 110]. Propane is introduced into a fluidized bed reactor (1) equipped at the top with cyclones for rapid product and catalyst separation. The catalyst is then stripped with a stripping gas to recover any product entrained with it. Then, the spent catalyst is transported to a regenerator (2) (also called a combustor), where it is reheated (using external heating fuel) and activated by coke burning at about 700 °C. The heat generated in the regenerator is used to carry out the endothermic dehydrogenation in the reactor. The regenerated catalyst is separated from the flue gas in cyclones and returns to the reactor, where it is activated with air. In the reactor the catalyst cools down again to temperatures below 560 °C. The catalyst residence time in the reactor and regenerator ranges from ~10 to 30 min [89].

The following advantages of the use of fluidized-bed technology can be pointed out [96]:
- continuous operation under stable operating conditions,
- catalyst replacement or filling-up while the plant is operating, and thus elimination of need to halt reactor operation,
- safety of the operation due to moderate process pressure (higher than the atmospheric pressure).

As it was mentioned before, the pressure during dehydrogenation significantly affects the maximum conversion of the paraffins into olefins. FCDh process operates between 0.16 and 0.25 MPa, which is slightly above atmospheric pressure, and is advantageous compared to the Catofin process due to reduction in compressor operating cost and capital [110]. Dow Chemicals declares that FCDh provides high conversion of propane, 43–53%, with propane to propylene selectivity of 92–96%. It is significantly higher than the competing commercial technologies (see Table 3.3). Additionally, the FCDh supplier declares emissions of NO_x lower than from alternative technologies, what may enhance potential customers to invest in this technology in the near future.

Table 3.3: Comparison of the commercialized dehydrogenation processes.

Process name	Catofin	Oleflex	STAR	Dow FCDh
Reactor, way of operation	Fixed-bed parallel reactors	3-4 moving-bed reactors, continuous catalyst regeneration (CCR) unit	2 reactors: fixed-bed tubular and oxyreactor	1 fluidized reactor, 1 fluidized regenerator
Conditions of reaction	550–650 °C, 0.016–0.1 MPa	525–705 °C, 0.1–0.6 MPa	500–600 °C, 0.6–0.9 MPa	550–600 °C, 0.16–0.25 MPa
Catalyst	Cr_2O_3/Al_2O_3, 2–4 years lifetime	Pt-Sn/Al_2O_3 1–3 years lifetime	Pt-$Sn/NaAlO_2$ up to 7 years	Pt-Ga/Al_2O_3 no data
Catalyst regeneration, providing heat to reaction step	Cyclic operation of reaction and regeneration steps, coke burned with air provides heat to catalyst having high heat capacity	Continuous regeneration in separate regenerator, coke burning, drying, Pt redispersion with chlorine addition, H_2 reduction; interstage external heating of reaction mixture	Cyclic coke regeneration in tubular reactor heated externally by an oven; additional source of reaction heat – oxyreactor	Coke burning in fluidized bed and activation with air, regenerated catalyst provides heat to reactor
Reaction catalyst residence time	7–15 min	approx. 7200–14,400 min (5–10 days)	420 min (7 h)	<2 min
Conversion	45% of propane, 50% of isobutane	25% of propane, 35% of isobutane	40% of propane	45% of propane
Selectivity	88% to propylene, 89% to isobutene	89–91% to propylene, 91–93% to butylene	89% to propylene	93% to propylene

Four dehydrogenation processes presented in this section are compared in Table 3.3 to show similarities and differences in conditions, technical solutions, and resulting conversion and selectivity.

3.6 Metathesis of olefins to propylene

Olefin methathesis involves two olefin substrates which form a four-membered ring intermediate (first proposed by Chauvin, the Nobel Prize winner in 2015) and then rearrange the substituents to form two new C-C double bonds. Olefins Conversion Technology (OCT), also called the Phillips Triolefin Process or the ethylene-to-propylene (ETP), is the industrial process in which ethylene and 2-butene are converted into two molecules of propylene according to the following reaction [113]:

$$CH_2{=}CH_2 + CH_3CH{=}CHCH_3 \rightarrow 2CH_2{=}CHCH_3 \qquad (3.7)$$

This is the only commercially demonstrated route to propylene using metathesis chemistry. Olefin metathesis was discovered in the 1960s by Standard Oil of Indiana and Phillips Petroleum. Up to now, several companies have developed this process at a demonstration or commercial scale, e. g. Lummus Technology (Olefin Conversion Technology, OCT), Sinopec (Olefin Metathesis Technology, OMT), and IFP-CPC (Meta-4 process) [113]. Metathesis was first commercially applied to propylene production at the BASF/FINA Port Arthur Cracker, USA, and several plants have been built in Asia since then [90]. In 2017 over 37 OCT plants (18 units in operation, and 19 planned or under construction), and in 2019 even 49, were licensed by Lummus Technology for an over 9 million tons of production per year, i. e. more than 10% of the world's propylene capacity. Unit capacity ranges from 60 000 t/year of polymer grade propylene to 800 000 t/year of propylene [113–115].

The components for methathesis and isomerization reactions are olefins, e. g. ethylene, butenes or iso- and n-pentenes from various refinery processing streams (from steam cracking, raffinate from MTBE or butadiene extraction and from FCC units) or the C_4 and C_5 stream from MTO units, that react to produce propylene. Polymer grade propylene is produced in a simple fixed bed reactor filled with a mixture of WO_3/SiO_2 (metathesis) and MgO (isomerization) catalysts, and without the use of super fractionators since no paraffins are formed in the metathesis or isomerization reactions. The process is carried out at temperatures higher than 260 °C and pressure in the range of 3–3.5 MPa [113, 114, 116].

As it is shown in Figure 3.12, fresh and recycled ethylene feeds are mixed with the C_4– C_5 fractions and heated prior to entering the fixed-bed metathesis reactor (3), where the catalytic reaction of ethylene and butene-2 is carried out resulting in formation propylene, and butylenes, and simultaneously isomerization of 1-butene to 2-butene. As a small amount of coke is produced, and deactivates the heterogeneous catalyst used in the reaction, so periodic regeneration with nitrogen-diluted air is necessary. The process uses swing reactors, namely when one fixed bed reactor is on stream, the other is being regenerated. The methathesis products are cooled and fractionated (4) to remove and recycle ethylene. A small portion of this recycle stream is purged to remove methane, ethane, and other light impurities from the process. The ethylene column bottoms is fed to the propylene column (5) where butylenes and pentenes are separated to be recycled to the reactor (3) and some is purged to remove unreacted butylenes, isobutenes, butanes, unreacted pentenes, isopentenes, pentanes and heavies from the process. The product – polymer-grade propylene – leaves the column overhead. Such an OCT unit can be integrated with any refining/petrochemical complex. As Lummus Technology emphasizes, in OCT process butene conversions are high – between 60% and 80%, with over 92% selectivity to propylene. The high selectivity limits feed consumption and by-product quantities [113].

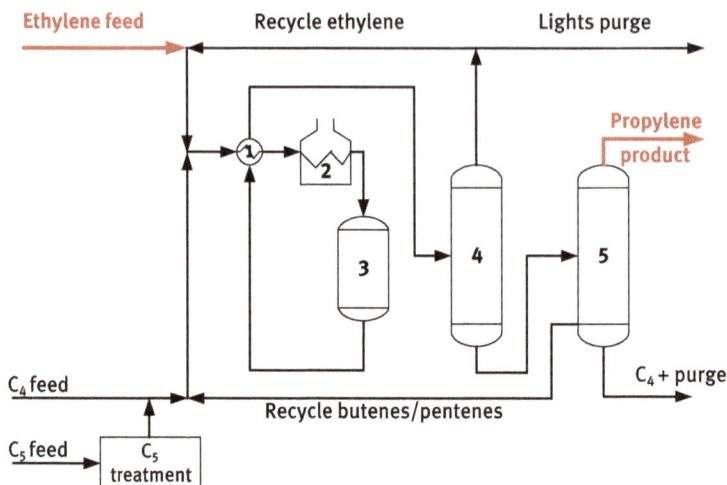

Figure 3.12: Scheme of Olefins Conversion Technology (OCT): 1 – heat exchanger, 2 – furnace, 3 – methathesis reactor, 4 – ethylene column, 5 – propylene column (according to [114]).

The metathesis reactions are mildly exothermic, so the process is considered nearly "energy neutral" because no energy input to the reaction step is required, which significantly suppresses greenhouse gas emission. OCT can be readily integrated with a steam cracker of naphta or ethane. When integrated with a naphtha cracker, the maximum propylene to ethylene ratio can be increased from 0.6 to 1.2, doubling the amount of propylene produced. When integrated with an ethane cracker, which as a standalone unit produces very little propylene, OCT combined with ethylene dimerization can be used to produce any desired quantity of polymer-grade propylene, making it the only route of ethane feed to propylene product [114].

OCT integrated with refinery streams can use as a feedstock ethylene recovered from refinery offgas streams. Integration of OCT and MTO units increases by up to 15% ethylene and propylene production.

3.7 Summary

In this chapter a great variety of current technological approaches to olefins production was shown. The continuous improvement and development of on-purpose processes is a response to dynamic changes on feedstock market of petrochemical raw materials, consumer needs, and environmental regulations. It was emphasized that boom in dehydrogenation processes (particularly, PDH) results from shift to ethylene production in steam crackers, a gap between supply of propylene, butylenes and higher olefins compared to the continuously growing demand for their derivatives. However, from time to time the positive trends happen to breakdown, and then a flexible and easy-to-modify units and technologies are able to adjust to the global market situation. It is clear that simple hydrocarbon intermediates (olefins) market is still changing, and ready for future challenges.

References

[1]   Plastics Europe, Association of Plastics Manufacturers, Belgium. Plastics – the Facts 2017, An analysis of European plastics production, demand and waste data, 2018.

[2]   Sinclair V 4 threats to the global ethylene boom. Wood Mackenzie, 2018. Available at: https://www.woodmac.com/news/opinion/4-threats-to-the-global-ethylene-boom/. Accessed: 15 Aug 2019.

[3]   ResearchMoz. H1 2019 global ethylene capacity and capital expenditure outlook-saudi aramco and exxon lead global capacity additions. Research Study, 2019. Available at: https://portnews24.com/global-ethylene-capacity-and-capital-expenditure-outlook-key-play ers-and-production-information-analysis-with-forecast-2030/93143/. Accessed: 15 Aug 2019.

[4]   Molenda J. Chemia w przemyśle. Surowce – procesy – produkty. (Chemistry in industry. Raw materials – processes – products.) WSiP, Warsaw 1996 (in Polish).

[5]   Grzywa E, Molenda J. Technologia podstawowych syntez organicznych (Technology of the basic organic syntheses). Warsaw: WNT, 1995. (in Polish).

[6]   Oviol L, Bruns M, Fridman V, Merriam J, Urbancic M Mind the gap. Catalysis & Energy (Süd-Chemie). Hydrocarbon Engineering. Clariant, September 2012.

[7]   Six revolutions. 1914–2014. A century of innovations in the oil and gas industry. UOP A Honeywell Company, USA, 2014. Available at : https://www.uop.com/wp-content/uploads/Six-Revolutions/. Accessed: 17 July 2019.

[8]   O'Reilly C CTO and MTO projects in China may decelerate. Hydrocarbon Engineering. May 2019. Available at: https://www.hydrocarbonengineering.com/petrochemicals/29052019/cto-and-mto-projects-in-china-may-decelerate/. Accessed: 17 July 2019.

[9]   Jasper S, El-Halwagi MM. A techno-economic comparison between two methanol-to-propylene processes. Processes. 2015;3:684–98. DOI:10.3390/pr3030684.

[10]   Amghizar I, Vandewalle LA, Van Geem KM, Marin GB. New trends in olefin production. Engineering. 2017;3:171–8. DOI:10.1016/J.ENG.2017.02.006.

[11]   Sattler JJHB, Ruiz-Martinez J, Santillan-Jimenez E, Weckhuysen BM. Catalytic dehydrogenation of light alkanes on metals and metal oxides. Chem Rev. 2014;114:10613–53. DOI:10.1021/cr5002436.

[12]   Sadrameli SM. Thermal/catalytic cracking of hydrocarbons for the production of olefins: A state-of-the-art review I: Thermal cracking review. Fuel. 2015;140:102–15. DOI:10.1016/j.fuel.2014.09.034.

[13]   Steam cracking – cracking furnaces. Metso application report 2015, 11/15, 1–6.

[14]   Emerson Process Management brochure. Ethylene Production. Chapter 1, Fisher Controls International LLC 2010.

[15]   SRT® Ethylene Furnaces. Available at: https://www.mcdermott.com/What-We-Do/Technology/Lummus/Heat-Transfer-Equipment/SRT-Ethylene-Furnaces. Accessed: 15 Aug 2019.

[16]   Barendregt S, Bowers Pitcher M, Den Uijl J. Cracking furnace. Patent No. US7964091B2, 2004.

[17]   Spoto M, Spoto B. Steam cracking furnace. Patent No. EP1716379B1, 2004.

[18] Van Cauwenberge DJ, Van Geem KM, Floré J, Marin GB, Laib H. Reactor for a cracking furnace. Patent No. WO2017178551A1, 2017.

[19] Spicer DB. Process for the on-stream decoking of a furnace for cracking a hydrocarbon feed. Patent No. EP2310472B1, 2008.

[20] Mukherjee S. KBR Olefins Technology – Technology options to meet uncertain market conditions. 4th Petrochem Conclave Delhi, 12 February 2015.

[21] Bruijnincx PCA, Weckhuysen BM. Shale gas revolution: an opportunity for the production of biobased chemicals? Angew Chem Int Ed. 2013;52:11980–7. DOI:10.1002/anie.201305058.

[22] Qatar petroleum partners with chevron phillips to build the largest ethane cracker in the middle east raising Qatar's polyethylene production capacity by 82%. 2019. Available at: https://www.euro-petrole.com/qatar-petroleum-partners-with-chevron-phillips-to-build-the-largest-ethane-cracker-in-the-middle-east-raising-qatars-polyethylene-production-capacity-by-82-n-i-18955. Accessed: 10 Aug 2019.

[23] Charlesworth R. Direct oil and gas to ethylene. MERTC Annual Meeting, Bahrain 2017.

[24] New ExxonMobil and Saudi Aramco Technologies produce ethylene directly from crude oil, cutting refining costs, IHS Says. 2016. Available at: https://news.ihsmarkit.com/press-release/new-exxonmobil-and-saudi-aramco-technologies-produce-ethylene-directly-crude-oil-cutti. Accessed: 10 Aug 2019.

[25] Kempf R Advanced MTO: breakthrough technology for the profitable production of light olefins. UOP LLC, A Honeywell Company, 2013. Available at: https://www.google.com/url?sa=t&rct=j&q=&esrc=s&source=web&cd=11&ved=2ahUKEwi13o_nwNPjAhXososKHY5dAPAQFjAKegQIBBAC&url=https%3A%2F%2Fwww.engerati.com%2Fsites%2Fdefault%2Ffiles%2Fday%25202%25201130%2520petchemRick%2520Kempf.pdf&usg=AOvVaw214A19_SXeN3-a-Hh9fsRv. Accessed: 26 July 2019.

[26] Chemicals Technology. 2001. Available at: https://www.chemicals-technology.com/projects/tjeldbergoddenethpro/. Accessed: 10 Aug 2019.

[27] Statoil testing process to convert natural gas into propylene. 2001. Available at: https://www.ogj.com/refining-processing/article/17261118/statoil-testing-process-to-convert-natural-gas-into-propylene. Accessed: 10 Aug 2019.

[28] Jenkins S. World's largest single-train methanol-to-olefins plant now operating. Chem Eng. 2018. Available at: https://www.chemengonline.com/worlds-largest-single-train-methanol-to-olefins-plant-now-operating/.

[29] UOP technology licensed to eurochem to convert methanol to olefins in new Nigerian petrochemicals plant. 2008. Available at: https://www.uop.com/?press_release=uop-technology-licensed-to-eurochem-to-convert-methanol-to-olefins-in-new-nigerian-petrochemicals-plant. Accessed: 10 Aug 2019.

[30] Otaraku IJ, Egun IL, Nyebuchi Q. Optimization of methanol to ethylene process in Nigeria using aspen HYSYS. Int J Latest Technol Eng Manage Appl Sci (IJLTEMAS). 2018;VII:15–18. ISSN 2278-2540.

[31] Mamudu OA, Igwe GJ, Okonkwo E. Performance evaluation of the nigerian petrochemical industry: reliability of the bag filter in carbon black production. Ewemen J Petrochem Res Innovation. 2016;1:7–17.

[32] Ren T, Patel MK, Blok K. Steam cracking and methane to olefins: Energy use, CO_2 emissions and production costs. Energy. 2008;33:817–33. DOI:10.1016/j.energy.2008.01.002.

[33] CCF Group. Methanol-to-olefins producers feel the pain. 2018. Available at: http://www.ccfgroup.com/newscenter/newsview.php?Class_ID=D00000&Info_ID=20180613038. Accessed: 10 Aug 2019.

[34] Environmental concerns are expected to delay new CTO/MTO projects in China. Hydrocarbons technology. 2019. Available at: https://www.hydrocarbons-technology.com/comment/olefin-china-demand/. Accessed: 17 July 2019.

[35] Gogate MR. Methanol-to-olefins process technology: current status and future prospects. Petroleum Sci Technol. 2019;37:559–65. DOI:10.1080/10916466.2018.1555589.

[36] Xu S, Zhi Y, Han J, Zhang W, Wu X, Sun T, et al. Advances in catalysis for methanol-to-olefins conversion. Advances in Catalysis. Vol. 61. Burlington: Academic Press, 2017:37–122.

[37] Barger P. Methanol to olefins (MTO) and beyond in: zeolites for cleaner technologies. Chapter 12 Catalytic Science Series, 2002:239–60. DOI:10.1142/9781860949555_0012.

[38] Eng CN, Arnold EC, Vora BV. Integration of the UOP/HYDRO MTO process into ethylene plants. Proceedings of 1998 AIChE Spring National Meeting. Session 16, Fundamental Topics in Ethlyene Production. New Orleans, LA, USA 1998; paper 16g.

[39] De Haan S. Methanol to olefins process. Patent No. US 2012/0240615 A1, 2012.

[40] Tian P, Wei Y, Ye M, Liu Z. Methanol to olefins (MTO): from fundamentals to commercialization. ACS Catal. 2015;5:1922–38. DOI:10.1021/acscatal.5b00007.

[41] Bozzano AG, Gregor JH, Senetar JJ. Integration of a methanol-to-olefin reaction system with a hydrocarbon pyrolysis system. Patent No. US8829259B2, 2011.

[42] Sun Q, Xie Z, Yu J. The state-of-the-art synthetic strategies for SAPO-34 zeolite catalysts in methanol-to-olefin conversion. National Science Rev. 2018;5:542–58. DOI:10.1093/nsr/nwx103.

[43] Boltz M, Losch P, Louis B. A General overview on the methanol to olefins reaction: recent catalyst developments. Adv Chem Lett. 2013;1:247–56. DOI:10.1166/acl.2013.1032.

[44] Sapre AV. Conversion of methanol to olefns in atubular reactor with light olefin co-feeding. Patent No. US4590320A, 1985.

[45] Chang CD, Lang WH, Silvestri AJ. Conversion of methanol to olefinic components. Patent No. US4052479A, 1977.

[46] Vora B, Pujado P Process for enhanced olefin production. Patent No. US7317133 B2, 2002.

[47] Fournier JF, Wagner ES, Laugier JP, Haag S, Wurzel T, Gorny M Recycling system and process of a methanol-to-propylene and steam cracker plant. Patent No. US20170253539A1, 2016.

[48] Honeywell's UOP and total petrochemicals successfully demonstrate technology to produce plastics from feedstocks other than oil. 2010. Available at: https://www.uop.com/?press_release=honeywells-uop-and-total-petrochemicals-successfully-demonstrate-technology-to-produce-plastics-from-feedstocks-other-than-oil. Accessed: 10 Aug 2019.

[49] Koempel H, Liebner W. Lurgi's methanol to propylene (MTP) report on a successful commercialisation. In: Proceedings of the 8th Natural Gas Conversion Symposium, Natal, Brazil, 2007:261–81.

[50] Teketel S, Erichsen MW, Lønstad Bleken F, Svelle S, Lillerud KP, Olsbye U. Shape selectivity in zeolite catalysis. The methanol to hydrocarbons (MTH) reaction. Catalysis. 2014;26:179–217. DOI:10.1039/9781782620037-00179.

[51] Chen JQ, Bozzano A, Glover B, Fuglerud T, Kvisle S. Recent advancements in ethylene and propylene production using the UOP/Hydro MTO process. Catal Today. 2005;106:103–7.

[52] Cao G, Shah MJ. Synthesis of silcoalumnophosphates. US Patent 6680278 B2, 2004.

[53] Barger PT, Vora BV. Methanol to olefin process with increased selectivity to ethylene and propylene. Patent No. US6534692B1, 2001.

[54] Chen D, Moljord K, Holmen A. A methanol to olefins review: diffusion, coke formation and deactivation on

[55] SAPO type catalysts. Microporous mesoporous mater. 2012;164: 239–50. DOI:10.1016/j.micromeso.2012.06.046.

[56] Fan D, Tian P, Su X, Yuan Y, Wang D, Wang C, et al. Aminothermal synthesis of CHA-type SAPO molecular sieves and their catalytic performance in methanol to olefins (MTO) reaction. J Mater Chem A. 2013;1:14206–13. DOI:10.1039/C3TA12829F.

[57] Gao B, Fan D, Sun L, An H, Fan F, Xu S, et al. Insights into the aminothermal crystallization process of SAPO-34 and its comparison with hydrothermal system. Micropor Mesopor Mater. 2017;248:204–13. DOI:10.1016/j.micromeso.2017.04.035.

[58] Wu P, Yang M, Tian P, Liu Z, Wang L. Preparation of SAPO-34 molecular sieve used for acid catalyst, involves carrying out hydrothermal synthesis in presence of templating agent containing piperazinyl organic compound e.g. organic compound containing hydroxyl structure. Patent No. WO2019113948-A1, 2019.

[59] Liu X, Wang T, Wang Z, Zhang P, Zhang B. Synthesizing SAPO-34 molecular sieve having CHA structure comprises adopting hydrothermal synthesis method for synthesizing SAPO-34-type molecular sieve, mixing phosphorus source with deionized water. Patent No. CN109678178-A, 2019.

[60] Li X, Li K, Ma H, Xu R, Tao S, Tian Z. Ionothermal synthesis of a CHA-type aluminophosphate molecular sieve membrane and its formation mechanism. Micropor Mesopor Mater. 2015;217:54–62. DOI:10.1016/j.micromeso.2015.06.005.

[61] Available at: https://www.uop.com/processing-solutions/petrochemicals/olefins/. Accessed: 10 Aug 2019.

[62] Tian P, Su X, Wang Y, Xia Q, Zhang Y, Fan D, et al. Phase-transformation synthesis of SAPO-34 and a novel SAPO molecular sieve with RHO framework type from a SAPO-5 precursor. Chem Mater. 2011;23:1406–13. DOI:10.1021/cm103512m.

[63] Zeeshan N. Light alkane dehydrogenation to light olefin technologies: a comprehensive review. Rev Chem Eng. 2015;31:413–36. DOI:10.1515/revce-2015-0012.

[64] Available at: https://www.tecnimont.it/en/what-we-do/chemical-petrochemical/pdh-plant. Accessed: 11 Aug 2019.

[65] Brelsford R. PetroLogistics plans US Gulf Coast PDH plant. Oil Gas J. 2019. Available at: https://www.ogj.com/refining-processing/petrochemicals/article/14036217/petrologistics-plans-us-gulf-coast-pdh-plant.

[66] Maddah HA. A comparative study between propane dehydrogenation (PDH) technologies and plants in Saudi Arabia. Am Sci Res J Eng Technol Sci (ASRJETS). 2018;45:49–63.

[67] AL WAHA petrochemicals company to use honeywell connected plant to improve plant performance. 2017. Availabel at: https://www.uop.com/?press_release=al-waha-petrochemi cals-company-to-use-honeywell-connected-plant-to-improve-plant-performance. Accessed: 11 Aug 2019.

[68] Al-waha starts up new propane dehydrogenation unit and largest spherizone polypropylene plant. 2009. Available at: https://www.lyondellbasell.com/en/news-events/products–tech nology-news/al-waha-starts-up-new-propane-dehydrogenation-unit-and-largest-spherizone-polypropylene-plant/. Accessed: 10 Aug 2019.

[69] Al-Duhaish IU, An example of a successful project & partnership leveraging all financing options/support. Saudi Polyolefins Company (SPC), 2013. Available at: https://www.enger ati.com/sites/default/files/day%202%200930%20Mr%20Al%20Duhaish%20slides%20day %202.pdf. Accessed: 11 Aug 2019.

[70] Stark A, South Korea: catalyst for PDH unit successful start-up. Process Worldwide 2016. Available at: https://www.process-worldwide.com/successful-start-up-a-538111/. Accessed: 11 Aug 2019.

[71] ARC Research Team, Saudi Petrochemicals Sector. Petrochemical–industria. 2012. Available at: http://www.alrajhi-capital.com/en/research/Petrochemicals/2Petro12May2012.pdf. Accessed: 11 Aug 2019.

[72] PDH Polska. Available at: https://pdhpolska.eu/en/news/. Accessed: 11 Aug 2019.

[73] Menachery M, McDermott announces successful start-up of world's largest catalytic dehydrogenation plant. Refining Petrochemicals, 2019. Available at: https://www.refinin gandpetrochemicalsme.com/petrochemicals/26147-mcdermott-announces-successful-start-up-of-worlds-largest-catalytic-dehydrogenation-plant. Accessed: 11 Aug 2019.

[74] Chinese operator lets contract for PDH unit. Oil Gas J 2018. Available at: https://www.ogj. com/refining-processing/refining/capacities/article/17297391/chinese-operator-lets-con tract-for-pdh-unit. Accessed: 11 Aug 2019.

[75] First of three new propylene/isobutylene plants in China starts operating. Processing 2015. Available at: https://www.processingmagazine.com/first-of-three-new-propyleneisobuty lene-plants-in-china-starts-operating/. Accessed: 11 Aug 2019.

[76] Liang C, Analysis: new PDH plants to drive China's LPG demand, trade war a concern. Petrochemicals 2019. Available at: https://www.spglobal.com/platts/en/market-insights/lat est-news/petrochemicals/022819-analysis-new-pdh-plants-to-drive-chinas-lpg-demand-trade-war-a-concern. Accessed: 11 Aug 2019.

[77] BASF Tarragona. Available at: https://chemicalparks.eu/parks/basf-tarragona. Accessed: 11 Aug 2019.

[78] BASF SONATRACH invests nearly 10 million euros in its PDH plant in Tarragona. 2019. Available at: https://chemicalparks.eu/news/basf-sonatrach-invests-nearly-10-million-euros-in-its-pdh-plant-in-tarragona. Accessed: 11 Aug 2019.

[79] EPCC Propylene/Polypropylene (PP) Complex. Chemicals Technology 2019. Available at: https://www.chemicals-technology.com/projects/eppc/. Accessed: 11 Aug 2019.

[80] Iran in talks with Uhde over propylene technology. Plastics New Europe 2017. Available at: https://www.plasticsnewseurope.com/article/20170202/PNE/170209983/iran-in-talks-with-uhde-over-propylene-technology. Accessed: 12 Aug 2019.

[81] Available at: https://www.thyssenkrupp-industrial-solutions.com/en/references/chemical-in dustry/organic-chemicals. Accessed: 12 Aug 2019.

[82] Dow's Oyster Creek PDH unit reaches full capacity. Oil Gas J 2016. Available at: https://www. ogj.com/refining-processing/article/17249355/dows-oyster-creek-pdh-unit-reaches-full-ca pacity. Accessed: 12 Aug 2019.

[83] PetroLogistics ll announces 500 KTA gulf coast PDH project using dow FCDh technology. Quantum Energy Partners 2019. Available at: https://www.globenewswire.com/news-re lease/2019/07/15/1882866/0/en/PetroLogistics-ll-Announces-500-KTA-Gulf-Coast-PDH-Project-Using-Dow-FCDh-Technology.html. Accessed: 12 Aug 2019.

[84] Honeywell UOP's Oleflex technology has been selected for 52 out of 64 propane projects. MENAFN - Daily News Egypt 2018. Available at: https://menafn.com/1097491343/Honeywell-UOPs-Oleflex-technology-has-been-selected-for-52-out-of-64-propane-projects. Accessed: 11 Aug 2019.

[85] Cosfeld M STAR process® technology. Robust on-purpose propylene production technology. MERTC, Bahrain 2017. Available at: http://www.wraconferences.com/wp-content/uploads/ 2017/01/20161216-TKIS-Presentation-MERTC-STAR-process-rev01.pdf. Accessed: 12 Aug 2019.

[86] Cosfeld M STAR process – state-of-the-art dehydrogenation technology. CIS Petrochemicals, Moscow 2019. Available at: https://globuc.com/wp-content/uploads/2019/04/Martin_ Cosfeld_ENG.pdf. Accessed: 12 Aug 2019.

[87] Volkova M PDH plant taken off-stream by Ningbo Haiyue. Market Report Company 2019. Available at: http://www.mrcplast.com/news-news_open-350044.html. Accessed: 12 Aug 2019.

[88] HMC polymers. Available at: https://www.hmcpolymers.com/production. Accessed: 13 Aug 2019.

[89] Plotkin JS. The propylene quandary. Industrial Chemistry & Engineering, ACS, 2016. Available at: https://www.acs.org/content/acs/en/pressroom/cutting-edge-chemistry/the-propylene-quandary.html. Accessed: 15 Aug 2019.

[90] ThyssenKrupp Uhde. The Uhde STAR process. Oxydehydrogenation of light paraffins to olefins. Dortmund: Uhde, 2011.

[91] Sanfilippo D. Dehydrogenations in fluidized bed: catalysis and reactor engineering. Catal Today. 2011;178:142–50. DOI:10.1016/j.cattod.2011.07.013.

[92] Heinritz-Adrian M, Wenzel A, Youssef F, Advanced propane dehydrogenation. Digital Refining 2008. Available at: https://www.digitalrefining.com/article/1000632,Advanced_propane_dehydrogenation.html. Accessed: 11 Aug 2019.

[93] McDermott brochure, CATOFIN Propane/Butane Dehydrogenation. Available at: https://www.mcdermott.com/What-We-Do/Technology/Lummus/Petrochemicals/Olefins/Propylene-Production/Propane-Butane-Dehydrogenation. Accessed: 10 Aug 2019.

[94] McDermott brochure, CATADIENE® n-Butane Dehydrogenation to Butadiene. Available at: https://www.mcdermott.com/What-We-Do/Technology/Lummus/Petrochemicals/Olefins/Butadiene-Butylene-Production/CATADIENE-n-Butane-Dehydrogenation-to-Butadiene. Accessed: 11 Aug 2019.

[95] Available at: https://www.clariant.com/en/Solutions/Products/2018/11/28/12/33/CATADIENE. Accessed: 24 July 2019.

[96] Schulze J, Homann M. C₄-Hydrocarbons and derivatives: resources, production, marketing. Berlin, Heidelberg: Springer Science & Business Media, 2012.

[97] Amadei C, editor. Encyclopaedia of hydrocarbons. refining and petrochemicals, vol. II. Rome: Instituto Della Enciclopedia Italiana Fondata Da Giovanni Treccani, 2006.

[98] Vora BV. Development of dehydrogenation catalysts and processes. Top Catal. 2012;55:1297–308. DOI:10.1007/s11244-012-9917-9.

[99] Houdry EJ. Process for producing diolefins. Patent No. US2423029, 1947.

[100] Houdry EJ. Catalytic dehydrogenation of aliphatic hydrocarbons. Patent No. US2419997, 1947.

[101] Urbancic MA, Fridman V. Improved endothermic hydrocarbon conversion process. Patent No. WO2009051767A1, 2007.

[102] Fridman V, Merriam J, Urbancic M. Catalytically inactive heat generator and improved dehydrogenation process. Patent No. US20070054801A1, 2007.

[103] Clariant. Clariant's breakthrough HGM concept for Propane dehydrogenation Unit - a proven success in China. Available at: https://www.clariant.com/en/Corporate/News/2015/03/Clariants-breakthrough-HGM-concept-for-Propane-dehydrogenation-Unit-a-proven-success-in-China. Accessed 12, 2019.

[104] Mondal KC, Chowdhury DR, Saravanan V Catalytic composite and improved process for dehydrogenation of hydrocarbons. Patent No. WO 2016/135615 Al, 2016.

[105] Imai T, Jan DY. Catalytic oxidative steam dehydrogenation process. Patent No. US4788371A, 1987.

[106] Schubert M, Müller U, Kiener C, Teich F, Crone S, Simon F, et al. Method for producing propene from propane. Patent No. US20090312591A1, 2007.

[107] UOP OleflexTM DeH-16 Catalyst, UOP Honeywell brochure. Des Plaines 2006.

[108] Hayes JC, Pollitzer EL. Dehydrogenation catalysts containing platinum rhenium a group VI transition metal and an alkali or alkaline earth metal. Patent No. US3649566A, 1970.

[109] Chaiyavech P. Commercialization of the world's first Oleflex unit. J Royal Inst Thailand. 2002;27:1–9.

[110] SMART SM Styrene Monomer Advanced Reheat Technology, CBI brochure. Chicago 2012.

[111] Pretz M, Fish B, Luo L, Stears B. Petrochemical developments. Shaping the future of on-purpose propylene production. Hydrocarbon Process. 2017:29–36.

[112] Pretz M Dow fluidized catalytic dehydrogenation (FCDh): the future of on-purpose propylene production. 2019. Available at: https://refiningcommunity.com/presentation/dow-fluidized-catalytic-dehydrogenation-fcdh-the-future-of-on-purpose-propylene-production/. Accessed: 12 Aug 2019.

[113] Pretz MT, Luo L. A process for catalytic dehydrogenation. Patent No. WO2017196602A1, 2017.

[114] Blay V, Epelde E, Miravalles R, Alvarado Perea L. Converting olefins to propene: ethene to propene and olefin cracking. Catal Rev. 2018;60:1–58. DOI:10.1080/01614940.2018.1432017. Available at: https://www.mcdermott.com/What-We-Do/Technology/Lummus/Petrochemicals/Olefins/Ethylene-Production/Complementary-Technologies-Ethylene-Production/Olefins-Conversion-OCT.

[115] Ghashghaee M. Heterogeneous catalysts for gas-phase conversion of ethylene to higher olefins. Rev Chem Eng. 2017;34:595–655. DOI:10.1515/revce-2017-0003.

[116] Mol JC. Industrial applications of olefin metathesis. J Molec Catal A: Chemical. 2004;213:39–45. DOI:10.1016/j.molcata.2003.10.049.

Martyna Rzelewska-Piekut and Magdalena Regel-Rosocka

4 Technology of large volume alcohols, carboxylic acids and esters

Abstract: Paper describes industrial synthesis of the most important alcohols (methanol and ethanol), organic acids (acetic and lactic), and fatty acid methyl esters (biodiesel). Also, current industrial solutions and global trends in manufacturing of these chemicals are presented. Moreover, several alternative production technologies of these chemical compounds are discussed, which might successfully replace current commercial methods in the future.

Keywords: methanol, ethanol, acetic and lactic acids, fatty acid methyl esters (biodiesel)

4.1 Introduction

Conventional industrial synthesis of alcohols, carboxylic acids and esters is still prevailing against bioprocesses (except from ethanol production). However, an increasing relevance of biotechnological processes or biomass treatment in production of important chemicals results from concept of biorefineries (first, second and third generation), and stringent environmental restrictions. In the future an integrated biorefinery (third generation, 3G) will produce various products, which include electricity produced from thermochemical processes and bioproducts from the combination of sugar and other existing conversion technology platforms [1]. Now, edible plants are used mainly as a feedstock for bioprocesses, however, an increasing number of installations is developed for agricultural, forestry and industrial wastes as sustainable raw materials. Development in biorefinery concept is a consequence of increasing awareness of environmental threats coming from huge emissions of greenhouse gases (mainly CO_2), and shortage in fossil fuels and raw materials, and switch to renewable resources.

Concept of biorefinery (Figure 4.1) takes into consideration complex industrial system based on renewable raw materials producing energy, fuels, chemical intermediates and final products [2]. Platform chemicals, among others produced compounds, are crucial substances produced in biorefineries, e. g. intermediates from biomass treatment, such as propionic, lactic, succinic, levulinic acids, furfural, 5-hydroxymethylfurfural, isoprene or ethanol [3]. Still is vivid the question if biomass is able to replace fossil raw materials in the organic technology [4, 5]. Biotechnology of

This article has previously been published in the journal Physical Sciences Reviews. Please cite as: Rzelewska-Piekut, M., Regel-Rosocka, M. Technology of large volume alcohols, carboxylic Acids and esters *Physical Sciences Reviews* [Online] 2020, 5. DOI: 10.1515/psr-2019-0034.

https://doi.org/10.1515/9783110656367-004

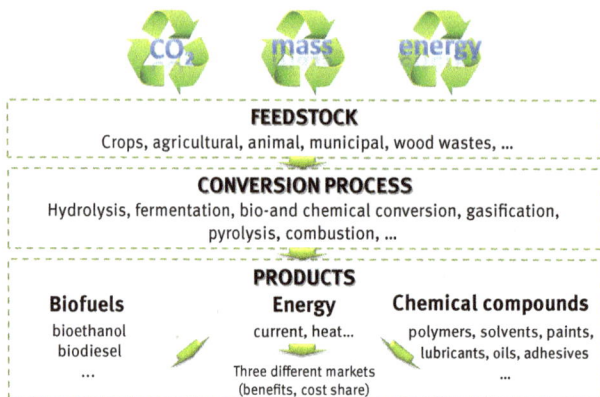

Figure 4.1: Concept of biorefinery (based on [2, 3]).

some high-volume products is advanced, e. g. ethanol, and could replace conventional methods in the future. However, bioprocessing is still more expensive than oil fractions or natural gas based syntheses, and, thus, it is not competitive and requires special legal regulations and financial support of governments.

In this chapter, various aspects of technology of the most important (from industrial point of view) alcohols, i. e. methanol and ethanol, organic acids (acetic and lactic), and fatty acid methyl esters (biodiesel) are presented. Current industrial solutions and global trends in manufacturing of these chemicals are described.

4.2 Technology of large volume alcohols

4.2.1 Methanol

4.2.1.1 Applications and production

Methanol has been one of the world's most widely used industrial chemicals since the 1800s. Formerly, methanol was obtained only in the process of thermal decomposition of wood. In this process, methanol was a by-product in the aqueous phase (containing acetic acid and acetone). From $1\,m^3$ of wood it was possible to obtain 2–6 kg of impure methanol [6].

From 1923, methanol is obtained synthetically by using synthesis gas (syngas) from various sources. It has also been discovered that syngas can be supplied by coal gasification. The process was initiated in Leuna (Germany) and it was a catalytic process that required difficult process conditions such as catalyst (zinc/chromium oxide), the high temperature (300–400 °C) and the high pressure (25–35 MPa). Since then, researchers have focused on improving process conditions; development of new catalysts to reduce temperature and pressure. In the 1940s, methanol began to be produced using from natural gas. In 1966, Imperial Chemical Industries (ICI, UK), presented a low-pressure methanol process (Cu/Zn/Cr catalyst, temperature 200–300 °C, pressure 3–12 MPa). The next process that improved the previous technology was Lurgi. The conditions in Lurgi process were lower temperature 230–250 °C, pressure 4–5 MPa than in ICI process [7, 8].

Despite technological progress, methanol is still mainly produced from fossil fuels and the production is based on natural gas reforming and coal gasification. Methanol production from coal is mainly used in China where natural gas is expensive and not easily available.

Nowadays global production of methanol reached 110 mln metric tons per year in more than 90 plants (North and South America, Asia, Africa, Europe) [9]. The world's largest producers are Methanex (Canada), QAFAC (Qatar Fuel Additives Company), Metafrax (Russia) and AMPCO (Atlantic Methanol Production Company, the National Gas Company of Equatorial Guinea).

High interest in methanol results from its relatively easy availability and low price, as well as the possibility of its processing into important chemicals, such as: formaldehyde (10 mln t of methanol is an intermediate for production of polymers and resins) acetic acid, methyl halides, methyl-t-butyl ether (MTBE), methylamines, methyl methacrylate (MMA). Methanol is also very important substrate or semipoduct for production of high-volume olefins (methanol-to-olefins process, *MTO*, described in the section *Simple hydrocarbon intermediates*) or gasoline (methanol-to-gasoline process, *MTG*), particularly in China [10]. Global demand for methanol in 2015 and world application of methanol as a substrate for further synthesis are presented in Figure 4.2.

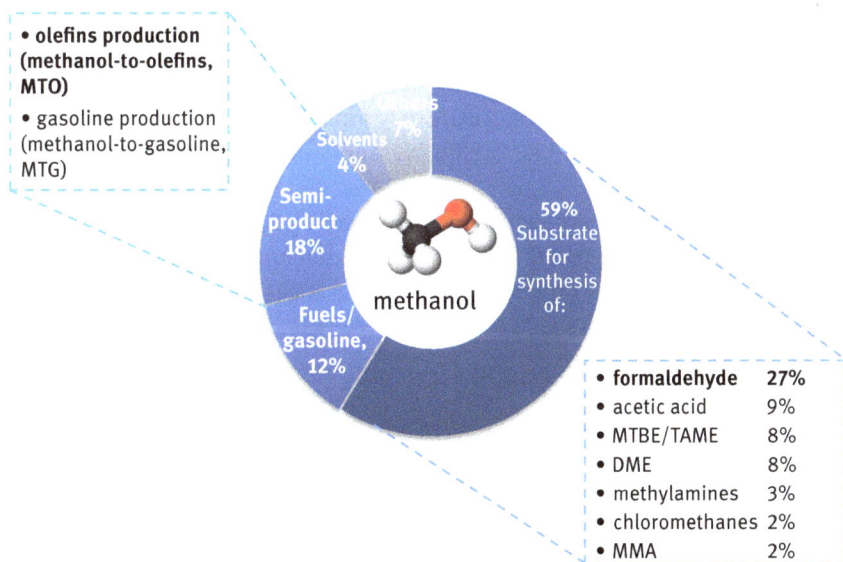

Figure 4.2: Global demand for methanol in 2015 [11].

More than half (59 %) of global demand for methanol is used as a substrate for synthesis. All commercial methanol production technologies focus on the three main sections [12]:
1. syngas preparation (accounts for about 60 % of the investment),
2. methanol synthesis,
3. methanol purification.

The three process sections may be considered and optimized independently. The carbon may be obtained from: coal (mainly in China), natural gas (almost 90 % of methanol is produced from natural gas) or heavy hydrocarbons. The oxygen and hydrogen may be obtained from electrolysis. Further, methanol may be converted into high molecular weight alternative liquid fuels [12–14].

4.2.1.2 High pressure process
On 16 January 1923, Mittasch and Pier during studying ammonia metal-catalyzed synthesis discovered that the hydrogenation of CO with a Fe-based catalyst, over 500 °C and 10 MPa, provided methanol. They tried to use waste gases derived from the ammonia synthesis to synthesize methanol using Fe-based catalyst. Unluckily, as a result of this reaction, not only methanol was formed but also long-chain alcohols, which can evolve in aldehydes and ketons, hydrocarbons C_nH_{2n+2}, dimethyl ether (DME). Moreover, the efficiency was low because of the presence of impurities (chlorine, hydrogen sulfide, methane) in the reactant gases. Due to these impurities, the catalyst was deactivated (a reaction between carbon monoxide and iron from catalyst) [13, 14].

Researchers carried out several methanol syntheses under the same conditions (320–450 °C, 25–30 MPa) but using different catalysts. The best results were obtained using two catalysts: ZnO/Cr_2O_3 and ZnO/CuO. The former catalyst (ZnO/Cr_2O_3) was highly stable to chlorine and sulfur compounds present in the syngas [13–15].

High pressure methanol production (HPM) began on 26 September 1923 at the Leuna site in BASF process based on zinc oxide/chromium oxide catalysts (ZnO/Cr_2O_3) [16, 17]. Methanol was produced from synthesis gas, by the following reactions (1–3) [8]:

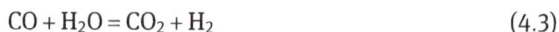

$$CO + 2H_2 = CH_3OH \tag{4.1}$$

$$CO_2 + 3H_2 = CH_3OH + H_2O \tag{4.2}$$

$$CO + H_2O = CO_2 + H_2 \tag{4.3}$$

The temperature and pressure ranged to 300–450 °C and 30 MPa, respectively. The reaction is exothermic and occurs with a decrease in volume of products (high pressure shifts the reaction equilibrium to the right). Hydrogen was added with a slight excess relative to the stoichiometry of the reaction. The BASF process was carried out to a small conversion (approx. 20 %). The unreacted mixture is recycled to the reactor [8, 15].

From 1927 BASF process has been used by both DuPont and the Commercial Solvents Corporation in the USA and for the next 45 years has been the dominant technology of methanol production [13]. Nowadays, the methanol production with zinc oxide/chromium oxide catalyst by the high pressure process is no longer profitable. In 1980s, the last methanol installation based on high pressure process was closed [15].

4.2.1.3 Low pressure process

The low or medium pressure processes of methanol synthesis (LPM) carried out in the world are similar. The reactions are carried out at temperatures of 220–260 °C and 5–10 MPa pressure. The differences occur in the synthesis reactor design and in the syngas preparation process. The composition of catalyst and method of obtaining the catalyst is also very important; a copper catalyst ($CuO/ZnO/Cr_2O_3$ or $CuO/ZnO/Al_2O_3$) are mainly used [6].

The low and medium pressure production of methanol is much less expensive than high pressure processes. The reduction in production costs results from [6]:
- pressure reduction – lower energy consumption for compression of syngas and recirculated gas in LPM than HPM,
- significant reduction of side reactions and reduction of consumption by about 10 % of synthesis gas,
- the possibility of using compressors powered by steam turbines, which are powered by steam from the syngas plant (which is integrated with the methanol synthesis installation).

Nowadays two LPM for methanol production from synthesis gas are widespread:
- *ICI method* (Imperial Chemical Industries) – selectivity over 99.5 %,
- *Lurgi method*.

4.2.1.3.1 Low pressure methanol production process by ICI method

The low pressure methanol synthesis by ICI method was carried out at 200–300 °C and 3–12 MPa, using a new type of catalyst: $Cu/ZnO/Al_2O_3$. The success of this method was associated not only with the invention of a new catalyst but also with the improvement in synthesis gas purification processes. The Cu/ZnO catalyst was thermally stabilized with Al_2O_3 and became more active, stable and selective than previously used catalysts. All commercially available catalysts are based on $Cu/ZnO/Al_2O_3$ or Cr_2O_3 with different promoters and additives. Promoters (e. g. Zr,

Cr, Mg and rare earth metals) have a significant influence on the Cu dispersion and also on their mobility on the catalyst surface [13–15].

In ICI process (Figure 4.3) a shelf reactor (4.1) is used, which usually contains five layers of catalyst. A fresh syngas is mixed with the recycled gas and then compressed (4.3). The gas mixture is heated by the heat from the exothermic reaction and (part of the stream) is delivered between layers, on several levels, by special distributors. The rest of the stream is used as the cooling factor (used to receive the heat of reaction). The products are cooled by heat exchange (4.3) with recirculating stream of feedstock and cooling water. Then, the gaseous phase under pressure is formed. The products are again cooled by air in the heat exchanger (4.6), in which methanol and water are condensed. An uncondensed gas is recycled to the process. The raw methanol is purified (4.8) by distillation in two columns connected in series [6]. The evolution of methanol production from the high pressure (BASF process) to the low pressure (ICI process) caused an increase in methanol yield to 105 tons per day [14].

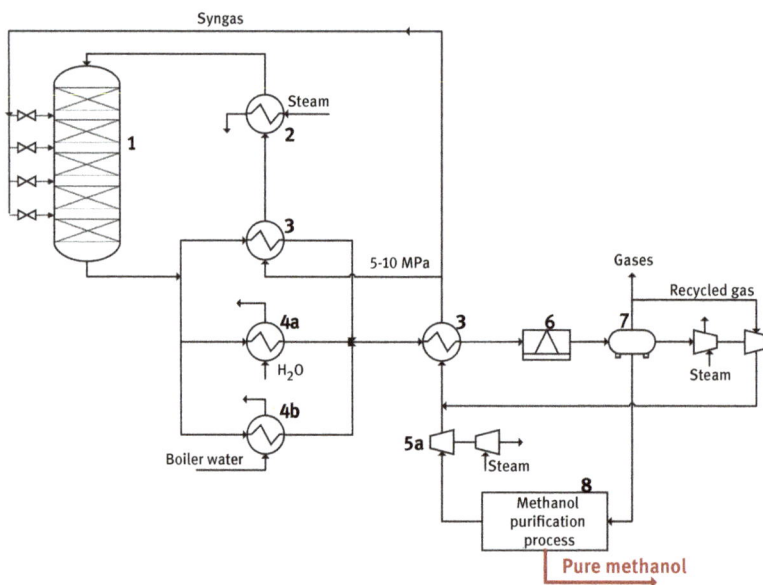

Figure 4.3: Scheme of the methanol production process by ICI [6]: 1 – shelf reactor, 2 – heater, 3 – heat exchangers, 4 – water heaters, 5 – compressors, 6 – air cooler, 7 – separator, 8 – methanol purification process.

4.2.1.3.2 Low pressure methanol production process by Lurgi method

The low pressure methanol synthesis by Lurgi method is carried out in shell-tube reactor (the catalyst is in the tubes). The process is carried out at 250–260 °C and 5–6 MPa. Water flows in the inter-pipe area/space and receives the heat generated during the exothermic reaction, thus, the high pressure steam is formed. The hot reaction gas is received from the bottom of the reactor and cooled. The heat from a hot reaction gas is used to heat the fresh synthesis gas. After re-cooling, the product is directed to a separator to recover raw methanol from the rest of reaction gases. The raw methanol is directed to the distillation columns for purification and the gas stream is recycled to the reactor [6].

An important stage in methanol production is its purification. In industrial processes, not only main reactions (reactions 1 and 2) occur but also side reactions:
-   methanization:

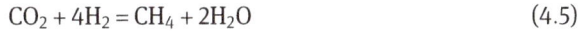

$$CO + 3H_2 = CH_4 + H_2O, \tag{4.4}$$

$$CO_2 + 4H_2 = CH_4 + 2H_2O \tag{4.5}$$

-   dimethyl ether (DME) formation:

$$2CH_3OH = CH_3OCH_3 + H_2O \tag{4.6}$$

-   formation of higher alcohols:

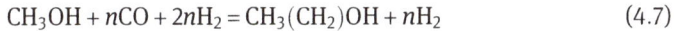

$$CH_3OH + nCO + 2nH_2 = CH_3(CH_2)OH + nH_2 \tag{4.7}$$

The purification process is carried out at low pressure in two or three distillation columns connected in a series. First, dissolved gases are removed along with low-boiling by-products. Then methanol is separated from ethanol and water is removed. Water can be reused in the steam system, and the mixture of ethanol and heavy organic compounds can be recycled back to the process. The Lurgi process produces about 1.4 t of steam at a pressure of 4 MPa per 1 t of methanol. This steam is used to drive the recirculating gas compressor, and then as a heat carrier in the raw methanol purification columns [6].

Scheme of methanol production using combined reforming by Lurgi method is presented in Figure 4.4.

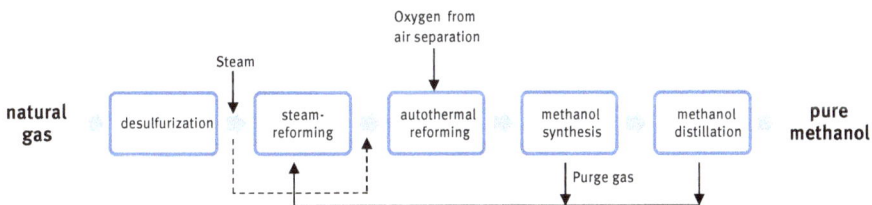

Figure 4.4: Methanol production using combined reforming by Lurgi method (Air Liquide Group) [18].

Air Liquide Group (ALG, France) is the world leader in gases, technologies and services for industry. The company was founded in 1902. In 2007, ALG acquired the engineering firm Lurgi, which was owned by Global Engineering Alliance (GEA Group AG, Germany). Nowadays, over 20 ALG plants can produce from 250 to 10,000 metric tons of methanol per day from any feedstock (e. g. natural gas, coal syngas). The availability of feedstock for methanol production is a key factor in choosing the proper installation. The company developed many new and innovative technologies [18]:

- low pressure methanol production (2500 ton per day),
- Lurgi Mega Methanol (5000 tpd),
- Lurgi Giga Methanol (10,000 tpd),
- Lurgi MTP methanol-to-propylene (1410 tpd),
- Lurgi GTP gas-to-propylene (1410 tpd) (combining Lurgi Mega Methanol and Lurgi MTP technology),
- G2G gas-to-gasoline (16,000 bbl/d),

and is also supplier of high-capacity methanol plants. The largest plants in the world constructed by ALG and using their technologies are shown in Table 4.1.

Table 4.1: Plants in the world using ALG technologies [18].

Company	Country	Feedstock	Process	Yield, mln tpy	Year
Atlas Methanol Company Point Lisa	Trinidad and Tobago	Natural gas	Lurgi Mega Methanol	1.65	2001
Zagros Petrochemicals	Iran	Natural gas	Lurgi Mega Methanol	1.65	2001, 2004
Methanex Punta Arenas	Chile	Natural gas	Lurgi Mega Methanol	0.80	2002
Petronas Labuan	Malaysia	Natural gas	Lurgi Mega Methanol	1.65	2005
Shenhua Ningxia Coal Industry Group	China	Coal syngas	Syngas to propylene	0.47	2006
Datang International Power Generation	China	Coal syngas	Syngas to propylene	0.47	2006
Shenhua Ningxia Coal Industry Group	China	Coal syngas	Lurgi MTP	0.47	2011
Zhong Tian Synergetic Energy	China	Coal syngas	Lurgi Mega Methanol	2 × 1.80	2011
Sinopec/Wanbei Group	China	Coal syngas	Lurgi Mega Methanol	1.80	2013
OCI Natgasoline	USA	Natural gas	Lurgi Mega Methanol	1.65	2014
Yuhuang Chemical	USA	Natural gas	Lurgi Mega Methanol	1.65	2014
ZeoGas	USA	Natural gas	Lurgi Mega Methanol	1.65	2014
BASF	USA	Natural gas	Gas to propylene	0.47	2015
Methanol Severny	Russia	Natural gas	Lurgi Mega Methanol	1.65	2015

tpy – tons per year

4.2.1.4 Catalysts

The synthesis of methanol from syngas is a high pressure exothermic reaction. At low temperatures the reactions take place very slowly and the catalysts used are not very active. Owing to that reason the reactions are carried out at temperatures above 200 °C [6].

Under the conditions of methanol synthesis using copper catalysts, copper in the catalyst is in the metallic form. It should be indicated that at temperature higher than 270 °C a recrystallization of the catalyst occurs and the catalyst quickly loses its activity [6].

Cu-based catalysts for methanol synthesis are very selective. In their presence, side reactions are very limited. As a result, the selectivity of methanol production is high (around 95 %) and crude methanol contains no more than 0.15 % of admixtures (reactions 4–7) which must be removed by distillation [6].

Figure 4.5 presents the conditions of nowadays and previous methanol production processes.

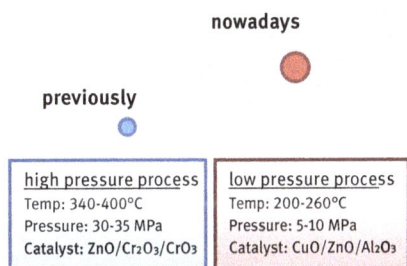

nowadays

previously

high pressure process	low pressure process
Temp: 340-400°C	Temp: 200-260°C
Pressure: 30-35 MPa	Pressure: 5-10 MPa
Catalyst: ZnO/Cr₂O₃/CrO₃	Catalyst: CuO/ZnO/Al₂O₃

Figure 4.5: Methanol production: nowadays and previously.

In the high pressure method ZnO/Cr_2O_3 was used as a catalyst. The temperature (340–400 °C) and pressure (3–3.5 MPa) of the process were high, which had an impact on the installation costs [13, 15].

One of the first patents for catalysts for industrial low temperature synthesis of methanol was proposed in Poland in 1950 by Błasiak [19]. The catalyst of the following composition: 62.5 % CuO, 25 % ZnO, 7.5 % Al_2O_3, 5 % H_2O was used for the first time in the world in Chemical Works at Oświęcim in Poland at the end of 1950s. However, the first full-scale trial of a copper catalyst 1963 was not successful [16].

In 1960s and 1970s, many types of low temperature/pressure catalysts (e. g. ICI and Lurgi companies) were discovered, and many of them have been widely used worldwide in industry till today. The compositions of these catalysts are similar to the Polish catalyst but in the production of these catalysts a variety of (protected by company secrets) additive components and various methods of preparation and processing are used [6].

Currently, all commercial methanol installations are based on $Cu/ZnO/Al_2O_3$ catalysts, which allow the process to be run in less restrictive conditions, in comparison to high pressure processes catalyzed with ZnO/Cr_2O_3. Composition of commercial catalysts used in low pressure methanol process is presented in Table 4.2. The pressure reduction significantly reduces installation costs due to lower energy consumption (necessary for compression of syngas and circulated gas) and reduction of side reactions (higher purity of the product) [31].

Table 4.2: Composition of LPM commercial catalysts.

Composition	Commercial name	Producer/licensor	Ref.
$CuO/ZnO/Al_2O_3$ $CuO/ZnO/Al_2O_3$ $Cu/$ $ZnO/MgO/Al_2O_3$ $CuO/ZnO/Cr_2O_3$ $CuO/ZnO/Al_2O_3$ $CuO/ZnO/MnO_2/$ Cr_2O_3 $CuO/ZnO/MnO_2/Al_2O_3/Cr_2O_3$ $CuO/ZnO/MnO_2/Al_2O_3$ $Cu/ZnO/Al_2O_3$ $CuO/ZnO/Al_2O_3$ promoted with MgO or ZrO_2	KATALCO KATALCO APICO TMC-3/1	Chemical Works Oświecim ICI ICI Lurgi BASF BASF BASF BASF Johnson Matthey INS Puławy, ZA Tarnów S.A	[20] [21] [22] [23] [24] [25] [26] [27] [28] [29] [30, 31]

The LPM catalysts produced by Johnson Matthey, named KATALCO 51-series, has been developed since 1960s [32]. The following features decide of their commercial success [8]:
- four-component: Mg-Cu-Zn-Al,
- life time: 4–6 years (even up to 8 years),
- sensitive to poisoning – necessity of desulfurization of syngas,
- very selective (about 95 %), limited side reactions,
- high purity of crude methanol – no more than 0.15 % admixtures.

Nowadays, KATALCOAPICO 51–100 is the successor of KATALCO 51-series and, as Johnson Matthey reports, "the premier methanol synthesis product now available to the methanol market" [32].

Hydrogenation of CO_2 to methanol has been studied using Ag, Zn, Cr or Pd as an active metal, instead of Cu. Nevertheless, the obtained results have shown that these catalysts are less active and less selective than Cu-based catalysts. The oxides: ZnO, Ga_2O_3 and Nb_2O_5 were used as catalyst promoters. The interest in using the carbon nanotube and carbon nanofibers as catalyst supports has increased recently. However, the cost of these supports is much higher than the commercial silica, alumina or zirconia supports [33].

The catalysts most commonly used in the synthesis of methanol from CO_2 and conditions of the process are shown in Table 4.3.

Table 4.3: The most commonly used catalysts in the synthesis of methanol from CO_2 [34].

Catalyst	T, °C	P, MPa	Selectivity to methanol, %
Cu/Zn/Ga/SiO$_2$	270	2.0	99.5
Cu/Ga/ZnO	270	2.0	88.0
Cu/ZrO$_2$	240	7.6	48.8
Cu/Ga/ZrO$_2$	250	2.0	75.5
Cu/B/Cr, Zr, Th/ZrO$_2$	170	5.0	72.9
Au/Zn/ZrO$_2$	220	8.0	100
Cu/Zn/ZrO$_2$	220	2.0	71.1
Cu/Zn/Al/ZrO$_2$	240	2.0	47.2
Pd/Zn/CNTs	250	2.0	99.6

4.2.1.5 Methanol production – perspectives

Nowadays, the climate change and the global warming are the serious threat for the Earth. CO_2 emissions into the atmosphere are a great problem that many scientists are struggled against. The promising way to decrease CO_2 emission to the atmosphere is catalytic conversion of CO_2 to methanol. Also, methanol production from biomass and charcoal as feedstocks can be competitive way for traditional industrial processes (synthesis process is similar to the process based on natural gas) [13, 34–36]. Below some perspectives for methanol production are presented.

4.2.1.5.1 Nippon Steel Corporation process

Nippon Steel Corporation (NSC, Japan) has proposed synthesis of methanol in liquid phase in a slurry reactor. Transferring the methanol synthesis reaction from the gaseous phase to the aqueous phase would significantly reduce production costs by reducing process pressure and temperature (150–200 °C, 5–10 MPa, the slurry Cu-based catalyst). The liquid phase method allows smaller reactor to be used than in the conventional methods. However, the low conversion of substrates hampers the application in a large-scale reactor [37].

4.2.1.5.2 CO_2 hydrogenation for methanol production

The presence of CO_2 in synthesis gas is advantageous, because the addition of 2 % CO_2 to syngas increases reaction rate a hundred times. However, higher carbon dioxide concentration than 2 % disturbs the progress of methanol synthesis [34].

A promising strategy for the future is using CO_2 as a raw material in methanol production. Carbon dioxide is inexpensive, available, non-toxic, non-corrosive and non-flammable and it can be easily stored. Application of CO_2 as a feedstock in production of methanol will not only contribute to reduce CO_2 emissions to the atmosphere but also to save the fossil fuels (the idea of sustainable development). The conversion of CO_2 to methanol runs according to reaction (4.2). The suggested temperature and pressure are 200 °C and 3.5–5.5 MPa, respectively, while $Cu/ZnO/Al_2O_3$ may be used as a catalyst (see also Table 4.3). Also, it is suggested that hydrogen may be generated by the electrolysis of water using electricity produced by renewable energy [34, 36].

4.2.1.5.3 Green freedom

"Green Freedom" (GF) is the concept of a large-scale production of gasoline and methanol from CO_2 captured from the air. The basis of this technology is the newly developed process of separation and capture of carbon dioxide from the air. GF is proposed to be an alternative methanol synthesis to the synthesis based on syngas (CO/H_2). However, this process requires the elaboration of new catalysts. The catalysts traditionally used in the industry for methanol synthesis $(CuO/ZnO/Al_2O_3)$, as mentioned before, lose their activity when the concentration of carbon dioxide in the reaction medium increases above 2 % [38]. Up to now it is rather laboratory concept not tested on a big scale.

4.2.1.5.4 Bioconversion of methane to methanol

In recent years, a concept for the process of converting methane to methanol using methanotrophic bacteria (*Methylomonas methanica*) has occurred. The traditional conversion of methane to methanol is an energy-consuming process (requires high temperature and pressure). Methanotrophic bacteria by means of the strong methane monooxygenase enzymes are able to bioconverse methane to methanol in ambient conditions. The potential of these bacteria is great but there are many technological barriers to overcome, i. e. relatively slow cell growth, methanol is not the final product in the oxidation of methane and methanol is toxic to most methanotroph (must be removed from the reactor) [39].

4.2.1.5.5 Self-sufficient hybrid installations

Pellegrini et al. [40] have proposed the concept of a hybrid process: traditional methanol production coupled to a power plant powered by waste gas. The produced energy can power a hybrid installation that may become self-sufficient in terms of energetic. Any excess energy generated may be sold. The investment cost of hybrid installation is higher than for a traditional installation but the calculated return investment time is about 5 years, which is profitable.

4.2.2 Ethanol

4.2.2.1 Application and production

Ethanol (ethyl alcohol) has a huge number of industrial applications, beginning with alcoholic beverages (fermentation to the beverages is known for more than 12,000 years [41], through automotive fuel (biofuel or, so called gasohol – mixture of petrol and alcohol) and solvent, to substrate/intermediate for various chemical synthesis. Ethanol can be converted by catalytic dehydration to diethyl ether or ethylene, by conversion with steam to acetone, by esterification to ethyl acetate, diethyl phthalate or other esters, by bacterial oxidation to acetic acid. Nowadays, ethyl alcohol is one of the main sustainable biofuels. Its sustainability lies in renewable feedstocks used for its production, and reduction in CO and CO_2 emission compared to combustion of fossil fuels. According to the US Environmental Agency it can reduce greenhouse gas emissions even up to 61 % compared to gasoline [42]. For example, in 2016 Japan government established its own sustainability standards for biofuels and only allowed import of ethanol with carbon emissions less than 50 % that of gasoline. A life cycle assessment (LCA) to calculate CO_2 emissions includes the entire supply chain, from cultivation of feedstocks to transportation of the final product to the consumer. The Japanese report indicated the only Brazilian sugarcane ethanol met the sustainability standards [43].

Bioethanol is an oxygenated additive (with blending octane rating of 114) to gasoline or diesel in order to increase octane number, reduce the dependency on petroleum, and limit green-house exhaust emissions. As an additive to gasoline anhydrous ethanol, i. e. 99.6 % of alcohol, is used, and the blend is called gasohol [44]. Due to the oxygen content, bioethanol has a more favorable combustion profile in spark engines and, consequently, enables the reduction of emissions of carbon monoxide, nitrogen oxide, hydrocarbons, carbon oxides and particulate matter into the atmosphere. Other reported advantages of ethanol blending are increased engine torque, brake power and thermal efficiency, broader ignition boundaries, higher combustion speed and higher evaporation temperature. However, bioethanol use as a fuel has some disadvantages such as its calorific value lower by 24 % relative to gasoline, corrosiveness on engine and combustion parts, and lower vapor pressure, which make cold starts (particularly in low temperatures), engine efficiency and engine durability more challenging compared to a pure gasoline-fuel based engine [45]. Noteworthy is the fact that ethanol is well soluble in water and biodegradable, which determines its low toxicity to the environment and minimizes the risk of ecological disasters in case of any leakage. In blends with gasoline, bioethanol is normally applied in the range of 5–20 % blends (E5–E20), in Europe typically 5 % or 10 %, and these fuels do not require special car engines. However, ethanol can also be used in higher concentrations. A mixture of between 65 % and 85 % (called E65 or E85) ethanol and the rest petrol or pure hydrous or anhydrous ethanol (E100) requires dedicated flex fuel vehicles (FFVs) or neat ethanol vehicles (used mainly in Brazil), which are able to run on E85, petrol, or any mixture of the two, without the need for separate fuel tanks.

In the USA, the revised Renewable Fuel Standard (RFS2) requires that primarily petroleum refiners blend 32 mln m^3 of cellulosic biofuels in 2019 to achieve 60 mln m^3 in 2022 [46]. In the European Union addition of biochemicals is regulated by Directive 2003/30/EC of the European Parliament and of the Council of 8 May 2003 on the promotion of the use of biofuels and other renewable fuels for transport [47]. However, for example, in Brasil also pure bioethanol is used as a biofuel. It is a consequence of ProAlcool program, a part of the Brazilian National Bio-Fuel Program commenced in 1975 as the answer to the first oil crisis, aiming to substitute gasoline for sugarcane alcohol in automobile use [44, 45, 48]. Today, most vehicles in Brazil (more than 90 %) are flex fuel cars, which are capable of running on gasoline or pure ethanol (a hydrated ethanol, i. e. 95.5 % of alcohol and 4.5 % of water) [44, 45, 48]. Though the Brasilian biofuel market depends on prices of sugar, or harvest quality (clearly visibly in recent 18 % increase in hydrous and anhydrous bioethanol prices due to rainfall disrupting the start of sugarcane harvesting) [49], Brasil is still the largest single consumer of the biofuel in the world.

Sugarcane-based bioethanol (first generation, 1 G) is used also as an intermediate for ethylene (bioethylene) production, e. g. in Brasil (the largest bioethylene producer, 200,000 t/y production capacity), India and China. However, such bioconversion installations still do not function without government subsidiaries (e. g. feedstock price support, capital cost support, income tax concessions or tax exemptions). In the USA, the Renewable Fuel Standard and California Low Carbon Fuel Standard, grants and loan guarantees from the US Department of Energy, Agriculture and Defense support production of cellulosic biofuels. In Europe, incentives within the Renewable Energy Directive and Research and Development programs support production of cellulosic and other biofuels [46, 50].

Ethanol production is carried out in two routes – chemical and microbiological [51]. The choice of the technological route depends on the feedstock available in a region of the world. In the regions rich in cheap biomass materials, production of bioethanol makes some economies independent of oil supplies [52]. However, it is noteworthy that most ethanol (93 %) is produced by fermentation, and only 7 % is produced by hydration of ethylene. LyondellBasell is the sole US producer of high purity synthetic ethanol [53]. African company SASOL [54] is the leader in production of synthetic ethanol by coal-to-liquids conversion; however, it produces also bioethanol in its Secunda plant in South Africa [55–57]. The trend to convert from synthetic ethanol to bioethanol production is visible also in Europe. Lillebonne plant in France, formerly SODES, after acquiring by Tereos BENP (Bio-Ethanol Nord Picardie) launched construction of the bioethanol installations and production began in 2007 [58]. The main producers of biobased or synthetic ethanol are listed in Table 4.4.

Table 4.4: Producers of biobased and synthetic ethanol and ethanol production capacities [53, 56, 57, 59–65].

Bioethanol	Synthetic ethanol
ADM, USA, 5.5 mln t/y (in 2016)	**INEOS Solvents**, UK, 900 000 t/y
Poet, USA, 4.7 mln t/y (in 2016)	(in 2018)
Green Plains Renewable Energy, USA, 4.3 mln t/y (in 2016)	**SADAF**, Saudi Arabia, 300 000 t/y
Valero Energy Corporation, USA, 3.55 mln t/y (in 2016)	(in 2005)
Abengoa Bioenergy, Spain, 2.5 mln t/y (in 2011)	**Sasol** (Secunda plant), RPA, 285
Jilin Fuel Ethanol, China, 600 000 t/y (in 2004)	000 t/y (in 2013)
CropEnergies, Germany, 550 000 t/y (in 2011)	**LyondellBasell**, Equistar
Tereos BENP, France, 430 000 t/y (in 2011)	Chemicals (Tuscola plant), USA,
GranBio, Brasil, 65 000 t/y 2 G (in 2014)	160 000 t/y (in 2016)
China New Energy, China (no data on capacity)	**Mitsubishi Chemical**, Japan (no
Bevap Biorenergia, Brasil (no data on capacity)	data on capacity)

The world's largest ethanol producers are Brazil and the United States, their share in worldwide production, with more than 90 mln m^3 of ethanol in 2018, accounts for 85 % (Figure 4.6) [66–68].

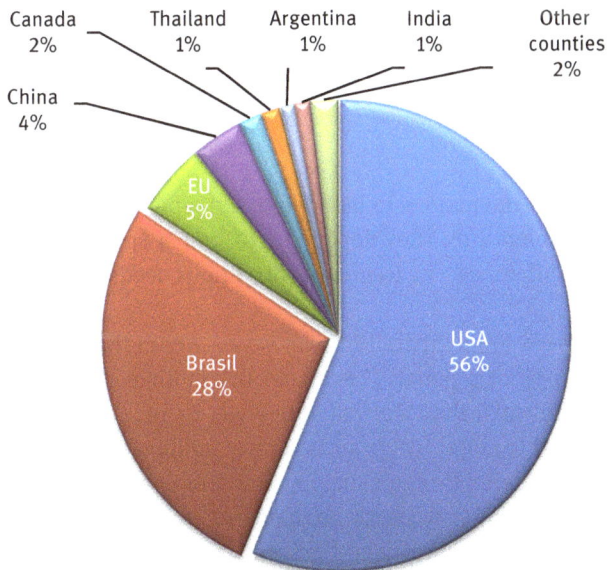

Figure 4.6: Bioethanol producers in 2018 according to [66].

4.2.3 Production of bioethanol

Despite the potential of various feedstocks including fermentable sugars (sugar cane, sugar beets, sweet sorghum), starches and fructoses (corn, potatoes, rice, wheat, agave), cellulosics (stover, grasses, corn cobs, wood, sugarcane bagasse), due to regional availability and technological challenges, almost 90 % of the 108 mln m^3 of bioethanol produced worldwide in 2018 was produced from corn and sugarcane [66]. In other regions, e. g. in China, sweet sorghum is applied in large quantities as a crop for such bio-based chemicals production as bioethanol or biobutanol [42].

Prior to alcoholic fermentation, polysaccharides must be hydrolyzed to simple sugars such as glucose or fructose. This reaction is called saccharification (Figure 4.7, reaction 8) and is driven by invertase enzyme. Afterwards, the simple sugars (hexoses or/and pentoses) are enzymatically fermented by zymase (Figure 4.7, reactions 9 and 10) produced by the yeast *Saccharomyces cerevisiae* (Melle-Boinot process in Brasil) [44]. During fermentation, sugars are converted into ethanol, energy, cellular biomass, CO_2 and other by-products by yeast cells.

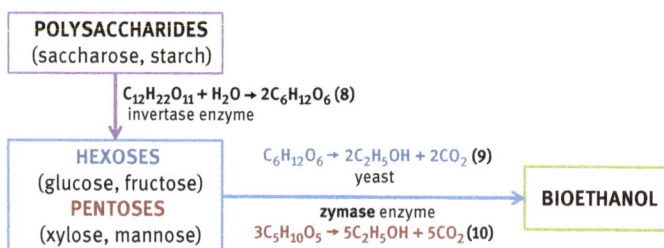

Figure 4.7: Reactions (8-10) of simple sugar fermentation to ethanol.

Many options of industrial sugar fermentation to bioethanol are reported in patent [69–71] and research [42, 44, 51] literature. They depend on technical opportunities or type of feedstock applied. Differences in treatment of various feedstocks are shown in Figure 4.8.

For example, corn is processed to bioethanol in two ways – dry milling (90 % of ethanol) and wet milling (10 % of ethanol) [73–75]. In dry milling, the entire grain kernel is first ground into "meal", then slurried with water to form a "mash", and afterwards starch is converted to sugar by enzymes. In wet milling, the grain is first separated into its basic components through soaking and grinding. Then, after separating the remaining fiber, gluten and starch components, the starch is fermented into ethanol (a process similar to the dry mill process). In the case of sugarcane processing, before the sugarcane juice is fermented to ethanol, sugar concentration is adjusted to desired level with molasses and the juice from

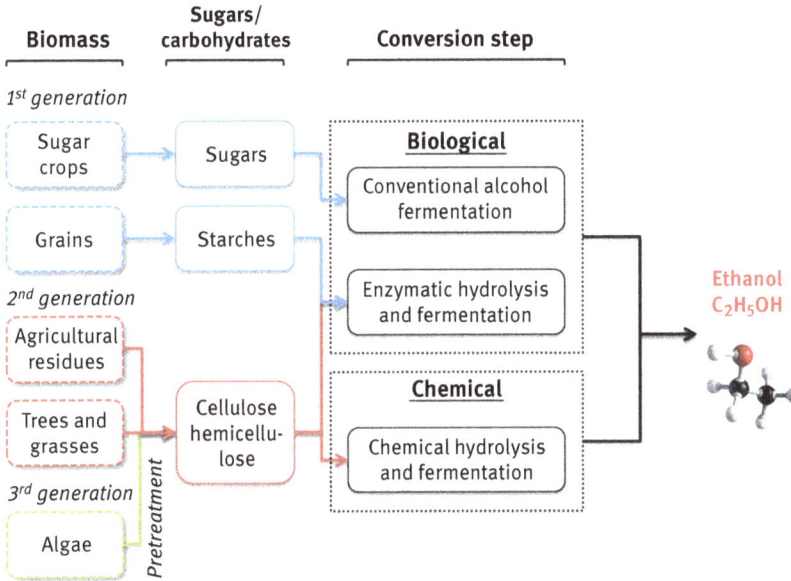

Figure 4.8: Scheme for bioethanol conventional fermentation from various feedstocks (based on [72]).

evaporators. It is noteworthy that before fermentation of sugarcane juice no saccharification is necessary [76]. After fermentation (at temperature around 25–34 °C, in time range from 6 h to even 72 h) the resulting broth (called "wine", in case of corn fermentation "beer") has about 6–12 vol% of alcohol. Yeast cells are separated from the wine by centrifugation, resulting in a concentrated yeast cell suspension (the yeast "cream" – 60–70 % of cells), and the yeast-free wine is sent to the distillation unity, where it is purified and dehydrated [42, 44, 73]. A general scheme of bioethanol production is shown in Figure 4.9.

Besides feedstock, the difference in corn and sugarcane processing lies also in the recycling of yeast cells. In the Brazilian distilleries, after the end of fermentation cycle, the yeast cells (cream) return to large-volume tanks (250–3000 dm^3) for a new fermentation cycle, while corn distilleries in the USA do not recycle yeast cells, and they direct the fermented mixture (also yeast) to distillation. Fermentation without yeast recycling is less effective.

The fermentation processes have been developed to increase yield of ethanol production, minimize the waste generated in each stage, and reduce investment costs. An example of such an improvement is cultivating optimized yeasts with better tolerance to alcohol, and capable to compete with wild strains contaminating the industrial processes. Also, introducing continuous or multistage fermentation systems improves flexibility and decreases investment costs [44]. However, among typical processes applied in distilleries (i. e. batch, fed-batch, and continuous), still fed-batch processes are most common (70–75 %) due to maintenance of the

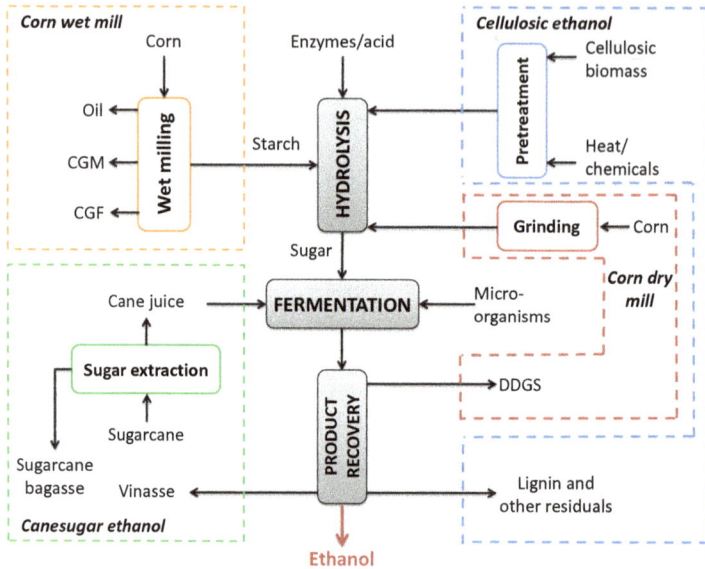

Figure 4.9: Scheme for bioethanol production by fermentation from various feedstocks; DDGS – distillers dried grains with solubles, CGM – corn gluten meal, CGF – corn gluten feed (based on [77]).

maximum concentration of viable cells, prolongation of cell lifetime, and less inhibition of yeast by the high substrate concentration [42].

To increase ethanol yields, new technologies focus on the reduction of vinasse volumes by increasing the ethanol concentrations in wine during fermentation above 9–10 vol%. However, it is difficult due to susceptibility of yeast for high alcohol concentration. It was discovered that some industrial strains are more robust and viable for fermentation with high ethanol concentrations in wine than other traditional yeasts. Thus, isolating these yeast strains or introducing some mutations to typical *Saccharomyces cerevisiae* open new opportunities for Brazilian distilleries to reduce vinasse volume by increase in ethanol concentration in wine. Also, treatment of cellulosic sugarcane or corn biomass and conversion into fermentable sugars for 2 G (second generation) ethanol production is a promising alternative to increase the alcohol production [68].

Approximately 4–8 % of the feedstock is converted into glycerol – one of the major by-products obtained during industrial fermentation to bioethanol. The yield of ethanol production can be increased by reducing the by-product formation by changes in genotype of the yeast strains and introduction of such mutants [42].

1 G ethanol is produced from edible plants that compete with food resources, thus, recently 2 G biomass processing has been developed and is quite advanced now. Bioethanol derived from lignocellulosic materials (2 G) takes account of use of inedible biomass resources such as forestry biomass (sawdust, wood crops),

agricultural wastes (crop residues, straws, sugarcane bagasse), dedicated energy crops (switchgrass, *Miscanthus*, biomass sorghum and energy cane), municipal solid wastes, industrial residues (wastes from paper and pulp industries) [74, 78, 79]. Production of bioethanol (more generally also biofuels) from lignocellulosic biomass is more complicated than from 1G feedstock. Processing of complex crystalline structure of the agricultural biomass is a challenging multi-step treatment including pretreatment (chemical/thermal), enzymatic hydrolysis, microbial fermentation, and purification of the product (Figure 4.9). The main components of lignocellulosic biomass such as cellulose (40–55 % of dry mass), hemicellulose (structural carbohydrates, 24–40 % of dry mass) and lignin (a complex heterogeneous phenolic polymer, 18–25 % of dry mass) must be pretreated by physicochemical methods (e. g. ozonolysis, steam treatment/explosion, wet oxidation, or ammonia fiber explosion, AFEX), chemical methods (acid hydrolysis, alkaline hydrolysis, and organic solvent treatment), electric methods (e. g. pulsed electric field treatment), biological methods (with bacteria or fungi producing ligninolytic enzymes), or a combination of these methods. The most complicated is lignin processing, as it gives variety of aliphatic and aromatic degradation products, employing surfactants to improve lignin-enzyme interaction. In short, the pretreatment step should reduce the degree of crystallinity of cellulose and hemicellulose to make the biomass more accessible to the hydrolytic step to achieve high yields of simple sugars for fermentation. Additionally, the biomass is delignificated to improve the porosity of lignocellulosic material prior to enzymatic hydrolysis. Hydrolysis and further fermentation can be realized as multi-step separate operations, however, also simultaneous saccharification and fermentation (SSF) is carried out combining hydrolysis and fermentation of cellulose in one step, thus, avoiding potential loss of sugar and reduced end-product inhibition [79]. Fermentation is followed by distillation and dehydration to obtain pure and concentrated ethanol.

Lignocellulosic biomass conversion to bioethanol seemed to be industrially advanced enough to construct the largest biorefinery of cellulosic ethanol in the world, commenced by DuPont in 2015, however, not profitable enough to be economically viable. Although, the biorefinery, powered by corn stover, produced 113,000 m^3 of bioethanol per annum, it generated losses, and finally was sold to VERBIO Vereinigte BioEnergie AG (a German biofuels company) in 2018. The new owner plans to convert the plant to produce renewable natural gas (RNG) [80]. Another company, Abengoa Bioenergy Biomass sold the Hugoton cellulosic ethanol plant (95,000 m^3 bioethanol capacity) to Synata Bio in 2016 but it is still unclear when cellulosic ethanol production will resume there. In the US, only three (DuPont, Abengoa, POET-DSM) out of nine planned cellulosic ethanol refineries were put into operation between 2013 and 2015, and only one has been working till today [46]. The only of the working cellulosic ethanol plants Project Liberty is a demonstration plant in Emmetsburg (planned bioethanol capacity 114,000 m^3, not yet reached), built and run by POET-DSM [46, 81]. Another working 2G ethanol plant is Raízen Energia's refinery in Piracicaba, Brazil, opened in 2014 with an 50,000 m^3 per annum nameplate capacity [50]. On the other hand, the world's first commercial-scale refinery (50,000 t/y bioethanol capacity) – the Crescentino cellulosic refinery (Italy) built and operated by Beta Renewables (Grupo M&G), was shut down in 2017 [82]. However, in 2018 Clariant confirmed construction of the large-scale commercial Sunliquid plant (in Romania) for the production of cellulosic ethanol from agricultural residues [83].

2G plants meet more obstacles in processing cellulosic biomass than grain-ethanol installations. 2G biorefineries face such challenges as increased investment cost (e. g. more resistant construction materials, more power, longer time of processing than in 1 G due to tough and abrasive feedstock), provision of steady supplies of feedstock, fluctuating markets (changes in crude oil and natural gas prices) and stalled policy decisions. They are also strongly dependent on government support, e. g. via mandated volumes of cellulosic ethanol, because prices of cellulosic ethanol are still higher than grain ethanol. For example, in 2018 Raízen Energia could sell cellulosic ethanol for $0.57 per dm^3, while corn ethanol in the US is sold even below $0.38 per dm^3 [46, 50, 81]. Some voices [50] indicate that "the illusion that cellulosic biofuel production has dramatically increased recently reflects a redefinition of 'cellulosic' to include transport fuels made from landfill gas, biogas and corn kernel fibre". Authors of the Biofuelwatch report express their concern for the development of genetically engineered trees, crops and microbes to provide biorefineries with better quality feedstock.

3G biofuels refer to biofuels derived from algae and has been established to emphasize that algae are capable of much higher yields with lower resource inputs than other feedstock. Algae feedstock is beneficial due to the fact that algae use a diverse array of carbon sources, and are proposed to be integrated directly to carbon emitting sources (power plants, industry) where they could directly convert emissions into usable fuel and reduce total emissions. Though, there are several new patents [84–86] on recovery of starch from algae (e. g. *Sargassum, Glacilaria, Prymnesium parvum* and *Euglena gracilis*) for biofuel production, it seems that in the near future no large-scale implementation of algae to produce biofuel will occur for a long time because of large amounts of water, nitrogen and phosphorous necessary to grow algae. It must be emphasized that production of fertilizer to grow biofuel algae would produce more greenhouse gas emissions than were saved by using algae-based biofuel [48, 51].

To overcome problems with fermentation of lignocellulosic biomass to ethanol, new methods are proposed, on the one hand, integrating biomass conversion and direct synthesis of ethanol from syngas [50, 78] or, on the other hand, applying hybrid conversion technology for the conversion of renewable feedstocks or gas waste streams containing CO, CO_2 and H_2 to biofuels and chemicals by fermentation with *Clostridium ragsdalei* bacteria [50, 87–89].

Ineos New Planet BioEnergy plant (30,000 m^3 of ethanol per annum) started in 2012 at full commercial scale of conversion of a variety of different lignocellulosic waste biomass feedstock to bioethanol and renewable electricity (up to 6 MW of electricity). The Ineos Bio technology combined gasification of biomass to syngas (CO, CO_2 and H_2 mixture), and syngas fermentation. After four years of processing, in 2016, Ineos Bio announced its intention to sell ethanol business, including the New Planet BioEnergy plant. Alliance Bio-Products declared to convert the Ineos process into cellulose-to-sugar (CTS) technology [90]. The CTS technology is based on the solid acid catalyzed (kaolin or bentonite) hydrolysis of cellulose-containing materials to soluble C_5 and C_6 sugars, with pure lignin as a by-product, in a CTS reactor (with no enzymes) [91]. Finally, Alliance Bio-Products withdrew its offer, and the plant was sold for other production.

The gasification-fermentation technology was sold to Chinese biofuel company Jupeng Bio [88, 92]. In 2018, Jupeng Bio and Lu'An Group signed an agreement to construct a 20,000 t/y demonstration plant, as a phase I of a 200,000 t/y bio-fermentation fuel ethanol project in Tunliu county, Shanxi province. Beginning of the plant operation was planned for the first quarter of 2019; however, up to now there has been no information on its commissioning. The demonstration plant project assumes integration of CO_2 reduction technology (industrial off-gas recycling) with the Jupeng Bio patented syngas bio-fermentation technology (Figure 4.10). The SARI CO_2-CH_4 reforming technology will convert CO_2 and CH_4 to syngas stream (CO and H_2), which is a feedstock for bio-fermentation plant [89]. The fermentation step is at the heart of the Jupeng Bio process technology. In a fermenter, at low temperature and pressure, naturally occurring anaerobic bacteria continuously convert the syngas to bioethanol with a high conversion rate and selectivity within minutes (10 min from waste input to bioethanol output). Jupeng Bio emphasizes that the planned process involves good tolerance to variations in syngas composition and to common catalyst poisons, fewer process steps than in conventional fermentation

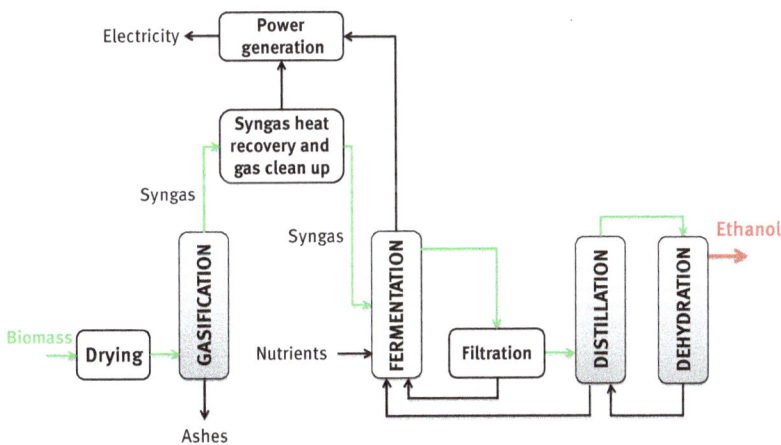

Figure 4.10: Jupeng Bio gasification-fermentation technology for bioethanol production [92].

are necessary to obtain high yield, high energy efficiency, low bioethanol production costs and attractive investment returns [88, 89].

4.2.3.1 Production of synthetic ethanol

As it was mentioned before, ethanol can be produced by catalytic direct or indirect hydration of ethylene, as well. Only 7 % of the global ethanol production comes from hydration of ethylene, mainly because of low economic viability of the process, as ethanol is a cheaper chemical than ethylene [57].

Reaction of hydration means that water molecule is bound to ethylene molecule. Mechanism of olefins hydration runs according to Markovnikov's rule which states that the hydrogen atom is added to the carbon with the greatest number of hydrogen atoms while the X component (OH^-, in the case of hydration) is added to the carbon with the least number of hydrogen atoms. Hydration is reversible and exothermic ($\Delta H = -46$ kJ/mol) reaction that can be generally written as:

$$CH_2 = CH_2 + H_2O = C_2H_5OH \tag{4.8}$$

However, actually, it is two-step electrophilic addition [51]:
1) carbocation formation (intermediate) – the slowest step:

ethylene:

$$CH_2 = CH_2 + H^+ = {}^+CH_2CH_3 \tag{4.9}$$

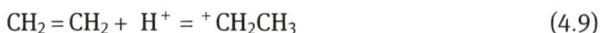

higher hydrocarbons:

$$RCH = CH_2 + H^+ = R^+CHCH_3 \tag{4.10}$$

2) the positively charged intermediate combines with an anion:

– indirect hydration

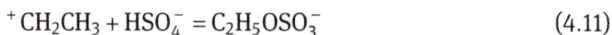

$$^+CH_2CH_3 + HSO_4^- = C_2H_5OSO_3^- \tag{4.11}$$

$$C_2H_5OSO_3^- + {}^+CH_2CH_3 = (C_2H_5O)_2SO_2 \tag{4.12}$$

– direct hydration

$$^+CH_2CH_3 + H_2O = C_2H_5OH + H^+CH_2{}^+ \tag{4.13}$$

As the hydration is exothermic and runs with decrease in product volume (one molecule is formed on the product side from two molecules of substrates), according to Le Chatelier's principle, a shift towards the high yield of ethanol (to the right) is favored if the temperature is lowered and pressure is increased in the equilibrium mixture. However, lowering the temperature results in a significant

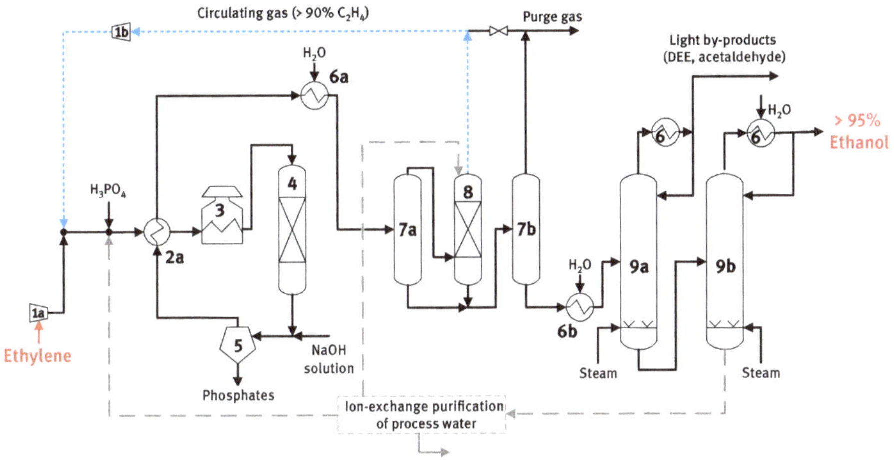

Figure 4.11: Direct synthesis of ethanol: 1 – compressors, 2 – heat exchangers, 3 – tubular furnace, 4 – hydration reactor, 5 – phosphate separator, 6 – coolers, 7 – separators (a – high-pressure, b – low-pressure), 8 – scrubber, 9 – rectification columns; DEE – diethyl ether (based on [6]).

decrease in reaction rate, therefore the reaction is carried out at 300 °C and catalyst is used to accelerate the reaction. At this temperature, approximately 5 % of ethylene is converted into ethanol at each pass over the catalyst in industrial process. The schematic representation of the direct synthesis of ethanol is shown in Figure 4.11.

The catalytic direct ethylene hydration (introduced by Shell in 1947) includes three steps, i. e. reaction, recovery and purification. Ethylene (very pure 97–99 %) is mixed with steam at a molar ratio of 0.4–0.9, preheated to 250–300 °C in tubular furnace (4.3), and then passes over an acidic catalyst (unhydrous phosphoric (V) acid on solid silica) in a fixed bed reactor (4.4) at 7–8 MPa. The steam-to-ethylene ratio should be less than one to avoid catalyst losses. The ethylene conversion in one run is about 5 %, thus, it is returned to the reactor several times. The ethanol selectivity is 98.5 %. Lost phosphoric acid is neutralized in the separator (4.5). Ethanol is separated from the vapor mixture of reactants, after cooling down, by dissolving in water in the scrubber (4.8). About 15–20 % solution of ethanol in water formed in scrubber is then purified and concentrated by distillation (9a and b) to at least 95 % ethanol. By-product such as acetaldehyde can be sold or hydrogenated to ethanol, other by-products are formed, namely diethyl ether, or butylenes and higher olefins (result of ethylene polymerization at high pressure). Vapor from the scrubber (4.8) – rich in ethylene – is recycled to the reactor. Generally, the direct hydration of ethylene is more common in industry than the indirect one. Most operating synthetic ethylene plants apply direct hydration process (Table 4.4) [6, 51, 57]. As an example, INEOS Solvents, Europe's largest producer of synthetic ethanol, can be given, who employs three stage process that ultimately produces a single grade of ethanol (DRAA – Double Rectified Absolute Alcohol) which is 99.9 wt% pure alcohol (anhydrous).

Ethanol synthesis by an indirect hydration (esterification-hydrolysis, Figure 4.12), the process known since 1930, can use as a feedstock not only pure ethylene but also an ethylene-ethane fraction derived from olefin pyrolysis, containing about 50–60 % ethylene. Application of ethylene-ethane feedstock is profitable because costly operation of separation of pyrogas fraction is avoided. As ethane does not participate in reactions, it is recycled to olefin pyrolysis. In the first step of the synthesis (4.2), ethylene is absorbed in 96–98 % sulfuric acid (acidic catalyst of hydration) resulting in mono and diethyl sulfates as intermediates. This step is carried out at 55–80 °C and 1–3.5 MPa. Gases from upper part of the reactor contain lost sulfuric acid which is neutralized in scrubbing and neutralization unit. Downstream from the reactor flows to hydrolysis (5) in 70–100 °C and 0.5 MPa to decompose ethyl esters of sulfuric acid with water. Hydrolysis is followed by distillation to separate ethanol from diluted sulfuric acid (6). Distillate is directed to neutralization column (7) to remove traces of sulfuric acid. The resulting crude product contains 25 % ethanol, 5 % diethyl ether (by-product) and water, and is distilled again to concentrate ethanol [6].

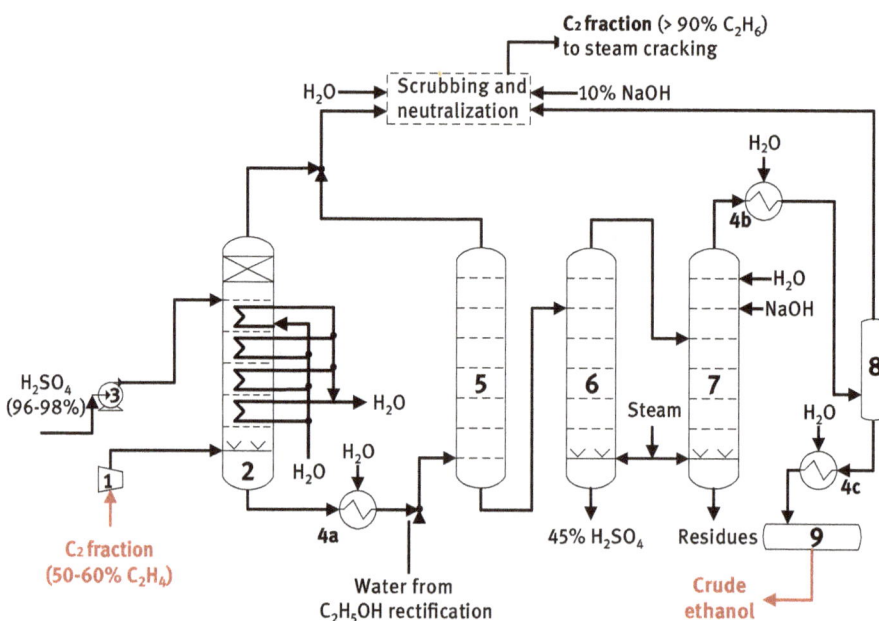

Figure 4.12: Indirect synthesis of ethanol: 1 – compressor, 2 – absorber, 3 – pump for sulfuric acid, 4 – coolers, 5 – hydrolyser, 6 – distillation column, 7 – neutralization column, 8 – separator, 9 – crude ethanol container (based on [6]).

4.2.3.2 Dehydration (drying) of ethanol

For the processes of ethanol production (either fermentation or chemical synthesis) final drying of alcohol is an important step affecting quality of the product and increasing investment and operation costs of the plant. Dehydration is commonly two-step process: in the first ethanol is rectified to concentration of 92.4 wt%, and, in the next step, it is further dehydrated by various methods [93]. Ethanol-water mixture forms an azeotrope at 95.6 wt% (89.5 mol%) ethanol content which needs special dehydration techniques to increase ethanol concentration, even up to an anhydrous alcohol (99.9 wt%); for blend with gasoline or biofuel itself, >99.3 wt% ethanol is required [76]. It cannot be done by simple distillation. Currently, three main dehydration (drying) methods are applied in ethanol industry (see Figure 4.13) [94]:

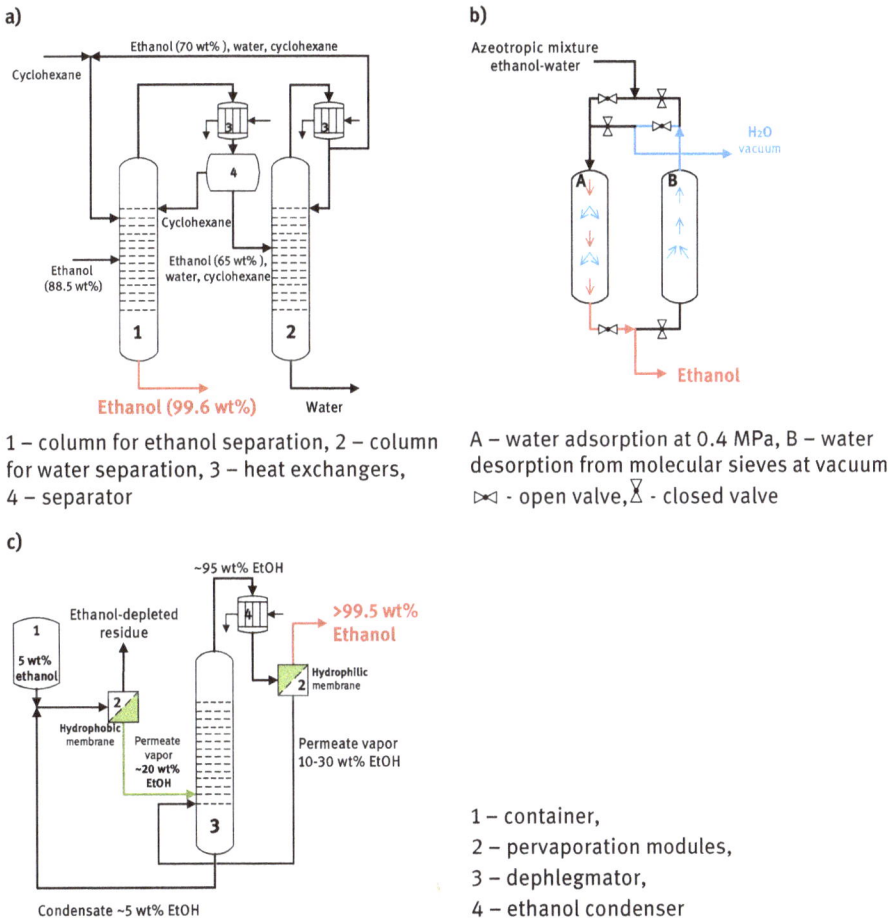

a)

Ethanol (70 wt%), water, cyclohexane
Cyclohexane
Cyclohexane
Ethanol (65 wt%), water, cyclohexane
Ethanol (88.5 wt%)
1   2
Ethanol (99.6 wt%)   Water

1 – column for ethanol separation, 2 – column for water separation, 3 – heat exchangers, 4 – separator

b)

Azeotropic mixture ethanol-water
H₂O vacuum
A   B
Ethanol

A – water adsorption at 0.4 MPa, B – water desorption from molecular sieves at vacuum - open valve, - closed valve

c)

~95 wt% EtOH
Ethanol-depleted residue
>99.5 wt% Ethanol
5 wt% ethanol
Hydrophilic membrane
Hydrophobic membrane   Permeate vapor ~20 wt% EtOH
Permeate vapor 10-30 wt% EtOH
Condensate ~5 wt% EtOH

1 – container,
2 – pervaporation modules,
3 – dephlegmator,
4 – ethanol condenser

Figure 4.13: Methods for ethanol dehydration (drying).

a) azeotropic distillation with cyclohexane,
b) adsorption with molecular sieves (zeolites),
c) pervaporation (membrane technique).

Azeotropic distillation with cyclohexane is used in 60 % distilleries in Brazil, and is effective for streams containing 10–85 wt% of ethanol. Additional solvent – cyclohexane (also benzene till the 90s) – changes azeotrope composition, and forms a new tertiary heterogeneous azeotrope with ethanol and water (Figure 4.13a). The azeotrope is removed at the top of the column, while anhydrous ethanol is produced at the bottom. In spite of high energy consumption, this method is the most common in sugarcane mills [76]. Similar to azeotropic distillation is, also popular in Brazilian distilleries, extractive distillation with monoethyleneglycol (MEG). It is less energy consuming and produces ethanol less contaminated with solvent than azeotropic distillation. Extractive distillation, also called homogeneous azeotropic distillation, assumes that addition of a third component (MEG) changes the relative volatility of ethanol and water, promoting separation of anhydrous ethanol in an extractive column (collected at the top). MEG-water mixture from the bottom of the extractive column is directed to the second column where the solvent is recovered.

Adsorption with molecular sieves (zeolites) is realized as pressure swing adsorption or temperature swing adsorption (Figure 4.13b). Zeolite molecular sieves are employed to remove water from ethanol. Vapors of ethanol and water pass through a column containing pellets of a molecular sieve of 3 Å (0.3 nm) pore size. Only water molecules (diameter 0.28 nm) are able to pass through the pores, where are trapped in the cages of the zeolite, as the diameter of ethanol molecules is too large (0.44 nm). Ethanol passes through the column and is collected at the bottom. Two or three columns are used in parallel, and operate in a cyclic way. One is regenerating the zeolite bed and the other one or two are removing water at about 150 °C and 0.4 MPa. The regeneration of molecular sieves applies lower pressure (pressure swing adsorption) or increased temperature approx. 230 °C (thermal or temperature swing adsorption) to remove water from zeolite pores. Pressure swing adsorption is more common than the thermal method. Adsorption is characterized by low energy consumption, and has the advantage of producing ethanol with no solvent contamination [76, 95].

Pervaporation (PV) or vapor permeation (depending if the separation is performed in liquid phase or gaseous phase) is an optimal process to break azeotropes, concentrate the product from any water level to very low water content, and solve problems at competitive costs. A liquid stream containing two or more miscible components (e. g. ethanol-water) is contacted with a semipermeable nonporous polymeric membrane or molecularly porous inorganic membrane (such as a zeolite membrane), while a vacuum or gas purge is applied to the other side. The desired compound from the liquid stream, e. g. water, sorbs into/onto the hydrophilic membrane, permeates through the membrane, desorbs in the permeate side, and evaporates into the vapor phase ("pervaporate" = permeate + evaporate). The resulting vapor, the permeate, is then condensed [51, 96]. As pervaporation is considered a low-energy process, and can be integrated with a fermentation unit, its application arouses interest of bioethanol producers. For example, in Finland St1 Biofuels Oy producer of cellulosic bioethanol from waste (from bakeries, breweries, factories producing enzymes, sweets or potato flakes) uses combined fermentation-evaporation hybrid processing to obtain 85 wt% ethanol, which is further transported to a dehydration plant, concentrated to 99.8 wt% in the currently world biggest ethanol vapor permeation unit (annual capacity of 88 000 m^3 of 99.8 wt% ethanol), and blended to the final fuel product [93, 97].

An example shown in Figure 4.13c, considers condensation of diluted ethanol, and includes two PV modules integrated with dephlegmator. Such a system enables the use of dehydration membranes with lower water-ethanol separation factors. In the first PV module with hydrophobic membrane, the permeate is enriched in the transported ethanol, which is further concentrated and recovered in the dephlegmator-condenser system. Final dehydration takes place in the second PV module with hydrophilic membrane (polyvinyl alcohol, PVA), where water is removed by pervaporation, and at least 99.5 wt% ethanol is obtained in the retentate [98].

4.3 Technology of large volume carboxylic acids

4.3.1 Acetic acid

4.3.1.1 Applications and production

Acetic acid (AA, CH_3COOH) is one of the first organic compounds obtained by humans. In ancient times acetic acid was produced by vinegar fermentation in the presence of aerobic acetic bacteria [99, 100]. The solutions containing sugar or ethyl alcohol were used as raw materials for the vinegar production (Figure 4.14). To the present day vinegar (10 % CH_3COOH) is obtained only by fermentation.

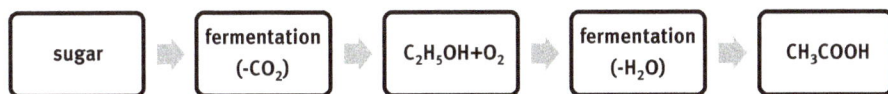

Figure 4.14: Scheme of the acetic acid (vinegar) production by fermentation (based on [99]).

In XIX century AA was obtained as a by-product in the process of thermal decomposition of wood [99]. Nowadays, acetic acid is produced synthetically using mainly methanol as a raw material but also acetaldehyde, n-butane, n-butene or the light gasoline fractions [101]. Important methods for AA synthesis are the following [100, 101]:

- methanol carbonylation (Monsanto process, Cativa process),
- ethylene and acetaldehyde oxidation,
- n-butane and light gasoline fractions (C_5-C_8) liquid phase oxidation.

In recent years, the demand for acetic acid is at a high level (13 mln tons in 2016), and still increases rapidly due to its wide application as a substrate or solvent in many chemical reactions [102, 103]. It is primarily used for production of acetic anhydride, purified terephthalic acid, acetate esters and vinyl acetate monomer. The main industrial directions of AA processing are shown in Table 4.5. Also, the low concentration acetic acid solutions (<5 %) are widely used as a food additives (vinegar) and preservatives in the food industry.

Table 4.5: The main applications of the AA derivatives [101, 103].

Acetic acid derivatives	Application
Acetic anhydride	– plastics – pharmaceuticals: aspirin, paracetamol, antibiotics – perfumes – dyes – explosives
Vinyl acetate	– latex emulsion resins and latex paint – paper coatings – adhesives – textile finishing
Cellulose acetate	– constituent of thermoplastics and fibers – cellulose acetate fiber – production of film, plastic sheets
Monochloro acetic acid	– carboxymethyl cellulose – herbicides – preservatives – bacteriostat – glycine

Worldwide, about 95 % of AA is conventionally produced by the Cativa or Monsanto processes, acetaldehyde oxidation and ethylene oxidation [104]. The world's producers of acetic acid are: the USA (Celanese, Lyondell Basell, Eastman, DuPont), the UK (BP Chemicals) and Saudi Arabia (Saudi International Petrochemical Company). The most important industrial acetic acid production processes and the conditions of these processes are compared in Table 4.6.

Table 4.6: Industrial AA production processes and conditions [99, 102].

Process	Feedstock	Catalyst	Temperature [°C]	Pressure [MPa]	Yield [%]
BASF AG	CH_3OH, CO	$Co_2(CO)_8$, CoI_2	250	50–60	90
Monsanto	CH_3OH, CO	$[Rh(CO)_2I_2]$-$AsPh_4$ + $RhCl_3 \bullet 3H_2O$, Rh_2O_3	150–200	2.0–3.5	95–99
Cativa	CH_3OH, CO	$Ir_2(CO)_8$	150–200	2.0–3.5	95–99
Acetaldehyde oxidation	CH_3CHO, O_2	$Co(C_2H_3O_2)_2$	150	5.5–6.0	95
Butane oxidation	C_4H_{10}, O_2	CrO_5, CoO_2	150–230	5.0–6.0	50

4.3.1.2 Acetic acid production by methanolcarbonylation

4.3.1.2.1 Monsanto and Cativa process
In 1913 BASF AG developed the AA production method by methanol carbonylation. In 1960 BASF built the first acetic acid production plant. The process was carried out at 250 °C and pressure 50–60 MPa with cobalt carbonyl promoted by cobalt iodide as a catalyst [99, 104]:

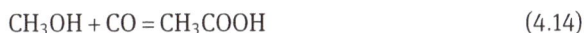

$$CH_3OH + CO = CH_3COOH \tag{4.14}$$

In 1966–1970, Monsanto Co. developed the carbonylation of methanol under milder conditions (180–200 °C, 2–3.5 MPa) than BASF, which was a great success. The reduction of pressure and temperature was possible due to the discovery of a new Rh-based catalyst with iodine as a promoter. The process took place in a liquid phase, in which the catalyst was dissolved. The new rhodium catalyst increased process efficiency from 90 % (BASF) to 99 % (Monsanto). Although the Monsanto method used a very expensive catalyst, high selectivity of the process, a limited number of operations and simple separation of the product from the post-reaction mixture caused it to be the most economical method of producing AA [104].

In 1986, Monsanto sold the developed method to British Petroleum Chemicals, which modernized the process. In 1996, BP Chemicals improved the methanol carbonylation process by developing a new iridium acetate catalyst with rhenium, ruthenium and osmium as promoters (Cativa), which significantly reduced the cost and the amount of impurities in the produced AA. Furthermore, the Ir-based catalyst was able to operate at reduced water levels: <5 % for Cativa versus ~15 % H_2O with the Monsanto method (water is needed to dissolve the Rh-based catalyst). Thus, in Cativa process highly concentrated methanol as a raw material (0.5 % water) can be used. A simplified scheme of the AA production by methanol carbonylation is presented in Figure 4.15 [105–107].

unreacted gases

CH₃OH + CO → **carbonylation** *Monsanto*: Rh-based catalyst *Cativa*: Ir-based catalyst → **catalyst separation** → **distillation** → CH₃COOH (95-99%)

recycled catalyst

propionic acid

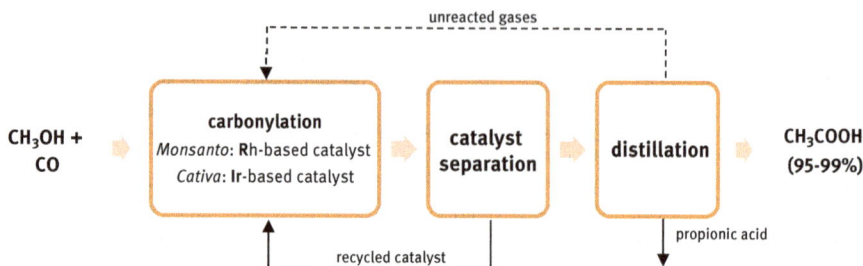

Figure 4.15: Scheme of the AA production by methanol carbonylation (based on [102, 105]).

4.3.1.2.2 Acid Optimization Plus technology

In 1980s, Celanese presented Acid Optimization Plus technology (AO Plus), which was an improved Monsanto process. In the AO Plus method the stability of the rhodium catalyst increased by adding inorganic iodide (mainly lithium iodide), which made possible reduce water concentration (to 4–5 %) in the reactor while maintaining high efficiency of the process. As a result, the costs of product purification was reduced [104, 108].

4.3.1.2.3 CT-ACETICA process

Chiyoda developed the CT-ACETICA process in which the rhodium catalyst was located on a thermostable vinylpyridine resin, however, there is no working industrial installation. The process is carried out in a bubble column loop reactor. The yield of acetic acid is more than 99 %. Advantages of using resin-supported rhodium catalyst are as follows [104, 109]:
- the amount of catalyst can be raised because there is no limit to the solubility of rhodium in water (carbonylation efficiency increases),
- the rhodium catalyst remains on the carrier in the reactor (no catalyst loss),
- low water content reduces the cost of purifying the final product,
- better contact of the reagents.

4.3.1.2.4 Haldor Topsoe method

The Danish company Haldor Topsoe developed a method for producing AA by carbonylation of a mixture of methanol and dimethyl ether or/and methyl acetate. The process is carried out at a relatively low pressure (3.5 MPa). The method is based on contacting a mixture of methanol and DME with CO and H_2 in a liquid phase reactor. The liquid phase contains a Rh-based catalyst, a methyl halide (a promoter), ruthenium acetate trimer and water. DME reacts with water in the reactor; methanol produced in this reaction (along with fresh methanol) is immediately carbonylated. Addition of Ru-compounds to the reaction mixture reduces the acetaldehyde concentration in the reactor (less than 400 ppm). Pilot studies were carried out and the method has been patented [6, 110].

4.3.1.3 Other methods for acetic acid production

4.3.1.3.1 Ethylene oxidation
In 1997, Showa Denko K.K. (Ōita, Japan) opened an ethylene oxidation plant (100,000 t/y capacity). AA is produced by direct oxidation of ethylene (4.14) [111]:

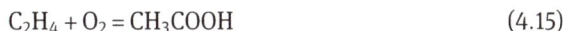

$$C_2H_4 + O_2 = CH_3COOH \tag{4.15}$$

and also, by-products are formed:

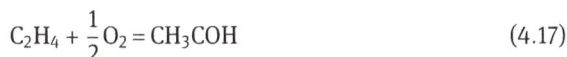

$$C_2H_4 + 3O_2 = 2CO_2 + H_2O \tag{4.16}$$

$$C_2H_4 + \frac{1}{2}O_2 = CH_3COH \tag{4.17}$$

The process is conducted in a fixed bed reactor at 160–240 °C in the presence of a Pd metal catalyst supported on a silicotungstic acid. The selectivity of this process towards AA is over 86 % [111].

4.3.1.3.2 N-butane and light gasoline oxidation
The AA production by n-butane oxidation can be carried out non-catalytically (Hüls AG, Germany) or with Co or Mn catalyst (Celanese, USA), according to the reaction (21):

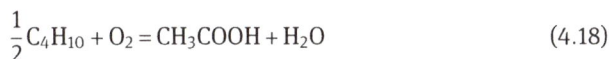

$$\frac{1}{2}C_4H_{10} + O_2 = CH_3COOH + H_2O \tag{4.18}$$

The non-catalytic process is carried out in two ways [104]:
- oxidation in a vapor phase, temperature: 350–400 °C, pressure 0.5–1 MPa,
- oxidation in a liquid phase, temperature 160–180 °C, pressure 0.45–1.0 MPa.

The catalytic process is carried out in a liquid phase at 160–180 °C and 4.5–5.5 MPa. The need to separate the multi-component reaction mixture is emphasized as a disadvantage of this process. The main product is acetic acid (85 %) but also methyl ethyl ketone, acetone, formic acid, methanol, ethanol and propionic acid are formed [6].

The autocatalytic oxidation of n-butane is carried out on a large scale only in the USA by the Celanese company [108]. In 1961 BP, as the first company in the world, commenced an installation for the industrial production of AA by the light gasoline oxidation method (production capacity about 200 000 tons CH_3COOH per year) [6, 104].

The light gasoline (C_5-C_8) is a cheaper raw material than n-butane. The oxidation process is carried in the liquid phase with air at 170–200 °C and 4–5 MPa at Co-based catalyst. The reaction products are the following: 70–80 % acetic acid, 15 % formic acid, 10 % propionic acid and others (alcohols, ketone, esters) [6, 106].

4.3.2 Lactic acid

Generally, lactic acid (LA, $CH_3CH(OH)COOH$), discovered in 1780 by Scheele in sour milk, and obtained by Fermi in fermentation in 1881, is applied in food and beverage, pharmaceuticals, personal care polymer industries [112]. In 2010, lactic acid was indicated as a potential building block for the future (so called "platform chemical") on a list reported by the US Department of Energy [113]. Building block chemicals can be made from sugars either by fermentation or synthetically, and can be converted to a number of high-value bio-based chemicals or materials. The environmental concerns have enhanced a development of recyclable and biodegradable plastics (bioplastics), such as polylactic acid (PLA), using renewable biomass as a feedstock. Two different processes are involved in producing PLA (condensation and polymerization) from LA-based monomers, i. e. optically pure L- and D-lactic acids (LA) or lactide (cyclic di-ester) [114, 115]. Lactic acid and its derivatives can be applied also for chemical foaming products used to produce polyurethane (PU) hard foams [116]. From 2010 to 2016, the global demand for LA and PLA were expected to increase from 482.7 to 1076.9 kilo tons (annual growth rate 14.2 %) and from 248.8 to 870.8 kilo tons (annual growth rate 20.8 %), respectively. Further LA annual growth rate (CAGR) is predicted to be around 19.5 % between 2019 and 2025 [117]. However, on industrial scale, compared with petrochemical raw materials, PLA production from LA is considered a relatively immature technology, mainly due to the high production cost of lactic acid [118].

Lactic acid can be produced, similarly to ethanol, by chemical synthesis or biotechnology (Figure 4.16). The rate of LA production by fermentation of sugars, mainly glucose, is low, and pH of the solution during the process must be kept in the range 5–7 to provide an environment suitable for lactic acid-fermenting bacteria. Although, various attempts have been made to develop catalytic (homogeneous with $PbCl_2$, $Pb(NO_3)_2$ and $ErCl_3$ or heterogeneous with $LaCoO_3$, NbF_5-AlF_3, AlW, ZrO_2) conversion of inedible cellulose to lactic acid, their use in industry remains still unrealized because of difficult separation of the products from the homogeneous catalysts, and their poor stability and recyclability, or, in the case of heterogeneous catalysts, due to metal species leach into reaction solution, high prices of catalytic metals, and low yields of conversion to LA [121, 122]. On the one hand, for the future biorefinery high-volume processes a chemocatalytic direct transformation of biomass feedstocks other than monosaccharides or oligosaccharides into chemicals in a one-pot process are necessary. Catalytic process provides

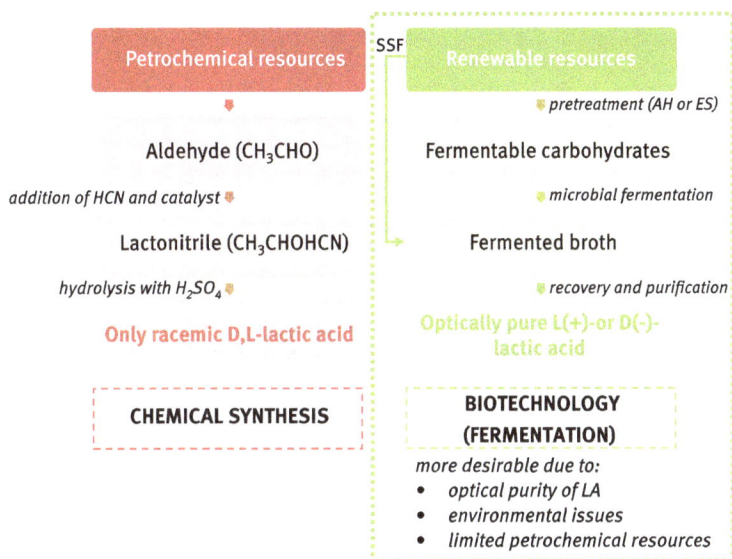

Figure 4.16: Chemical synthesis and biotechnology (microbial fermentation) routes of LA production (SSF – simultaneous saccharification and fermentation, AH – acidic hydrolysis, ES – enzymatic saccharification) based on [119, 120].

opportunity to increase the LA yield, ease process optimization and extends available biomass feedstock (e. g. lignocellulosic biomass) that can be processed. On the other hand, for pharmaceutical and food industry use only one L-lactic acid isomer is acceptable (D-lactic acid is considered harmful to humans), so fermentation route is more attractive instead of chemocatalytic or chemical synthesis to obtain high enantiomeric purity of LA (Figure 4.16) [119, 120]. It can be expected that, in the future, both fermentation and chemical processes will work complementary to produce LA for distinctive applications.

An exemplary process for the production of lactic acid by fermentation and for the separation and/or recovery of lactic acid from a lactate feed solution is shown in Figure 4.17. The biomass (plants) is put through a milling process extracting the starch (glucose). Enzymes are added to convert the glucose to dextrose via hydrolysis. Different bacteria strains, e. g. *Sporolactobacillus sp. CASD, Bacilus coagulans, Streptococcus thermophilus, Lactobacillus bulgaricus, Lactobacillus delbruckii leichmanni, Lactobacillus casei, Lactobacillus plantarum, Enterococcus faecalis,* are employed to ferment the dextrose into LA from various raw materials [112, 119, 120, 126, 127]. pH of the fermentation broth must be controlled (e. g. with Na_2CO_3) to be stable in the range 6–6.5. Prior to extraction step, biomass and other solids are physically removed from the broth by filtration or ultrafiltration, and concentrated by water evaporation to 40–70 wt% of LA. LA feed solution obtained from the fermentation broth is combined with and extracted by a water immiscible trialkyl amine (used as an organic

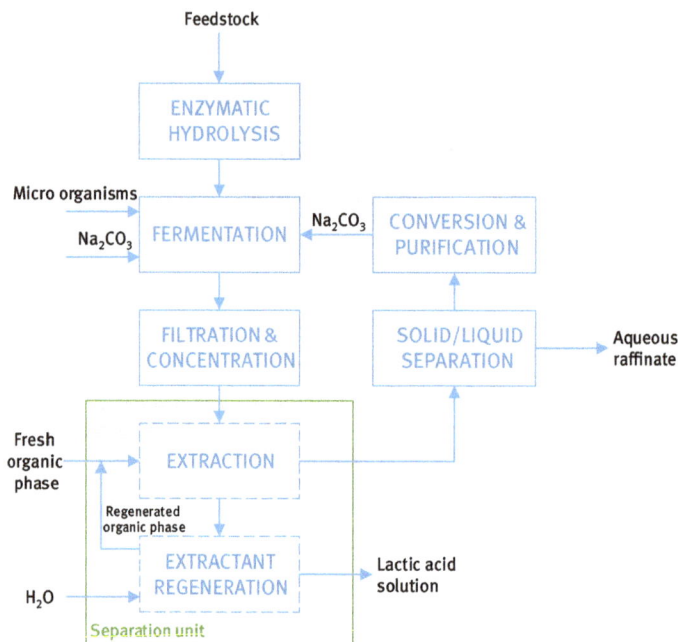

Figure 4.17: Scheme of fermentation production of lactic acid (based on [123–125]).

extractant) in the presence of carbon dioxide. Lactic acid is recovered from the result-ing organic phase by stripping with water in separation unit. Recovered carbonate from the aqueous phase is recycled to the fermentor, and regenerated extractant is re-cycled for use in the extraction [123]. Among other methods that can be used in separa-tion unit are proposed distillation, adsorption, ion exchange, electrodialysis. In the conventional LA recovery process, the acid is precipitated with $Ca(OH)_2$ to form cal-cium lactate, which is separated from the liquid by a filtration process. Finally, sulfuric acid is used to release the LA from the precipitation. Large amount of generated $CaSO_4$ is a problematic waste that must be managed. The diluted LA product is then purified sequentially using activated carbon, evaporation and crystallization. It must be em-phasized that costs of recovery and purification of biorefinery products reaches from 20 % to even 50 % of the operating costs of the biotechnological process [120].

In 1989, a Cargill started an innovative project for uses of carbohydrates from plants as feedstock for more sustainable plastics that resulted in such innovative materials as biopolymers, i. e. polylactic acid [123, 124]. In 2003, NatureWorks company owned by PTT Global Chemical and Cargill, built the world's largest lac-tic acid manufacturing facility (based on first generation feedstock – industrially sourced corn) to feed polymer plants. In 2016, a new fermentation laboratory to develop commercial-scale fermentation technology to turn CO_2 or CH_4 directly into green building blocks (e. g. lactic acid) – bypassing the agricultural step –

was opened [127]. Although biogas, waste or methane conversion into lactic acid is of great interest contemporary, the feedstock most commonly used for industrial production of LA and its derivatives is raw sugar from sugarcane; it helds approximately 40 % of the global revenue in 2017 [128, 129]. Also sugar beet, molasses, whey and barley malt are valuable sources of such carbohydrates as glucose, lactose, starch, maltose and sucrose for the LA industry [120]. However, other biomass substrates are tested to replace these plants used for human and animal feeding. Hence, the lignocellulosic biomass from agricultural, agroindustrial and forestry sources seems to be the most promising feedstock for LA production due to its low prices, availability, abundance, renewability and high polysaccharide content [120]. Additionally, the following carbon sources were investigated on laboratory scale: algae biomass, food waste, glycerol from biodiesel production [120, 130].

Nature Works LLC (140,000 t/y, USA), Corbion Purac (100,000 t/y, Netherlands), Pyramid Bioplastics Guben GmBH (60,000 t/y, Germany), Galactic (Belgium), Archer Daniels Midland Company (USA) and Henan Jindan (China) CCA (Changzhou) Biochemical (China), and Musashino Chemical (7,000 t/y, Japan) are the leading companies for LA and PLA production [114, 116, 131]. Companies have been working on development of LA biotechnology, e. g. since 1994 Galactic (Belgium) has been improving its LA production, recently participating in building a LA and a PLA factory in China (two joint-ventures with BBCA Biochemical) [132], or Polish petrochemical company Orlen has been working on innovative biotechnology for conversion of biorefinery by-products and biomass into LA since 2017, and plans to commence an installation in 2020 [133].

In 2013, ThyssenKrupp inaugurated multi-purpose pilot plant Uhde Inventa-Fischers in Guben (Germany), where fermentation and processing technologies were applied for the production of a variety of biochemicals, also LA and PLA. Based on this experience the company started to build the first commercial plant based on its patented "PLA neo" technology in Changchun, China, in 2017 [134, 135].

4.4 Technology of large volume esters

4.4.1 Fatty acid methyl esters – biodiesel

Mixture of fatty acid methyl esters is used as a biofuel, called biodiesel. In different regions of the world development both of bioethanol and biodiesel as engine fuels was response to oil crises in the past, and to global climate changes due to high CO_2 emissions nowadays. Compared to diesel, biodiesel produces no sulfur, no net CO_2, less carbon monoxide, zero particulate matters, no smoke and no hydrocarbons [136]. On the contrary, biodiesel produces more oxygen, which leads to complete combustion and reduced emissions. Legal acts regulate addition of biodiesel to conventional diesel fuel in the USA or Europe. In Europe the net quota of net greenhouse gas (GHG) reduction for biofuels should increase from a rate of 3.5 % in 2015, to 4 % in 2017 and to 6 % in and post 2020 [47]. In the US, the Renewable Fuel Standard is a program developed by the Environmental Protection Agency to comply with the Clean Air Act and the Energy Independence and Security Act (EISA). The second Renewable Fuel Standard (RFS2), established in 2010, mandates the inclusion of biofuels such as biodiesel into the country's petroleum fuel supply. RFS2 also mandates that biofuels used under the program cause less greenhouse gas emissions to enter the atmosphere compared to the petroleum fuels they replaced.

Besides ProAlcool program (Section 4.2.2.1), the Brazilian National Bio-Fuel Program comprises also Biodiesel program (conceived in 1983) of vegetable oil application as a fuel [48]. Biodiesel is promoted by governments of various countries which support and subsidize production and consumption of the biofuel.

Biodiesel has lubricating properties and cetane ratings compared to low sulfur diesel fuels. Its calorific value (37.27 MJ/kg) is 9 % lower than of regular diesel. Biodiesel quality is more related to the feedstock used in its production than in case of petrochemical fuels [136]. A drawback of this biofuel is low oxidation resistance due to presence of unsaturated bonds in ester molecules, which negatively influences stability of the fuel.

The major feedstock for biodiesel is animal fat or oil from plants (more than 350 oil-bearing crops can be used). The raw materials are divided in four categories, and, similarly to bioethanol, they are considered as 1 G (edible) or 2 G (inedible, waste) [136]: (a) edible vegetable oils such as rapeseed, soybean, peanut, sunflower, palm and coconut oil, (b) inedible vegetable oils such as jatropha, karanja, sea mango, algae and halophytes, (c) waste or recycled oil, (d) animal fats such as beef tallow, yellow grease, chicken fat and by-products from fish oils. The raw materials contain triglycerides that are chemically conversed by transesterification reaction with short-chain monohydric alcohols (in the presence of alkali or acidic catalyst or enzyme) at elevated temperatures into fatty acid alkyl esters (FAAE) and glycerol. Transesterification is the easiest and most cost-effective process to produce biodiesel [136].

In most cases, methanol is used as the reacting alcohol, thus, the main products are called fatty acid methyl esters (FAME). One molecule of triglyceride reacts stepwise with three molecules of methanol according to reactions (22–24). The basic catalyst provides OH⁻ anions, which are a strong nucleophilic agent. In the first stage of the reaction, the metoxide anion attacks a carbon atom of the carboxyl group in the triglyceride molecule. Next, the alcohol reacts with the ion produced in the first stage to regenerate the catalyst. In the last stage, the resulting ion breaks down to form a methyl ester and diglyceride, monoglyceride or glycerin depending on stage of the main reaction.

$$\underset{\text{triglycerides (fat or oil)}}{\begin{array}{l} H_2C-O-\overset{\overset{\displaystyle O}{||}}{C}-R \\ HC-O-\overset{\overset{\displaystyle O}{||}}{C}-R \\ H_2C-O-\overset{\overset{\displaystyle O}{||}}{C}-R \end{array}} + CH_3OH \rightleftharpoons \underset{\text{diglycerides}}{\begin{array}{l} H_2C-O-\overset{\overset{\displaystyle O}{||}}{C}-R \\ HC-O-\overset{\overset{\displaystyle O}{||}}{C}-R \\ H_2C-OH \end{array}} + \underset{\text{FAME}}{H_3C-O-\overset{\overset{\displaystyle O}{||}}{C}-R} \tag{4.19}$$

$$\underset{\text{diglycerides}}{\begin{array}{l} H_2C-O-\overset{\overset{\displaystyle O}{||}}{C}-R \\ HC-O-\overset{\overset{\displaystyle O}{||}}{C}-R \\ H_2C-OH \end{array}} + CH_3OH \rightleftharpoons \underset{\text{monoglycerides}}{\begin{array}{l} H_2C-O-\overset{\overset{\displaystyle O}{||}}{C}-R \\ HC-OH \\ H_2C-OH \end{array}} + \underset{\text{FAME}}{H_3C-O-\overset{\overset{\displaystyle O}{||}}{C}-R} \tag{4.20}$$

$$\underset{\text{monoglycerides}}{\begin{array}{l} H_2C-O-\overset{\overset{\displaystyle O}{||}}{C}-R \\ HC-OH \\ H_2C-OH \end{array}} + CH_3OH \rightleftharpoons \underset{\text{glycerol}}{\begin{array}{l} H_2C-OH \\ HC-OH \\ H_2C-OH \end{array}} + \underset{\text{FAME}}{H_3C-O-\overset{\overset{\displaystyle O}{||}}{C}-R} \tag{4.21}$$

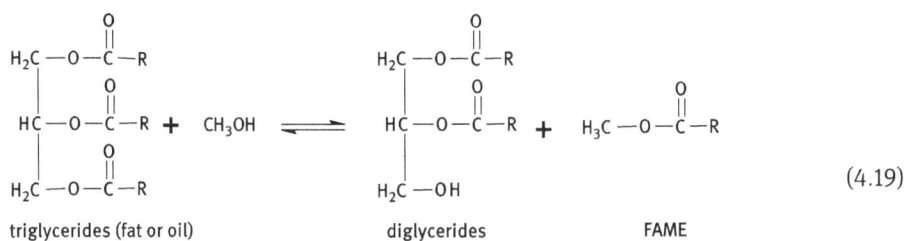

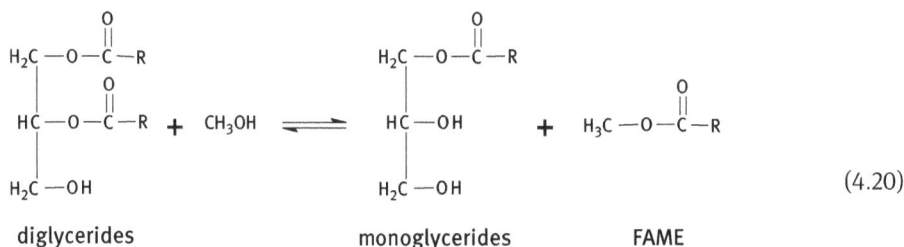

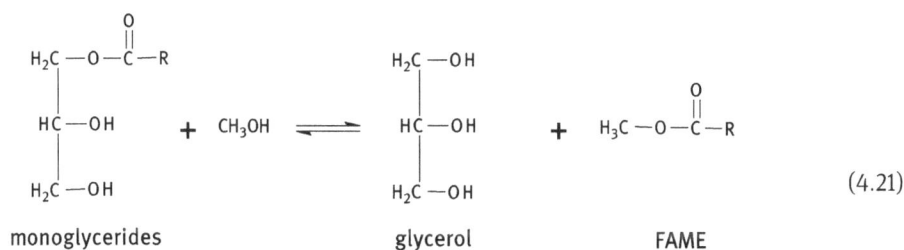

Biodiesel production increases constantly, for example in 2012 worldwide production amounted to 24.6 mln tons, and has increased to 33 mln tons in 2016 [137]. It is noteworthy that Europe is the main supplier of biodiesel – 42 % of the global production (12.4 mln t in 2016) – mainly from rapseeds, also the USA (5.9 mln t) and Brazil (3.4 mln t) are among the leaders. Since for each ton of biodiesel 90–110 kg of glycerol is formed (reaction (18)) as a by-product, surplus of this polyol is reported. Due to the unique structure of glycerol, its properties and renewability, new applications for glycerol and new processes to convert it to value-added products have been developed in recent years.

The catalysts for the transesterification of triglycerides are classified as alkali, acid, enzyme. They can be homogeneous or heterogeneous. The most effective and applied in industry are alkali catalysts such as sodium hydroxide, sodium methoxide, potassium hydroxide and potassium methoxide. Also, metoxides of other metals are patented as transesterification catalysts [138]. Acidic catalysts, such as sulfuric, hydrochloric and sulfonic acid, are less useful because the reaction must be carried out in higher temperature compared to alkali catalysts. Enzyme lipase

[139] or microbial biocatalysts [140] (a mixture of *Bacillus* and *Lactobacillus* bacteria) are proposed to process waste triglycerides. Heterogeneous catalysts include enzymes, titanium-silicates, alkaline-earth metal compounds (e. g. cement kiln dust with CaO [141]), anion exchange resins and guanidines heterogenized on organic polymers, zirconium sulfate or sulfated zirconia on support [142]. However, the most effective and well established industrial system is based on methanolysis, i. e. transesterification with methanol catalyzed by sodium or potassium hydroxide, or sodium or potassium methoxide.

General scheme of biodiesel production includes three main stages: transesterification of triglycerides, methanol recovery and glycerin (glycerol) purification (Figure 4.18). At first, mixture of alcohol (in excess) and alkaline catalyst is prepared, and than it is mixed with oil in a reactor for approximately an hour at 60–70 °C at atmospheric pressure [136, 145]. Smaller plants often use batch reactors, however, most larger plants (>4 mln dm^3/y) use continuous flow processes involving continuous stirred-tank reactors (CSTR) or plug flow reactors. After reaction the glycerol is removed from the methyl esters, and is followed by methanol recovery and recycling. As the glycerol stream leaving the separator is only about 50 % glycerol (contains the excess methanol and the catalyst and soaps), it must be purified by splitting soaps, neutralization of free fatty acids, and vacuum distillation or ion exchange [145].

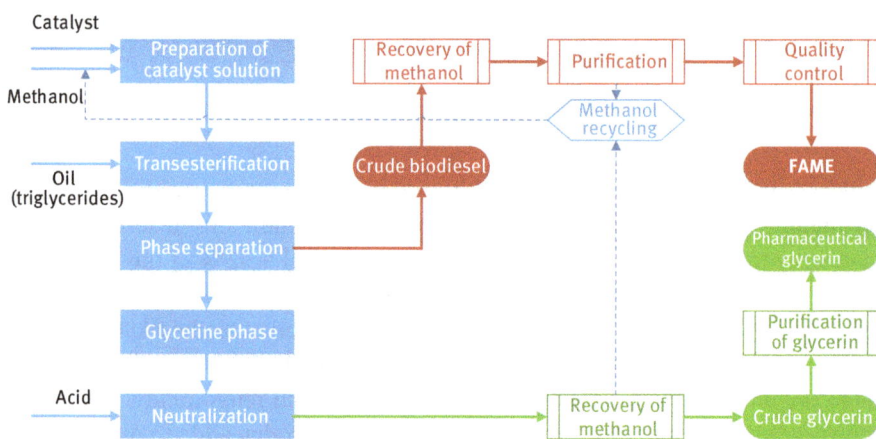

Figure 4.18: Scheme of biodiesel (FAME) production by transesterification (based on [143–145]).

Only in Poland there are many FAME producers connected to the petrochemical industry and fuel production, e. g. Orlen Group, Polmos Wrocław, Elstar Oils Malbork, J&S Energy, Zakłady Azotowe Puławy, Oleochemia, Solvent Dwory and many more. Among the largest global producers of biodiesel are Renewable Energy Group (USA) 212 mln t/y (in 2012) [146], ADM (USA) 145 mln t/y (in 2012), Avril

(France) – Europe's largest biodiesel producer 1.3 mln t/y (in 2018) [147], Biopetrol (Switzerland) 750 000 t/y (in 2010) [148], Envian Group (Slovakia) 280 000 t/y (in 2018) [149], UnitecBio (Argentina) 240 000 t/y [150], China Biodiesel (leader in China) 100 000 t/y [151].

4.5 Summary

The chemical industry plays a very important role in the global economy. It is strongly associated with other industries, and its products are used in all areas of the economy. Despite the well-known commercial processes of production of alcohols (except ethanol), carboxylic acids and esters, alternative bioproduction methods are still elaborated and developed. It seems that contribution of bioprocesses in industrial production of important chemicals will increase. However, in the nearest future biorefineries will be rather accompanied by conventional petrochemical or chemical processes, and do not replace them. Bioprocesses (e. g. production of cellulosic ethanol) still are less economic than traditional processes, and still need special legal regulations and financial support of governments.

References

[1]   Naik SN, Goud VV, Rout PK, Dalai AK. Production of first and second generation biofuels: a comprehensive review. Renewable Sus Energy Rev. 2010;14:578–97. DOI:10.1016/j.rser.2009.10.003.

[2]   Melero JA, Iglesias J, Garcia A. Biomass as renewable feedstock in standard refinery units. Feasibility, opportunities and challenges. Energy Environ Sci. 2012;5:7393–420. DOI:10.1039/c2ee21231e.

[3]   Jang YS, Kim B, Shin JH, Choi YJ, Choi S, Song CW, et al. Bio-based production of C2–C6 platform chemicals. Biotechnol Bioeng. 2012;109:2437–59. DOI:10.1002/bit.24599.

[4]   Bruijnincx PC, Weckhuysen BM. Shale gas revolution: an opportunity for the production of biobased chemicals? Angew Chem Int Ed. 2013;52:11980–7. DOI:10.1002/anie.201305058.

[5]   Hatti-Kaul R, Törnvall U, Gustafsson L, Börjesson P. Industrial biotechnology for the production of bio-based chemicals – a cradle-to-grave perspective. Trends Biotechnol. 2007;25:120–4. DOI:10.1016/j.tibtech.2007.01.001.

[6]   Grzywa E, Molenda J. Technologia podstawowych syntez organicznych (Technology of basic organic syntheses). Warsaw: WNT, 2000. (in Polish).

[7]   Ghasemzadeh K, Sadati Tilebon SM, Nasirinezhad M, Basile A. Economic assessment of methanol production (Chapter 23) in methanol – science and engineering. John Fedor, Elsevier, 2018:613–32. ISBN: 978-0-444-63903-5.

[8]   Johnson Matthey Technology Review. Methanol production – A technical history. Johnson Matthey Technol Rev. 2017;61:172–82.

[9]   The methanol industry. Available at: https://www.methanol.org/the-methanol-industry/. Accessed: 10 Aug 2019.

[10] Iaquaniello G, Centi G, Salladini A, Palo E. Methanol economy: environment, demand, and marketing with a focus on the waste-to-methanol process (Chapter 22) in Methanol – Science and Engineering. John Fedor, Elsevier, 2018:595–612. ISBN: 978-0-444-63903-5.

[11] Alvarado M. Methanol. IHS Markit Report. 2016. Available at: http://www.methanol.org/wp-content/uploads/2016/07/Marc-Alvarado-Global-Methanol-February-2016-IMPCA-for-upload-to-website.pdf. Accessed: 31 Aug 2019.

[12] Aasberg-Petersen K, Stub Nielsen C, Dybkjær I, Perregaard J. Large scale methanol production from natural gas. Haldor Topsoe, 1–14.

[13] Dalena F, Senatore A, Basile M, Knani S, Basile A, Iulianelli A. Advances in methanol production and utilization, with particular emphasis toward hydrogen generation via membrane reactor technology. Membranes. 2018;8:1–27. DOI:10.3390/membranes8040098.

[14] Dalena F, Senatore A, Marino A, Gordano A, Basile M, Basile A. Methanol production and applications: an overview (Chapter 1) in Methanol – Science and Engineering. Amsterdam, Netherlands: John Fedor, Elsevier, 2018:3–28.

[15] Ott J, Gronemann V, Pontzen F, Fiedler E, Grossmann G, Kersebohm DB, et al. Methanol. ULLMANN'S encyclopedia of industrial chemistry. Wiley-VCH, 2012.

[16] Lloyd L. Handbook of industrial catalysts. United Kingdom: Springer Science & Business Media, LLC; 2011.

[17] Mittasch A, Pier M, Winkler K. BASF AG, Ausführung Organischer Katalysen. German Patent No. 415686, 1925.

[18] Air Liquide Group. Air Liquide Engineering & Construction, Methanol and derivatives proven technologies for optimal production. Available at: www.engineering-airliquide.com. Accessed: 18 Aug 2019.

[19] Błasiak E. Way of manufacturing of highly active catalyst for synthesis of methanol (Sposób wytwarzania wysoko aktywnego katalizatora do syntezy metanolu). RP Patent, No. 34000.

[20] Blasiak E. Polish Patent No. PRL 34000, 1947.

[21] Davies P, Snowdon FF, Bridger GW, Hughes DO, Young PW. Water-gas conversion and catalysts therefor Patent No. GB1010871A, 1962.

[22] Chinchen GC, Jennings JR. Copper catalyst for carbon oxide conversion. Patent No. EP0296734A1, 1987.

[23] Bechtholdt G, Bratzler H, Liebgott K, Ehrhard Erwin H. A process for the production of methanol. Patent No. DE1300917B, 1967.

[24] Brücker FJ, Marosi L, Schröder W, Schwarzmann M. Patent No. 2056612, 1972.

[25] Brücker FJ, Engelbach H, Schroeder W. A process for the production of methanol. Patent No. 1930003A1, 1969.

[26] Catalyst for methanol prepn-contg copper, zinc, manganese, - aluminium and chromium. Patent No. 2026165A1, 1970.

[27] Catalyst for methanol prepn-contng copper zinc manganese - and aluminium. Patent No. 2026165, 1970.

[28] Johnson Matthey Process Technologies. Delivering world class methanol plant performance. Billingham, 2014.

[29] Ledakowicz S, Nowicki L, Petera J, Nizioł J, Kowalik P, Gołębiowski A. Kinetic characterisation of catalysts for methanol synthesis. Chem Proc Eng. 2013;34:497–506.

[30] LTS catalysts: TMC-3/1, TMC-3/1-K, TMC-3/1-Cs. Available at: http://www.ins.pulawy.pl/index.php/en/products/catalysts-and-sorbents/water-gas-shift. Accessed: 07 July 2019.

[31] Czardybon A, Więcław-Solny L, Ściążko M. Technological perspectives of carbon dioxide utilization. Energetyka: Instytut Chemicznej Przeróbki Węgla w Zabrzu, 2014:25–33. (in Polish).

[32] Methanol synthesis catalysts. Available at: https://matthey.com/products-and-services/
chemical-processes/chemical-catalysts/methanol-synthesis-catalysts. Accessed: 07 July
2019.

[33] Dina IU, Shaharunb MS, Alotaibia MA, Alharthia AI, Naeemc A. Recent developments on
heterogeneous catalytic CO2 reduction to methanol. J CO2 Util. 2019;34:20–33.

[34] Wilk A, Więcław-Solny L, Spietz T, Tatarczuk A. CO2-to-methanol conversion – an alternative
energy storage solution. CHEMIK. 2016;70:626–33.

[35] Tursunov O, Kustov L, Kustov A. A brief review of carbon dioxide hydrogenation to methanol
over copper and iron based catalysts. Oil Gas Sci Technol-Revue d'IFP Energies nouvelles,
Institut Français du Pétrole. 2017;72:30–9.

[36] Bukhtiyarova M, Lunkenbein T, Kähler K, Schlögl R. Methanol synthesis from industrial CO2
sources: a contribution to chemical energy conversion. Catal Lett. 2017;147:416–27.

[37] Development of highly efficient methanol synthesis process with new catalysts. Available at:
https://www.nipponsteel.com. Accessed: 01 Aug 2019.

[38] Martin FJ, Kubic WL. Green freedom. A concept for producing carbon-neutral synthetic fuelas
and chemicals (patent pending). Nov 2007.

[39] Bjorck CE, Dobson PD, Pandhall J. Biotechnological conversion of methane to methanol:
evaluation of progress and potential. AIMS Bioeng. 2018;5:1–38.

[40] Pellegrini LA, Soave G, Gamba S, Langè S. Economic analysis of a combined energy–
methanol production plant. Appl Energy. 2011;88:4891–7.

[41] Nathan PE, Conrad M, Skinstad AH. History of the concept of addiction. Annu Rev Clin
Psychol. 2016;12:29–51. DOI:10.1146/annurev-clinpsy-021815-093546.

[42] Bergmann JC, Trichez D, Sallet LP, de Paula E Silva FC, Almeida JR. Technological
advancements in 1G ethanol production and recovery of by-products based on the biorefinery
concept. In: Chandel AK, Silveira MH, editors. Advances in sugarcane biorefinery, Chapter 4.
The Netherlands: Elsevier, 2018. DOI:10.1016/B978-0-12-804534-3.00004-5.

[43] Voegele E. Japan's ethanol imports expected to increase slightly. 2016. Available at: http://
ethanolproducer.com/articles/13713/japanundefineds-ethanol-imports-expected-to-in
crease-slightly. Accessed: 22 Aug 2019.

[44] Soccol CR, Vandenberghe LP, Costa B, Woiciechowski AL, de Carvalho JC, Medeiros AB, et al.
Brazilian biofuel program: an overview. J Sci Ind Res. 2005;64:897–904.

[45] Manzetti S, Andersen O. A review of emission products from bioethanol and its blends with
gasoline. Background for new guidelines for emission control. Fuel. 2015;140:293–301.
DOI:10.1016/j.fuel.2014.09.1010016-2361/.

[46] Brown TB. Why the cellulosic biofuels mandate fell short: a markets and policy perspective.
Biofuels Bioprod Bioref. 2019. DOI:10.1002/bbb.

[47] Directive 2003/30/EC of the European Parliament and of the Council of 8 May 2003 on the
promotion of the use of biofuels and other renewable fuels for transport. Available at:
https://eur-lex.europa.eu/legal-content/EN/TXT/PDF/?uri=CELEX:32003L0030&from=en.
Accessed: 22 Aug 2019.

[48] Biofuel. National Fuel Alcohol Program (Proalcool). Available at: http://biofuel.org.uk/
National-Fuel-Alcohol-Program.html. Accessed: 22 Aug 2019.

[49] Herring P. Ethanol prices in brazil jump over 18 percent. The Rio Times. 2019. Available at:
https://riotimesonline.com/brazil-news/rio-business/ethanol-prices-in-brazil-jump-over-18-
percent/. Accessed: 22 Aug 2019.

[50] Ernsting A, Smolker R. Dead End Road: the false promises of cellulosic biofuels. Biofuelwatch
report. 2018. Available at: https://www.biofuelwatch.org.uk/2018/dead-end-road/.
Accessed: 22 Aug 2019.

[51]   Küüt A, Ritslaid K, Küüt K, Ilves R, Olt J. State of the art on the conventional processes for ethanol production. In: Basile A, Iulianelli A, Dalena F, Veziroğlu TN, editors. Ethanol. The Netherlands: Elsevier, 2019:61–101. DOI:10.1016/b978-0-12-811458-2.00003.

[52]   Leja K, Lewandowicz G, Grajek W. Produkcja bioetanolu z surowców celulozowych. Biotechnologia. 2009;4:88–101. (in Polish).

[53]   Tuscola Plant. LyondellBasell brochure. 2018. Available at: https://www.lyondellbasell.com/en/tuscola-plant/. Accessed: 22 Aug 2019.

[54]   Sasol brochure. Technology in motion fuels technology. Sasol: RPA, 2002.

[55]   ISSUU. Mpumalanga Bussiness 2018/19. Available at: https://issuu.com/globalafricanetwork/docs/mpumalangacompanies/43. Accessed: 22 Aug 2019.

[56]   Ethanol. Producers Association Of Southern Africa. Available at: http://www.epasa.org.za/about-us.html. Accessed: 22 Aug 2019.

[57]   Mohsenzadeh A, Zamani A, Taherzadeh MJ. Bioethylene production from ethanol: a review and techno-economical evaluation. Chem Bio Eng Rev. 2017;4:75–91. DOI:10.1002/cben.201600025.

[58]   Lillebone, France. Available at: https://www.tereos-starchsweeteners.com/en/contact/our-sites/lillebonne. Accessed: 24 Aug 2019.

[59]   ICIS. Sadaf ups project capacity. 2005. Available at: https://www.icis.com/explore/resources/news/2005/05/10/674897/sadaf-ups-project-capacity/. Accessed: 22 Aug 2019.

[60]   INEOS Solvents. Europe's largest producer of synthetic ethanol.

[61]   Ethanol Fact Sheet. Available at: http://www.etipbioenergy.eu/fact-sheets/ethanol-fact-sheet#prod. Accessed: 24 Aug 2019.

[62]   The 5 largest ethanol producers. 2016. Available at: https://www.farmprogress.com/ethanol/5-largest-ethanol-producers. Accessed: 24 Aug 2019.

[63]   Modl J. Jilin fuel ethanol plant. Report 2004. Available at: https://www.vogelbusch-biocommodities.com/assets/4-References/JilinCN/isj2004.pdf. Accessed: 24 Aug 2019.

[64]   Bevap Biorenergia. Available at: https://www.bevapbioenergia.com.br/?lang=en. Accessed: 24 Aug 2019.

[65]   FDT Group. Bioethanol producer employs advanced automation technology to optimize integration, operation and lifecycle management of critical assets. Available at: https://www.fdtgroup.org/case-study/bioethanol-producer-granbio/. Accessed: 24 Aug 2019.

[66]   RFA (Renewable Fuel Association). Available at: https://ethanolrfa.org/statistics/annual-ethanol-production/. Accessed: 24 Aug 2019.

[67]   Olszewska-Widdrat A, Alexandri M, López-Gómez JP, Schneider R, Mandl M, Venus J. Production and purification of L-lactic acid in lab and pilot scales using sweet sorghum juice. Fermentation. 2019;5:36. DOI:10.3390/fermentation5020036.

[68]   Lopes ML, de Lima Paulillo SC, Godoy A, Cherubin RA, Lorenzi MS, Carvalho Giometti FH, et al. Ethanol production in Brazil: a bridge between science and industry. Brazilian J Microbiol. 2016;47S:64–76. DOI:10.1016/j.bjm.2016.10.003.

[69]   Hayes FW. Production of ethanol from sugar cane. Patent No. US4326036A, 1980.

[70]   Asturias CER. Ethanol production from fermentation of sugar cane. Patent No. US4560659A, 1983.

[71]   Jain T. Ethanolic fermentation with immobilized yeast. Patent No. US20140273133A1, 2014.

[72]   Swain MR, Natarajan V, Krishnan C. Marine enzymes and microorganisms for bioethanol production. In: Advances in food and nutrition research, vol. 80, Chapter 9. The Netherlands: Elsevier, 2017. DOI:10.1016/bs.afnr.2016.12.003.

[73]   RFA report. Ethanol strong – 2018 ethanol industry outlook. USA: Renewable Fuels Association, 2018.

[74] Saville BA, Griffin WM, MacLean HL. Ethanol Production Technologies in the US: Status and Future Developments. In: SLM S-F, Barbosa CL, Ferreira Jardim da Silveira JM, SC T, Derengowski FM, editors. Global bioethanol, Chapter 7. The Netherlands: Elsevier, 2016. DOI:/10.1016/B978-0-12-803141-4.00007-1.

[75] Küüt A, Ritslaid K, Küüt K, Ilves R, Olt J. State of the art on the conventional processes for ethanol production, Chapter 3. In: Basile A, Iulianelli A, Dalena F, TN V, editors. Ethanol. The Netherlands: Elsevier, 2019. DOI:10.1016/B978-0-12-811458-2.00003-1.

[76] MOdS D, Filho RM, Mantelatto PE, Cavalett O, Rossell CE, Bonomi A, et al. Sugarcane processing for ethanol and sugar in Brazil. Environ Develop. 2015;15:35–51. DOI:10.1016/j.envdev.2015.03.004.

[77] Vohra M, Manwar J, Manmode R, Padgilwar S, Patil S. Bioethanol production: feedstock and current technologies. J Environ Chem Eng. 2014;2:573–84. DOI:10.1016/j.jece.2013.10.013.

[78] Gwak YR, Kim YB, Gwak IS, Lee SH. Economic evaluation of synthetic ethanol production by using domestic biowastes and coal mixture. Fuel. 2018;213:115–22. DOI:10.1016/j.fuel.2017.10.101.

[79] Gomes J, Batra J, Chopda VR, Kathiresan P, Rathore AS. Monitoring and Control of Bioethanol Production From Lignocellulosic Biomass. In: Bhaskar T, Pandey A, Mohan SV, Lee DJ, Khanal SK, editors. Waste biorefinery, Chapter 25. The Netherlands: Elsevier, 2018. DOI:10.1016/B978-0-444-63992-9.00025-2.

[80] Eller D. DuPont sells Iowa ethanol plant to German company; it will soon make renewable natural gas. 2018. Available at: https://eu.desmoinesregister.com/story/money/agriculture/2018/11/08/dupont-cellulosic-ethanol-plant-nevada-sold-german-company-verbio-north-america-claus-sauter/1938321002/. Accessed: 22 Aug 2019.

[81] Mayer A. Cellulosic Ethanol Push Stalls In The Midwest Amid Financial, Technical Challenges. 2018. Available at: https://www.kcur.org/post/cellulosic-ethanol-push-stalls-midwest-amid-financial-technical-challenges#stream/0. Accessed: 23 Aug 2019.

[82] McCoy M. DuPont seeks to sell cellulosic ethanol plant. 2017. Available at: https://cen.acs.org/content/cen/articles/95/web/2017/11/DuPont-seeks-sell-cellulosic-ethanol.html. Accessed: 22 Aug 2019.

[83] Clariant starts the construction of its cellulosic ethanol plant in Romania. 2018. Available at: https://biorrefineria.blogspot.com/2018/09/clariant-starts-construction-of-its-cellulosic-ethanol-plant-in-romania-podari-craiova-sunliquid.html. Accessed: 22 Aug 2019.

[84] Ogaki M, Tanaka S, Haruhiko K, Ishii T. Method of producing bio-ethanol. Patent No. US20090075353A1, 2008.

[85] Kang DH, Lee HY, Han JG, Park HS, Lee HS, Kang RS. Liquefied extract of marine algae for producing bio-ethanol under high pressure and method for producing the same. Patent No. US20100041926A1, 2008.

[86] Ferrero VM. Method for the production of petrochemical, agri-foodstuff or other products using bioethanol obtained at a multifunctional biorefinery. Patent No. WO2009125037A1, 2009.

[87] Devarapalli M, Lewis RS, Atiyeh HK. Continuous ethanol production from synthesis gas by Clostridium ragsdalei in a trickle-bed reactor. Fermentation. 2017;3:23. DOI:10.3390/fermentation3020023.

[88] Bell PS, Ko CW. Process for fermentation of syngas. Patent No. US20140045246A1, 2012.

[89] Jupeng Bio, Lu'An Group and Shanghai Advanced Research Institute sign Research and Development collaboration agreement. 2018. Available at: http://www.jupengbio.com/blog/jupeng-bio-lu-an-group-and-shanghai-advanced-research-institute-sign. Accessed: 23 Aug 2019.

[90] Alliance Bio-Products to demonstrate its CTS technology revamping closed ethanol plant. Available at: https://biorrefineria.blogspot.com/2017/07/alliance-bio-products-to-demonstrate-CTS-technology-former-INEOS-ethanol-plant.html. Accessed: 18 Aug 2019.

[91]   Blair RG, Hick SM, Truitt JH. Solid acid catalyzed hydrolysis of cellulosic materials. Patent No. US8062428B2, 2007.

[92]   Jupeng Bio website. Available at: http://www.jupengbio.com/. Accessed: 23 Aug 2019.

[93]   Niemistö J, Pasanen A, Hirvelä K, Myllykoski L, Muurinen E, Keiski RL. Pilot study of bioethanol dehydration with polyvinyl alcohol membranes. J Membr Sci. 2013;447:119–27. DOI:10.1016/j.memsci.2013.06.048.

[94]   Bastidas PA, Gil ID, Rodríguez G. Comparison of the main ethanol dehydration technologies through process simulation. In: Pierucci S, Buzzi Ferraris G, editors. 20th European symposium on computer aided process engineering – ESCAPE20. Oxford, Great Britain:Elsevier, 2010:1–7.

[95]   Ethanol Available at: http://www.essentialchemicalindustry.org/chemicals/ethanol.html. Accessed: 18 Aug 2019.

[96]   Kujawski W, Zieliński Ł. Bioethanol – one of the renewable energy sources. Environ Prot Eng. 2006;32:143–9.

[97]   Raiser T. Turning waste into energy. Sulzer Tech Rev. 2012;3:1–7. Available at: https://www. sulzer.com/-/media/files/products/separation-technology/distillation-and-absorption/tech nicalarticles/str_2012_3_04_07_raiser_e_lo_einzl.ashx. Accessed: 24 Aug 2019.

[98]   Vane LM. A review of pervaporation for product recovery from biomass fermentation processes. J Chem Technol Biotechnol. 2005;80:603–29. DOI:10.1002/jctb.1265.

[99]   Bogoczek R, Kociołek-Balawajder E. Technologia chemiczna organiczna. Surowce i półprodukty (Organic chemical technology. Raw materials and semi-products). WAE, Wrocław, 1992. (in Polish).

[100] Prudhvi Raj T. Plant design for manufacturing acetic acid. Department Of Petroleum Engineering & Petrochemical Engineering University College Of Engineering (A) Kakinada, 2010-4.

[101] Cheung H, Tanke RS, Torrence GP. Acetic Acid. ULLMANN'S Encyclopedia of Industrial Chemistry. Wiley-VCH, 2012.

[102] Pal P, Nayak J. Acetic acid production and purification: critical review towards process intensification. Sep Purif Rev. 2017;46:44–61.

[103] Acetic acid prices, markets & analysis. Independent chemical information service ICIS. Available at: https://www.icis.com/explore/commodities/chemicals/acetic-acid/ Accessed: 20 July 2019.

[104] Acetic Acid Production and Manufacturing Process. Independent Chemical Information Service ICIS. Available at: https://www.icis.com/explore/resources/news/2007/10/31/ 9074780/acetic-acid-production-and-manufacturing-process/. Accessed: 04 Aug 2019.

[105] Sunley GJ, Watson DJ. High productivity methanol carbonylation catalysis using iridium - The Cativa process for the manufacture of acetic acid. Catal Today. 2000;58:293–307.

[106] Acetic acid – Manufacturing process for Acetic acid. Guichon Valves. Available at: https://gui chon-valves.com/faqs/acetic-acid-manufacturing-process-for-acetic-acid/. Accessed: 04 Aug 2019.

[107] Jones JH. The cativa process for the manufacture plant of acetic acid. Platinum Metals Rev. 2000;44:94–105.

[108] Celanese Corporation. The chemistry inside innovation. Available at: https://www.celanese. com. Accessed: 04 Aug 2019.

[109] Chiyoda Corporation. Acetic Acid Production Process (CT-ACETICA). Available at: https:// www.chiyodacorp.com/en/service/gtl/acetica/. Accessed: 04 Aug 2019.

[110] Process for production of acetic production. Haldor Topsoe A/S, Lyngby (DK), Patent No. US6573403, 2000.

[111] Sano K, Uchida H, Wakabayashi S. A new process for acetic acid production by direct oxidation of ethylene. Catal Surv Jpn. 1999;3:55–60.

[112] Castillo Martinez FA, Balciunas EM, Salgado JM, Domínguez González JM, Converti A, Oliveira R. Lactic acid properties, applications and production: a review. Trends Food Sci Technol. 2013;30:70–83. DOI:10.1016/j.tifs.2012.11.007.

[113] Werpy T, Petersen G, Aden A, Bozell J, Holladay J, White J, et al. Top value added chemicals from biomass, vol. I: results of screening for potential candidates from sugars and synthesis gas. Report for the U.S. Department of Energy (DOE), 2004. Available at: https://www.en ergy.gov/eere/bioenergy/downloads/top-value-added-chemicals-biomass-volume-i-results-screening-potential. Accessed: 18 Aug 2019.

[114] Juturu V, Wu JC. Microbial production of lactic acid the latest development. Crit Rev Biotechnol. 2016;36:967–77. DOI:10.3109/07388551.2015.1066305.

[115] Polylactic Acid Properties, Production, Price, Market and Uses. Plastics Insight. Available at: https://www.plasticsinsight.com/resin-intelligence/resin-prices/polylactic-acid/. Accessed: 18 Aug 2019.

[116] Corbion, Polymers. Available at: http://www.corbion.com/biochemicals/chemicals/applica tions/polymers. Accessed: 18 Aug 2019.

[117] Zion Market Research. Global Lactic Acid Market Will Reach USD 10.06 Billion by 2025: Zion Market Research, 2019. Available at: https://www.globenewswire.com/news-re lease/2019/02/04/1709550/0/en/Global-Lactic-Acid-Market-Will-Reach-USD-10-06-Billion-by-2025-Zion-Market-Research.html. Accessed: 20 Aug 2019.

[118] Eş I, Mousavi Khaneghah A, Barba FJ, Saraiva JA, Sant'Ana AS, Hashemi SM. Recent advancements in lactic acid production – a review. Food Res Int. 2018;107:763–70. DOI:10.1016/j.foodres.2018.01.001.

[119] Wee YJ, Kim JN, Ryu HW. Biotechnological production of acid. Food Technol Biotechnol. 2006;44:163–72.

[120] de Oliveira RA, Komesu A, Vaz Rossell CE, Maciel Filho R. Challenges and opportunities in acid bioprocess design – from economic to production aspects. Biochem Eng J. 2018;133:219–39. DOI:10.1016/j.bej.2018.03.003.

[121] Wattanapaphawong P, Sato O, Sato K, Mimura N, Reubroycharoen P, Yamaguchi A. Conversion of cellulose to lactic acid by using ZrO2–Al2O3 catalysts. Catalysts. 2017;7:221. DOI:10.3390/catal7070221.

[122] Wang Y, Deng W, Wang B, Zhang Q, Wan X, Tang Z, et al. Chemical synthesis of lactic acid from cellulose catalysed by lead(II) ions in water. Nature Commun. 2013;4:2141. DOI:10.1038/ncomms3141.

[123] Baniel AM, Eyal AM, Mizrahi J, Hazan B, Fisher RR, Kolstad JJ, et al. Lactic acid production, separation and/or recovery process. Patent No. US5510526A, 1996.

[124] Baniel AM, Eyal AM, Mizrahi J, Hazan B, Fisher RR, Kolstad JJ, et al. Lactic acid production, separation and/or recovery process. Patent No. US5892109A, 1999.

[125] Miller C, Fosmer A, Rush B, McMullin T, Beacom D, Suominen P. Industrial production of lactic acid. Compr Biotechnol. 2011;179–88. DOI:10.1016/b978-0-08-088504-9.00177-x.

[126] Needle HC. Fermentation process for producing lactic acid and lactates. Patent No. US2143360A, 1937.

[127] Nature Works. Available at: https://www.natureworksllc.com/About-NatureWorks. Accessed: 18 Aug 2019.

[128] Subbian E. Production of lactic acid from organic waste or biogas or methane using recombinant methanotrophic bacteria. Patent No. EP3129513A2, 2014.

[129] Market Research Report. Lactic Acid Market Size, Share & Trends Analysis Report By Raw Material (Sugarcane, Corn, Cassava), By Application (Industrial, F&B, Pharmaceuticals, Personal Care, PLA), And Segment Forecasts, 2018 – 2025, 2018. Available at: https://www.

grandviewresearch.com/industry-analysis/lactic-acid-and-poly-lactic-acid-market. Accessed: 20 Aug 2019.

[130] Pleissner D, Lau KY, Schneider R, Venus J, Lin CS. Fatty acid feedstock preparation and lactic acid production as integrated processes in mixed restaurant food and bakery wastes treatment. Food Res Int. 2015;73:52–61. DOI:10.1016/j.foodres.2014.11.048.

[131] Sherman LM. Global lactic acid producer plans to enter PLA business. Plastics Technology, 2014. Available at: https://www.ptonline.com/blog/post/global-lactic-acid-producer-plans-to-enter-pla-business. Accessed: 18 Aug 2019.

[132] Galactic. Available at: https://www.lactic.com/en-us/aboutus/ourhistory.aspx. Accessed: 20 Aug 2019.

[133] Dotacje INNOCHEM. Available at: https://www.orlenpoludnie.pl/PL/OFirmie/Strony/Dotacje-INNOCHEM.aspx. Accessed: 18 Aug 2019.

[134] Laird K. Green matter: ThyssenKrupp Uhde inaugurates Europe's first multi-purpose fermentation plant. Plastics Today, 2013. Available at: https://www.plasticstoday.com/content/green-matter-thyssenkrupp-uhde-inaugurates-europe-s-first-multi-purpose-fermentation-plant/12167412219155. Accessed: 18 Aug 2019.

[135] PLA-production process – Commercial debut for a new PLA-production process. 2017. Available at: https://www.polyestertime.com/pla-production-process/. Accessed: 18 Aug 2019.

[136] Mishra VK, Goswami R. A review of production, properties and advantages of biodiesel. Biofuels. 2017;1–17. DOI:/10.1080/17597269.2017.1336350.

[137] Newsy (in Polish). 2017. Available at: https://nawozy.eu/aktualnosci/newsy/rosliny-oleiste/ue-najwiekszym-producentem-biodiesla-na-swiecie.html. Accessed: 24 Aug 2019.

[138] Yang SC, Chang JR, Lee MT, Lin TB, Lee FM, Hong CT, et al. Homogeneous catalysts for biodiesel production. Patent No. WO2014123711A1, 2013.

[139] Rahman Talukder MM, Wu J. Biodiesel production via enzymatic hydrolysis followed by chemical/enzymatic esterification. WO2010005391A1, 2008.

[140] Carpenter RS, Hartantio I. Compositions and methods for biodiesel production from waste triglycerides. Patent No. WO2016172698A1, 2016.

[141] Lin VS, Cai Y, Kern C, Dulebohn JI, Nieweg JA. Method for biodiesel production. Patent No. EP2205354B1, 2008.

[142] Koch M, Thewissen S, Rosso-Vasic M, Nuberg E, Van Der Griend H, Steenwinkel EE. Zirconium-based catalyst compositions and their use for biodiesel production. Patent No. WO2011089253A1, 2011.

[143] Perrier MJ. Process and system for producing biodiesel fuel. Patent No. US20090071063A1, 2007.

[144] Slade DA, Ellens CJ, Albin BN, Winkel DJ, Downey JA. Methods and devices for producing biodiesel and products obtained therefrom. Patent No. US20160145536A1, 2014.

[145] Farm Energy. Commercial and Large Scale Biodiesel Production Systems. 2019. Available at: https://farm-energy.extension.org/commercial-and-large-scale-biodiesel-production-systems/. Accessed: 26 Aug 2019.

[146] Renewable Energy Group. Available at: https://www.regi.com/. Accessed: 24 Aug 2019.

[147] Biofuels International. Avril biodiesel at full capacity, for now. Available at: https://biofuels-news.com/news/avril-biodiesel-at-full-capacity-for-now/. Accessed: 24 Aug 2019.

[148] Ecoreporter. 2010. Available at: https://www.ecoreporter.de/artikel/biopetrol-industries-ag-extraordinary-shareholders-meeting-20-12-2010/. Accessed: 24 Aug 2019.

[149] Envien Group. Available at: https://www.enviengroup.eu/en/products/biodiesel-en. Accessed: 24 Aug 2019.

[150] UnitecBio. Available at: http://www.unitecbio.com.ar/html/. Accessed: 24 Aug 2019.

[151] China Biosource. Available at: https://www.chinabiodiesel.com/. Accessed: 24 Aug 2019.

Katarzyna Materna

5 Trends in technology of oxygen containing hydrocarbons: aldehydes, ketones, ethers

Abstract: The structural element of many organic compounds in which the carbon atom combines with the oxygen atom in a double bond is called the carbonyl group. Both atoms forming such a double bond are characterized by hybridization of the sp2 type. Spatially, these two atoms and two other atoms directly related to the carbon atom lie in the same plane. One of the carbon–oxygen bonds is of σ type, formed by two overlapping sp2 orbital bonds, while the other is of π type, realized through the commonalization of non-hybridized electrons from orbital π. The carbonyl group "–C=O" is a common element in the structure of aldehydes and ketones and to a large extent determines their chemical properties. Aldehydes are organic compounds in which the carbonyl group is connected by one single bond to the hydrogen atom and the other to the rest of the hydrocarbon molecule. In aldehydes, the C=O group occurs at the end of the carbon chain; it is connected with a single C–C bond; the fourth bond is saturated with hydrogen atom. Ketones are organic compounds containing a carbonyl group which is combined with two hydrocarbon groups. In ketones, the C=O group occurs within the carbon chain of molecules; the carbon atom of the carbonyl group is secondary. Ethers are organic compounds in which C–O–C bonds are present, where none of the carbon atoms is bound to more than one oxygen atom.

Keywords: aldehydes, ketones, ethers, oxidation, dehydrogenation, primary and secondary alcohols, reduction, dehydration

5.1 Introduction

The structural element of many organic compounds in which the carbon atom combines with the oxygen atom in a double bond is called the carbonyl group. Both atoms forming such a double bond are characterized by hybridization of the sp^2 type. Spatially, these two atoms and two other atoms directly related to the carbon atom lie in the same plane. One of the carbon–oxygen bonds is of σ type, formed by two overlapping sp^2 orbital bonds, while the other is of π type, realized through the commonalization of non-hybridized electrons from orbital π.

This article has previously been published in the journal Physical Sciences Reviews. Please cite as: Materna, K. Trends in technology of oxygen containing hydrocarbons: aldehydes, ketones, ethers *Physical Sciences Reviews* [Online] 2020, 5. DOI: 10.1515/psr-2019-0035.

https://doi.org/10.1515/9783110656367-005

The carbonyl group "–C=O" is a common element in the structure of aldehydes and ketones and to a large extent determines their chemical properties.

Aldehydes are organic compounds in which the carbonyl group is connected by one single bond to the hydrogen atom and the other to the rest of the hydrocarbon molecule. In aldehydes, the C=O group occurs at the end of the carbon chain; it is connected with a single C–C bond, and the fourth bond is saturated with hydrogen atom.

Ketones are organic compounds containing a carbonyl group which is combined with two hydrocarbon groups. In ketones, the C=O group occurs within the carbon chain of molecules; the carbon atom of the carbonyl group is secondary.

Ethers are organic compounds in which C–O–C bonds are present, where none of the carbon atoms is bound to more than one oxygen atom [1, 2].

5.2 Aldehydes and ketones

5.2.1 Nomenclature

Aldehydes are carbonyl compounds in which carbonyl carbon is bonded to two hydrogen atoms (HC(=O)H) or to a hydrogen atom and carbon atom of the hydrocarbon substituent R–(RC(=O)H).

The acyclic (chain) aldehyde names are formed by adding the suffix -al (for monoaldehyde) or -dial (for dialdehyde) to the hydrocarbon name. The aldehyde group has priority in the name over the groups: C=C, OH.

The name of aldehyde, in which the aldehyde group is directly connected to a ring system (aliphatic or aromatic), is formed by adding the carboaldehyde suffix (dicarboaldehyde) to the name of the cyclic system.

The aldehyde group as a substituent is called by adding the prefix -formylo (or -metanoyl) to the name of the parent system.

The common names of aldehydes are formed by: replacement of the word acid in the common name of the carboxylic acid concerned by aldehyde; or by replacement of the -oil (or -yl) tip in the name of the acyl group by the aldehyde suffix (Table 5.1) [1].

Aldehydes are characteristic fragrant and toxic chemical compounds with fungicidal and bactericidal properties – the shorter the carbon chain, the less pleasant the smell and the higher the toxicity. Formalin and acrolein (burnt fat) have a very unpleasant smell, but vanillin (vanilla) or citronella (lemon) is pleasant. Formaldehyde solution (HCHO) in water (so-called formalin) is used for the preservation of biological preparations [2].

Ketones are compounds with a carbonyl group to which hydrocarbon groups are attached on both sides. The substituent names of ketones are formed by adding

Table 5.1: Formulas and names of sample aldehydes.

Aldehyde		Aldehyde	
	hexanal (capron aldehyde)		3-methoxy-2-methylbutane
	formylocyclopentane (cyclopentyl methanal)		propanedial (malonic aldehyde)
	3-chlorobenzoic aldehyde		phenylethanol (phenylacetaldehyde)

the suffix -on to the parent hydrocarbon name (acyclic or cyclic) or -dion (two ketone groups, etc.).

The carbon atom of the carbonyl group shall be denoted by the lowest possible locus.

Group-functional names of acyclic ketones are created by naming groups (in adjective form and in alphabetical order) after the word "ketone"; the names of both groups are separated by a hyphen (a line).

The names of cyclic ketones, in which the group C=O is a segment of a ring, are created by adding the suffix -on (or -dion, -trion, etc.) to the name of the cyclic system.

The carbonyl group as a substituent is called by adding the prefix -oxo to the name of the parent system [1].

5.2.2 Properties and application

Aldehydes are compounds that are mostly solids (except for: phenyldehyde, acetaldehyde and unsaturated aldehydes – they are gases that are relatively well soluble in water), poorly soluble in water, with a characteristic fragrance. They have strong reducing properties, which makes them different from ketones. This property is the basis for the detection and determination of these compounds, e. g. by means of a test: Tollens, Fehling, Benedict or Trommer.

Both aldehydes and ketones are highly polar molecules. They do not form hydrogen bonds with each other because they do not have hydrogen atoms directly connected to the oxygen atom of the carbonyl group. Simple aldehydes and ketones are water-soluble. The solubility of larger aldehydes and ketones decreases with the length of the carbon chain.

The carbonyl group of aldehydes is usually more reactive than the carbonyl group of ketones. Ketones undergo dehydration, oxidation and reduction reactions under specific reaction conditions that produce the desired products. That is, Clemmensen reduction or reduction using hydrogen anion gives the product a form of alcoholic anion, which after protonation transforms into suitable alcohol. Aldehydes are reduced to primary alcohols, while ketones are reduced to secondary alcohols [1–4].

The reaction with the Tollens reagent allows the presence of aldehyde to be detected. A positive result occurs in the form of a characteristic "silver mirror" which is deposited on the inner surface of the reaction vessel.

Aldehydes and ketones are widely used in chemical synthesis as starting substances for further compounding organic. Aldehydes and ketones are often found in everyday life.

Aldehydes are used for organic syntheses (plastics, dyes), in food and cosmetic industry (ingredients of fragrance compositions and food aromas), tanning (glutaric aldehyde). Formaldehyde solution (HCHO) in water (so-called formalin) is used for the preservation of biological preparations.

Ketones belong to one of the most widely dispersed chemical compounds both in nature and in the chemical industry. In the chemical industry, they are often used as substrates to obtain a wide range of chemical compounds or as solvents (e. g. acetone, etc.) [3, 4].

5.2.3 Preparation

The general method of aldehyde synthesis is the reaction of oxidation of primary alcohols [1]:

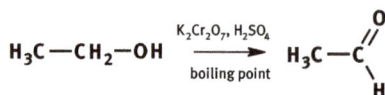

$$H_3C-CH_2-OH \quad \xrightarrow[\text{boiling point}]{K_2Cr_2O_7,\ H_2SO_4} \quad H_3C-C\begin{smallmatrix}O\\ \\H\end{smallmatrix}$$

The basic method of obtaining aldehydes in industry is catalytic oxidation of primary alcohols by oxygen from air or dehydrogenation over copper The oxidation should be carried out very carefully, as they are very reactive and easily oxidize to carboxylic acids [1]:

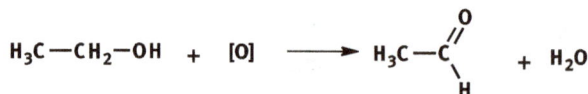

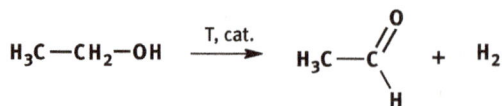

$$H_3C-CH_2-OH \ + \ [O] \quad \longrightarrow \quad H_3C-C\begin{smallmatrix}O\\ \\H\end{smallmatrix} \ + \ H_2O$$

$$H_3C-CH_2-OH \quad \xrightarrow{T,\ \text{cat.}} \quad H_3C-C\begin{smallmatrix}O\\ \\H\end{smallmatrix} \ + \ H_2$$

Another method of obtaining aldehydes is adding water to alkines using catalysts: $HgSO_4$ and H_2SO_4. This method, called catalytic hydration, is used on an industrial scale, among others, to obtain acetaldehyde [2]:

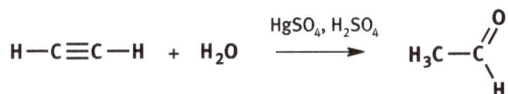

$$H-C\equiv C-H \ + \ H_2O \ \xrightarrow{\ HgSO_4,\ H_2SO_4\ } \ H_3C-C\overset{O}{\underset{H}{\diagup}}$$

The next method of aldehyde preparation is the reduction of carboxylic acids [2]:

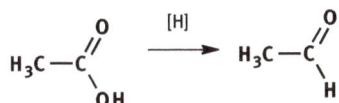

$$H_3C-C\overset{O}{\underset{OH}{\diagup}} \ \xrightarrow{\ [H]\ } \ H_3C-C\overset{O}{\underset{H}{\diagup}}$$

A specific method of ethanal preparation is the reaction of ethylene with oxygen using an appropriate catalyst [2]:

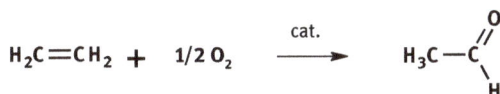

$$H_2C\!=\!CH_2 \ + \ 1/2\,O_2 \ \xrightarrow{\ cat.\ } \ H_3C-C\overset{O}{\underset{H}{\diagup}}$$

The general method of ketones synthesis is the dehydrogenation reaction of second-row alcohols – passing through the alcohol vapour, at a temperature of about 300 °C, over copper [3]:

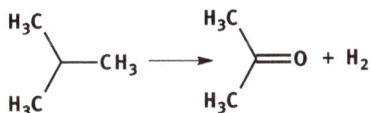

$$\underset{H_3C}{\overset{H_3C}{>}}\!\!<\!\!-CH_3 \ \longrightarrow \ \underset{H_3C}{\overset{H_3C}{>}}\!\!=\!\!O \ + \ H_2$$

Ketones can also be obtained by careful reduction of carboxylic acids – or careful oxidation of alcohols, although oxidation of secondary alcohols is often very difficult [1, 4].

Ketones are less reactive than aldehydes and the reactions to their addition to the carbonyl group require stricter conditions.

Ketones with small alkyl groups are liquids that mix well with both water and organic solvents. Ketones are very polar compounds and at the same time quite unreactive, so they are often used as solvents and additives in paint removers [4].

5.2.4 The most important aldehydes and ketones

Benzoic aldehyde, benzaldehyde or benzenecarboaldehyde is the simplest aromatic aldehyde with formula C_6H_5CHO [3]:

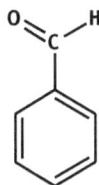

Properties: yellowish, volatile, oily liquid, density 1.04 g cm^{-3}, melting point −26 °C, boiling point of 178.1 °C.

It is obtained from toluene (methylbenzene) $C_6H_5CH_3$ by conversion into dichloromethylbenzene (benzylidene chloride) $C_6H_5CHCl_2$ and hydrolysis of this compound. It occurs in almonds and has almond smell, but unlike nitrobenzene and hydrogen cyanide, which smell the same, it is not toxic. It is used in the perfumery industry, in spices and in the production of dyes.

Formaldehyde (according to IUPAC methanal nomenclature) is the simplest organic compound from the group of aldehydes with HCHO formula (other − H_2CO) [3]:

It was discovered by Alexander Butlerrov in 1859. Under normal conditions, formic aldehyde is a gas with a characteristic suffocating smell and is a strong poison. It has a melting point of −113 °C and a boiling point of −21 °C. It is freely soluble in water, up to approximately 40% by weight. The solution has a weak acidic pH. In trade, the most common solution is 35–40% formaldehyde solution in water − formalin. Formaldehyde is formed during incomplete combustion of carbon-containing substances. It is obtained by oxidation and dehydrogenation of methanol on a copper or silver catalyst.

Acetaldehyde (according to the IUPAC nomenclature: ethanal) is an organic compound from the group of aldehydes with the CH_3CHO formula [3].

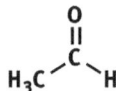

Properties: at room temperature a colourless liquid with a pungent odour, melting point −123 °C, a boiling point of 21 °C, density 0.78 g cm^{-3}, soluble in water and organic solvents, under the influence of small amounts of sulphuric acid it polymerizes with the formation of paraldehyde (3 monomers, formed at lower temperatures) or metaldehyde (4 or more monomers).

Occurrence and preparation: Acetaldehyde occurs in nature in ripe fruits and coffee. It is obtained by catalytic oxidation of ethanol, hydration of acetylene or oxidation of ethylene with oxygen in the presence of an aqueous solution of palladium chloride and copper salts:

Application: In the industry, acetaldehyde is used to produce acetic acid, acetic anhydride and many other compounds. Like formic aldehyde, it reacts with phenol and amines to form synthetic resins.

One of the best-known ketone is acetone [3]:

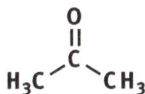

$$\underset{H_3C}{\overset{\displaystyle \overset{O}{\|}}{}}\overset{C}{}\underset{CH_3}{}$$

It is the simplest representative of a group of ketones with formula C_3H_6O [other records are $CO(CH_3)_2$ or CH_3COCH_3]. Acetone is a colourless liquid with a boiling point of 56.53 °C and a melting point of −95.4 °C. It has a relatively low density of 0.819 g cm^{-3}. It has a sharp, as some people claim, associated with fruit fragrance. It is mixed in all proportions with water, ethanol, ethers and other low molecular weight ketones. Acetone is contained in small amounts in blood and urine. Its concentration is higher than normal in the body when diabetes is advanced and untreated. Acetone is naturally present in the tissues of many plants, volcanic and exhaust gases. It is formed in large quantities during the dry distillation of wood, which used to be the main production method.

Nowadays, in industry, it is most often produced by the cumen method in the process of obtaining phenol. In the laboratory, it can be obtained by dry distillation of calcium acetate. Acetone is a widely used organic solvent with high polarity. It is used in the production of medicines, dyes, paints, varnishes and cleaning agents. Acetone is quite toxic. Particularly toxic are its vapours, easily absorbed by the lungs into the blood, which distributes this compound throughout the body. If the concentration of acetone in the blood is not too high, it is effectively metabolized by the liver. At higher concentrations, acetone begins to cause acute irritation of the mucous membranes of the nose and mouth, as well as tearing of the eyes and headache. A high concentration of acetone in the air causes unconsciousness and coma. Small amounts of acetone administered orally do not cause any major effects except irritation of the entire gastrointestinal tract and possible vomiting. Large amounts of acetone cause similar effects as when inhaled. Acetone irritates the skin, causing its redness. In frequent contact, it causes permanent damage to the skin structure similar to burns [4].

5.3 Ethers

Ethers in their structure contain an oxygen atom combined with two hydrocarbon groups: R−O−R [1].

The two carbon atoms (α-carbon) attached to the more electronegative oxygen atom are more acidic than the carbon atoms in hydrocarbons and less acidic than the hydrogen atoms in α positions relative to the carbonyl group.

Due to the bent structure of the C–O–C bond (angle 104°), the ethers show a weak dipole moment. However, this poor polarity does not significantly affect the boiling temperatures of ethers, which are similar to those of a similar molecular weight and significantly lower than isomeric boiling temperatures (e. g. 98 °C for n-heptane, 100 °C for methylpentyl ether and 157 °C for hexanol). Ethers are relatively unreactive. The ether bond is resistant to alkalis as well as to oxidizing and reducing agents. The bond can be cleaved under the action of strong acids [2].

They are soluble in water and their solubility is similar to that of the corresponding alcohols. Dimethyl ether and methyl ethyl ether are gases. Higher ethers are liquids. They are commonly used in paints, perfumes, oils or waxes [3].

Ethers are obtained by the reaction of suitable alcohols with sulphuric acid [1]:

$$R-O-R \xrightarrow{\text{H}_2\text{SO}_4, \text{ heating}} R-O-R + H_2O$$

Each pair of alcohol molecules loses a molecule of water. The course of the dehydration reaction towards the formation of ethers, not alkenes, is regulated by the selection of appropriate reaction conditions.

Many symmetrical ethers containing shorter alkyl groups are obtained on a large scale. They are mainly used as solvents. The most important is diethyl ether – known as an anaesthetic and a solvent used for extraction and for obtaining Grignard compounds.

An interesting example of ether are epoxides [1]:

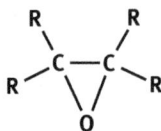

Epoxides are a class of organic compounds that contain an oxirane ring in their structure (ethylene oxide), which gives them exceptional properties. Ethylene oxide is obtained on an industrial scale from the catalytic oxidation of ethylene with molecular oxygen in air. The C–O bond in epoxy is more readily broken down under the influence of alkalis and catalytic acids than the C–O bond present in the structure of larger cyclic ethers. This is due to the presence of stresses in the three-part ring. Epoxy-specific reactions are those where the product is a compound containing two functional groups. The reaction of epoxies with water produces diols, while the reaction with alcohols produces compounds that are both ethers and alcohols. Epoxides also react with Grignard compounds [2].

Another interesting example of ethers are crown ethers [3]. Crown ethers are molecules in which the oxygen atom combines with two carbon atoms to form molecular rings resembling the shape of a crown, which play a very important role in the so-called supramolecular chemistry, where there are host–guest molecular complexes. In such complexes, guest molecules and ions can be placed inside the host molecules (in this case, in the rings of ethers) [5].

Recent research on the properties of crown ethers has led to a radical increase in the strength of bonds and the selectivity of corona ethers molecules, placing them in a flat graphene structure. Valuable properties such as high selectivity and solubility in almost all solvents and the ability to "curl" and "develop" many times make crown ethers widely used, e. g. in accelerating chemical processes involving ions, purifying water from heavy metal ions, transporting ions through cell membranes, etc. [5].

Crown ethers are also highly flexible and adapt to the shapes and ionic dimensions of guests, but they reduce their selectivity and the strength of molecular bonds, which means that they cannot bind to all ions. Scientists decided to insert ethers into the flat and rigid structure of graphene. Forcing the rings of crown ethers to form a two-dimensional rigid structure significantly increased their selectivity and multiplied the forces of bonds with guests' ions [5].

The most popular oxygenate is methyl tert-butyl ether (MTBE), which in the 1970s was added to fuel in small amounts as a component increasing the octane number (anti-knock agent) [4]. In the 1990s, its concentration in fuel was significantly increased (up to 15%) – it started to be considered as a component supporting total hydrocarbon combustion and widely used in the United States.

Methyl tert-butyl ether (MTBE) is a colourless liquid with an unpleasant ether odour, mainly used in the petrochemical industry as a component of gasoline for increasing the number of octanes, as well as in some medical procedures, e. g. for dissolving gallstones and in chemical reactions and analyses, e. g. as a solvent in polymerization reactions and as an eluent for chromatography [3].

5.3.1 MTBE production methods

Methyl tert-butyl ether (MTBE, with a structural formula $(CH_3)_3COCH_3$) is an ether compound made by combining methanol and isobutylene. The methanol typically is derived from natural gas; isobutylene can be derived as a byproduct of the petroleum refinery process. The synthesis of methyl tert-butyl ether is carried out in the liquid phase at a pressure of about 1.2 MPa at 90–100 °C, where only isobutylene reacts selectively with methanol. This makes it possible to use it as a raw material in the process of C_4 fraction from catalytic cracking or olefin pyrolysis. Butadiene must be removed beforehand. Butanes are also often removed and a fraction consisting only of butenes is processed [4].

MTBE production technologies differ, among others, in the solution of the synthesis node – instead of tubular reactors, for example, reactors with several layers of catalyst and cooling by cold streams of raw materials introduced between these layers are used.

The electricity and steam consumption of MTBE fusion plants depends on the degree of isobutylene reactivation and product purity. MTBE installations are integrated with installations carrying out other technological processes.

As a result of the high demand for MTBE, an unleaded petrol component, its production is constantly increasing, which makes it possible to combine the production of methanol from various raw materials with the refinery production of high octane petrol, aromatic hydrocarbons and olefins [4].

5.4 Green modifications for the oxidation of alcohols into carbonyl-containing products

5.4.1 Oxydation without solvent

Selective oxidation of alcohols to suitable aldehydes and ketones is one of the most important reactions in organic synthesis. For years it has aroused growing interest. Chromium trioxide (CrO_3), an oxidizing agent used in organic chemistry for years, is one of the most versatile and energetic oxidants used in these transformations [5]. Unfortunately, the subsequent oxidation of aldehydes to carboxylic acids, as well as the degradation of unsaturated substrates, constitute a serious problem when using ordinary chromium trioxide. These undesirable side effects occur in acidic aqueous solutions, where the aldehydes formed are hydrated to geminal diols and further oxidized. In addition, the formed aldehyde can react with unreacted alcohols to hemiacetals, which are then easily oxidized to esters. In recent years, chromium trioxide has been modified to overcome the fundamental problems associated with subsequent oxidation in particular [6]. Reactions conducted with the use of immobilized raw materials are of great interest. Immobilization of raw material on or in organic polymeric material or in inorganic porous material or in the form of a carrier film not only modifies the reactivity of the raw material and the selectivity of the product formation, but also greatly facilitates its release. A method of oxidation of aromatic alcohols to aldehydes and ketones by grinding the alcohols with chromium trioxide deposited on aluminium silicate without the use of solvent at room temperature was developed. All the reactions were completed after 2 min. Their efficiency was 86–96% [6].

Another method of oxidation of oximes to aldehydes and ketones by grinding the oximes with manganese dioxide deposited on aluminium silicate without the use of solvent and at room temperature was developed. All the reactions were carried out within 3 min. Their efficiency was 82–92% [7].

Carbonyl compounds are often used in organic synthesis to protect the carbonyl group (C=O) and are widely used for the separation, purification and characterization of carbonyl compounds. Oximes are also intermediate products in various chemical syntheses and have proven their usefulness as important and convenient raw materials in organic synthesis. Reconstitution of carbonyl compounds (aldehydes and ketones) from their oximes is an important process in organic chemistry and can be based, among others, on hydrolysis, reduction or oxidation reactions. However, the number of methods and substrates for converting oximes to suitable carbonyl compounds under mild conditions is small. Therefore, it is important to develop methods of oxime decomposition to ensure that the process is carried out under mild conditions. Oxidative decomposition is one of the most important ways of recovering carbonyl compounds from their oximes. So far, potassium permanganate ($KMnO_4$) and a few oximes-based ones have been used for the decomposition of oximes.

A method of oxidative decomposition of carbonyl compounds oximates to initial aldehydes and ketones by grinding the oximes with potassium permanganate deposited on graphite at room temperature without the use of solvent was developed. All the reactions were completed after 5 min. Their efficiency was 80–94% [8].

5.4.2 Catalytic systems

Dichromates and permanganates generating stoichiometric amounts of waste have been and still are the most commonly used oxidants. However, publications describing the oxidation of alcohols with oxygen in the liquid phase in the presence of various catalytic systems can be found. First, the method using the catalytic system $RuCl_2(PPh_3)_3$/TEMPO (TEMPO – radical (2,2,6,6-tetramethylpiperidin-1-yl)oxyl) seems to be of great importance [9, 10]. The reaction was carried out in an autoclave without solvent or in chlorobenzene or toluene. The use of a mixture of oxygen and nitrogen causes that the composition of the gas phase (including reagent pairs) exceeds the explosive limits of the mixture, while increased pressure improves the solubility of oxygen.

Primary alcohols were oxidized to aldehydes with high selectivity – for octanol-1 99% with a conversion rate of 85%. It is worth noting that the obtaining aldehydes are not further oxidized to carboxylic acids, which is attributed to the presence of TEMPO in the system [9, 10].

Oxidation of alcohols with oxygen was also carried out with the use of another ruthenium catalyst – ruthenium deposited on aluminium oxide, under heterogeneous catalysis conditions [11]. The advantage of this process is the possibility of oxidation of alcohols containing also sulphur atoms, nitrogen or C=C bonds in the molecule, which was not successful in the process with the use of $RuCl_2(PPh_3)_3$/TEMPO system. After application of oxygen at 0.1 MPa pressure, at 83 °C and (trifluoromethyl)benzene as a reaction medium, suitable aldehydes with 99% selectivity were obtained from benzyl alcohol and its analogues with substituents in a ring (4-OMe, 4-Cl, 4-NO$_2$) at a similar degree of conversion. Equally high selectivity was obtained by oxidation of 2-hydroxymethylthiophene and 3-hydroxy-2-methylpyridine to appropriate aldehydes.

Palladium catalysts are also effective in oxidation of alcohols. The catalyst used by Brinks et al. to oxidize olefins in the two-phase system has also proved useful in the oxidation of alcohols. No organic solvents need to be used and sodium acetate, present in the environment, stabilizes the catalyst [12]. Carboxylic acids with selectivity and degrees of conversion equal to 100% are obtained from primary alcohols, but the addition of TEMPO to the reaction system allows to obtain aldehydes (e. g. pentanal from pentanol-1 with selectivity 97%). The second-order alcohols are oxidized to ketones with almost theoretical efficiency.

Similarly high yields (85–99%) of aldehydes and ketones were obtained from alcohols in the presence of a palladium catalyst [13]. The use of heterocyclic carbene in the Pd(OAc)$_2$ complex eliminates the necessity of using this ligand in excess.

The palladium(0) nanoparticles deposited on dispersed, functionalized resin – poly(styrene)-poly(oxyethylene glycols) – also proved to be a convenient catalyst for the discussed reactions [14]. From the first-order of aliphatic alcohols, oxidized in 0.2 M solution K$_2$CO$_3$, carboxylic acids with the efficiency above 90% were obtained. From benzyl alcohol benzoic aldehyde was formed, and from second row alcohols suitable ketones were obtained with the efficiency of 85–99%. Palladium nanoparticles Pd(0) applied on the hydroxyapatite catalysed very effectively the oxidation with air oxygen of the first order of alcohols [15]. Also, Mg(Pd(0)) deposited on MgO, added in the amount of 1% to alcohol in trifluoromethylbenzene, effectively catalyses the oxidation of alcohols to appropriate carbonyl compounds with molecular oxygen [16].

Complexes of copper halogens with aromatic amines constitute another group of catalysts for oxidation of alcohols with oxygen to appropriate carbonyl compounds. Sheldon and others used the complex [CuBr$_2$(2,2'-bipyridine)] in the presence of TEMPO and base (e. g. potassium tert-butoxide, t-BuOK) as a cocatalyst, obtaining from primary alcohols suitable aldehydes under mild conditions of reaction (25 °C, air oxygen, 5% mol CuBr$_2$ and 5% mol bipyridine), with the selectivity exceeding 99% [17]. Marko $et\,al.$ used a complex [CuCl,10-phenanthroline/di-tert-butyl-azodicarboxylate] in the presence of 1-methylimidazole and in the alkaline medium (t-BuOK). Aldehydes were obtained from the primary alcohols with the efficiency of 82–97% [18]. The presence of 1-methylimidazole prevents the formation of carboxylic acids and the neutral reaction environment allows to oxidize alcoholic substrates sensitive to acids and bases.

The oxidation of benzyl alcohol to benzaldehyde was also carried out in the presence of heteropoly acid H$_5$PV$_2$Mo$_{10}$O$_{40}$ as a catalyst in the environment of polyoxyethylene glycol (PEG 200), at 100 °C, for 16 h [19]. The reaction is carried out with 100% selectivity and a similar degree of conversion. The advantage of the proposed method is the use of polyoxyethylene glycol as a non-toxic, non-volatile and cheap solvent. The used catalyst does not undergo oxidation, which often happens in ligand in metal complexes, and the catalyst-solvent system can be recycled to further reactions. However, the disadvantage of this process is the loss of about 5% of the solvent after each cycle.

Gandhari and coworkers [20] described a greener oxidation of aromatic aldehydes to carboxylic acids. The procedure is based on the use of Oxone (a 2:1:1 molar mixture of $KHSO_5$, $KHSO_4$, and K_2SO_4) in water to oxidize benzaldehyde to benzoic acid. Several derivatives of benzaldehyde can also be oxidized in this way by using a 4:1 water:ethanol mixture as the solvent. It seems that Oxone is safer and environmentally friendly alternative to $KMnO_4$ or $K_2Cr_2O_7$.

5.4.3 Oxidations using sodium hypochlorite (bleach)

The oxidation of secondary alcohols into ketones by sodium hypochlorite was studied by Stevens *et al.* For a variety of substrates, 85–96% yields were obtained [21]. Zuczek and Furth [22] described the modification of the oxidation of cyclohexanol to cyclohexanone. This method utilizes above 11% aqueous sodium hypochlorite and mild heat with an excellent yield. Likewise, the room temperature conversion of isopropanol to acetone via reaction with NaOCI was improved [23]. Bleach oxidation of secondary alcohols to ketones attends to green principles. Use of sodium hypochlorite instead of sodium or potassium dichromate eliminates carcinogenic reagents and the production of chromium salt waste. Furthermore, sodium hypochlorite decomposes in the environment to sodium chloride, water and oxygen.

Mohrig *et al.* also studied the use of household bleach for the oxidation of secondary alcohols to ketones [24]. In this case, the oxidation procedure was altered in two other important ways. Firstly, the amount of acetic acid was reduced from 7.5 to 0.5 mL per gram of alcohol reactant, and secondly each ketone product was purified by distillation rather than extraction. This change in the purification method eliminates hazardous solvents and reduces the amount of generated waste. This method can also be used for oxidation of borneol to camphor [25].

Straub noticed that the system does not oxidize benzoin to benzil [26]. However, this conversion can be resulted using household bleach, 2,2,6,6-tetramethyl-piperidine-l-oxyl (TEMPO), sodium bicarbonate and potassium bromide in dichloromethane [26].

Additionally, sodium hypochlorite can be used to oxidize acetophenone to benzoic acid [27].

5.5 Conclusions

This chapter demonstrates the oxygen-containing hydrocarbons: aldehydes, ketones and ethers. Aldehydes are important reagents used for the manufacture of plastics and dyes, in the food industry (as flavourings) and cosmetics (composition ingredients), in the tanning industry. 40% aqueous solution of formaldehyde (formalin) is used for the maintenance of biological preparations. Whereas ketones are mainly solvents for paint and varnish production, in the cosmetics industry (ingredients of fragrance compositions), semi-finished products in various organic syntheses. Finally, ethers (most often diethyl ether) are used as low-polar solvents, e. g. to obtain Grignard compounds. Other commonly used ethers are cyclic tetrahydrofuran (THF), used as an organic solvent with moderate polarity and a more polar cyclic dieter, dioxane. Each analysed group of compounds is characterized by original properties, which results in wide application possibilities. In addition, new methods of obtaining them are being developed, looking for alternatives and possibilities to meet the principles of green chemistry.

References

[1]   Morrison RT, Boyd RN. Chemia organiczna. Warszawa: PWN. t. I., 2012.
[2]   McMurry J. Chemia organiczna t. 3–4. Warszawa: PWN, 2005.
[3]   Winterberg M, Schulte Körne E, Peters U, Nierlich F. Ullmann's encyclopedia of industrial chemistry. Weinheim: Wiley VCH, 2002.
[4]   Grzywa E, Molenda J. Technologia podstawowych syntez organicznych t. II. Warszawa: WNT, 2009.
[5]   Guo J, Lee J, Contescu CI, Gallego NC, Pantelides S, Pennycook SJ, et al. Crown ethers in graphene. Nat Commun. 2014;5:5389.
[6]   Zhu LY, Gao YS, Qian JQ. Synthesis of aldehydes and ketones under solvent-free conditions. Part 1, efficient oxidation of alcohols to aldehydes and ketones with chromium trioxide on aluminum silicate. Przem Chem. 2017;96:2454.
[7]   Zhu LY, Gao YS, Qian JQ. Synthesis of aldehydes and ketones under solvent-free conditions. Part. 2, an efficient method for recovering aldehydes and ketones by oxidation of their oximes with manganese dioxide on aluminum silicate. Przem Chem. 2017;96:2457.
[8]   Zhu LY, Gao YS, Qian JQ. Synthesis of aldehydes and ketones under solvent-free conditions. Part 3, a method for recovering aldehydes and ketones by oxidation of their oximes with potassium permanganate on graphite. Przem Chem. 2017;96:2461.
[9]   Lou JD, Wu YY. Chromic acid adsorbed on aluminum silicate: a new reagent for oxidation of primary alcohols to aldehydes. Synth Commun. 1987;17:1717.
[10]   Lou JL, Zhu LY, Wang L. Efficient oxidation of alcohols with potassium permanganate adsorbed on aluminum silicate reagent. Monatsh Chem. 2004;135:31.
[11]   Huang LH, Lou JD, Zhu L, Ping L, Fu Y. An efficient and selective solvent-free oxidation of alcohols by shaking with chromium trioxide supported on aluminium silicate. Molecules. 2005;10:794.
[12]   Lou JD, Gao C, Li L, Fang Z. An efficient selective oxidation of alcohols with potassium permanganate adsorbed on aluminum silicate under solvent-free conditions and shaking. Monatsh Chem. 2006;137:1071.
[13]   Lou JD, Huang LH, Marrogi AJ, Li F, Li L, Gao C. Selective oxidation of alcohols with a new reagent: iron(III) nitrate supported on aluminum silicate. Synth Comm. 2008;38:428.
[14]   Lou JD, Li L, Vatanian N, Lu X, Yu X. Jones reagent supported on aluminum silicate: a new reagent for oxidation of alcohols. Synth React Inorg Metal Org Nano- Metal Chem. 2008;38:370.

[15] Lou JD, Ma YC, Vatanian N, Wang Q, Zhang C. Oxidation of benzoins with ferric(III) nitrate supported on aluminum silicate. Synth React Inorg Metal Org Nano-Metal Chem. 2010;40:157.

[16] Lou JD, Ma YC, Vatanian N, Wang Q, Zhang C. Selective oxidation of benzoins with chromic acid supported on aluminum silicate under viscous conditions. Synth React Inorg Metal Org Nano-Metal Chem. 2010;40:495.

[17] Lou JD, Ge J, Zou XN, Zhang C, Wang Q, Ma YC. Preparation of benzils by oxidation of benzoins with chromic acid supported on aluminum silicate. Oxid Comm. 2011;34:309.

[18] Lou JD, Ma YC, Wang Q, Natanian N, Zhang C. An efficient oxidation of benzoins to benzils with the jones reagent supported on aluminum silicate. Oxid Comm. 2011;34:349.

[19] Lou JD, Zou XN, Zhang C, Wang Q, Ma YC. An efficient oxidation of benzoins to benzils with manganese dioxide supported on aluminum silicate under heterogeneous conditions. Oxid Comm. 2011;34:355.

[20] Gandhari R, Maddukuri PP, Vinod TK. Oxidation of aromatic aldehydes using oxone. J Chem Educ. 2007;84:852.

[21] Stevens RV, Chapman KT, Weller HN. Convenient and inexpensive procedure for oxidation of secondary alcohols to ketones. J Org Chem. 1980;45:2030.

[22] Zuczek NM, Furth PS. Oxidation of cyclohexanol to cyclohexanone by sodium hypochlorite. J Chem Educ. 1981;58:824.

[23] Kauffman JM, McKee JR. Student preparation of acetone from 2-propanol. J Chem Educ. 1982;59:862.

[24] Mohrig JR, Nienhuis DM, Linck CF, Van Zoeren C, Fox BG, Mahaffy PG. The design of laboratory experiments in the 1980's: a case study on the oxidation of alcohols with household bleach. J Chem Educ. 1985;62:519–21.

[25] Dos Santos APB, Gongalves IRC, Pais KC, Martinez ST, Lachter ER, Pinto AC. Oxidation of borneol to camphor with bleach: a simple, green chemistry and inexpensive experiment. Quim Nova. 2009;32:1667–9.

[26] Straub TS. A mild and convenient oxidation of alcohols: benzoin to benzil and borneol to camphor. J Chem Educ. 1991;68:1048–9.

[27] Blunt SB, Hoffman VF. A multi-step synthesis of triphenylmethanol. Chem Educator. 2004;9:370–3.

Karolina Wieszczycka

6 Novel technologies of nitrogen-based compounds

Abstract: This paper discusses the main technological solutions used in the production of key nitrogen derivatives such as nitrobenzene, aniline, ethanolamine, and methylene diphenyl diisocyanate. The technologies presented are not only already functioning technologies, but also the newest installations that are at the testing stage.

Keywords: Nitrogen derivatives, Amine technology, Aniline, Nitrobenzene, MDI

6.1 Amines

Amines are very important chemicals that are used in the manufacturing of a wide range of intermediates or even final consumer products of e. g. agrochemical, pharmaceutical, or polymer industry [1, 2]. For example, cyclohexylamine (Figure 6.1) is applied as an intermediate for producing sulfenamide-based vulcanization accelerators and corrosion inhibitors [3], ethanolamine is used as a corrosion inhibitor, gaseous sulphur contaminants absorbent, and it is a key component used in a cocamide MEA production, diethylenetriamine is used mainly as a building block for wet strength agents in the paper industry, but also in the chelating agent synthesis, and as a lubricant oil additive [4]. From the group of aromatic amines, the most important is aniline, which, in the global market on 2019, has been estimated on 8.1 million tons [5], and is mainly resulted of a huge demand on production of methylene diphenyl diisocyanate (MDI). Aniline is also necessary in manufacturing of pharmaceuticals and rubber-processing chemicals. This compound is also used as an anti-knock additive and solvent in the improvement of reformat gasoline.

The variety of applications associated with the amines structures results in a variety of synthesis method. In the case of aliphatic amines, there are many approaches for the synthesis of primary, secondary, and tertiary amines; however, methods such as (I) halogenated hydrocarbon ammonolysis, (II) hydroamination of unsaturated carbon–carbon bonds, (III) aldehyde amination hydrogenation, (IV) hydroaminometylation (V) nitrile and nitro compounds reduction, (VI) alcohol amination, and (VI) nitroarene *N*-alkylation with alcohol have been developed in the last century. Most of these methods are also recommended to produce amines with benzyl moiety, various alkyl chains, and also having a complexed unsaturated

This article has previously been published in the journal Physical Sciences Reviews. Please cite as: Wieszczycka, K. Novel technologies of nitrogen-based compounds *Physical Sciences Reviews* [Online] 2020, 5. DOI: 10.1515/psr-2019-0101.

https://doi.org/10.1515/9783110656367-006

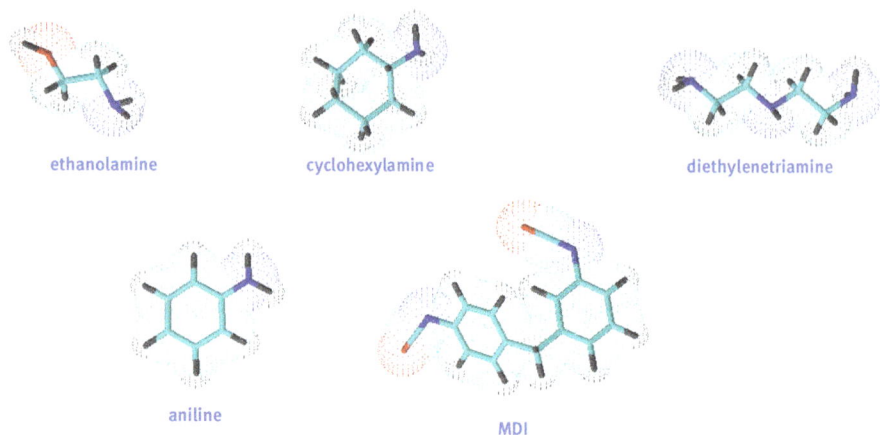

ethanolamine   cyclohexylamine   diethylenetriamine

aniline

MDI

Figure 6.1: Industrially relevant nitrogen derivatives.

structures. In the case of aromatic amines the most important is nitrile and nitro compounds hydrogenation. The main types of the methods are listed in Table 6.1.

6.1.1 Amination of alcohols

Today more than 1 million tons of methylamines are produced according to this method. Various modifications of the reaction of an alcohol with ammonia provide the most common commercial routes to alkylamines. The advantage of this methods over other is due to the availability of various alkyl alcohols (straight-chain and branched alcohol) by industrial processes such as direct production from synthesis gas (methanol), catalytic hydration of olefins (ethanol, propan-2-ol), fermentation of sugars or starch (methanol, ethanol, butanol), hydroformylation and subsequent hydrogenation of olefins (butanol, butan-2-ol, 2-methylbutan-1-ol, 3-methylbutan-1-ol). Butanol can also be produced using aldol condensation of acetaldehyde with subsequent hydrogenation of the crotonaldehyde [6, 7]. Another advantage is much less problematic by-product (water) than that observed in alkylhalogene amination. Conditions of the process depend on substrates, catalysts, and which product is desired.

Direct amination of aliphatic alcohols with ammonia in the presence of acidic solid catalysts such as aluminia, aluminosilicates, and aluminophosphates is manufactured on a multi-thousand-tonne scale. In this method, methanol and excess ammonia react at 300 °C to 500 °C and 15 to 30 bar [8]. Other heterogeneous catalysts for alcohol amination are based on tungsten, chromium, nickel, cobalt, iron, and copper [9–15]. The vapour phase amination of methanol with ammonia over alumina, silica-alumina or aluminophosphates leads to the products, which are an equilibrium composition of monomethylamine, dimethylamine, and trimethylamine [16–18]. Using

Table 6.1: The most important synthesis of amines.

I.   halogenated hydrocarbon ammonolysis

$NH_3 \xrightarrow{R-X} RNH_2 \xrightarrow{R-X} R_2NH \xrightarrow{R-X} R_3N \xrightarrow{R-X} R_4N^+X^-$

II. hydroamination of alkene, alkyne

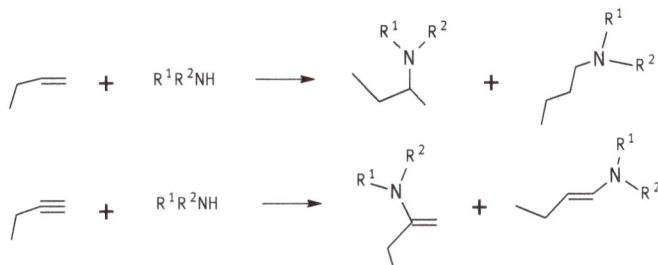

III. aldehyde amination hydrogenation

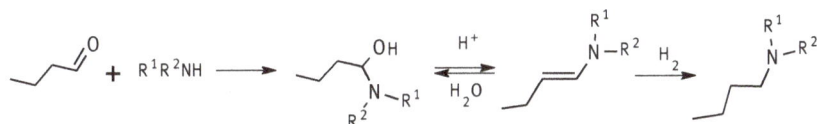

IV. hydroaminomethylation

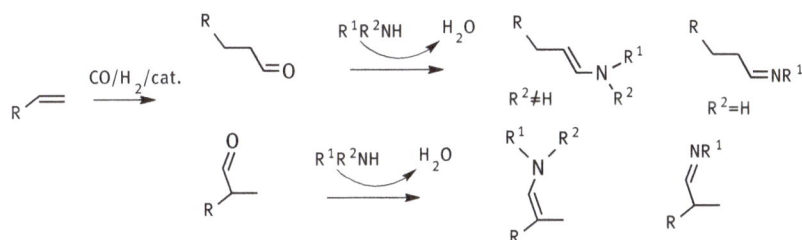

V.   nitrile and nitro compounds reduction

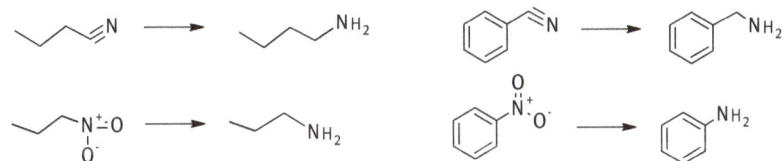

VI. alcohol amination

$NH_3 + ROH \xrightarrow{cat.} RNH_2 + R_2NH + R_3N$

VII. nitroarene N-alkylation with alcohol

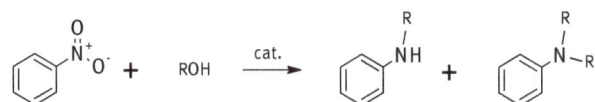

this type of catalyst, the reaction products formation is a result of the following equilibrium reactions [19–21]:

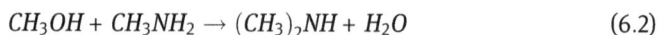

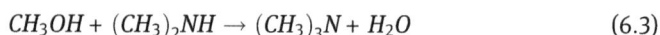

$$CH_3OH + NH_3 \rightarrow CH_3NH_2 + H_2O \tag{6.1}$$

$$CH_3OH + CH_3NH_2 \rightarrow (CH_3)_2NH + H_2O \tag{6.2}$$

$$CH_3OH + (CH_3)_2NH \rightarrow (CH_3)_3N + H_2O \tag{6.3}$$

The content of each methylamines also influences another type reaction known as disproportionation reaction:

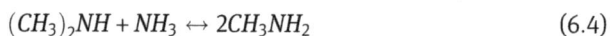

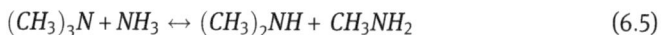

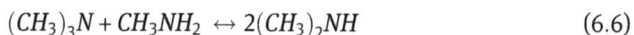

$$(CH_3)_2NH + NH_3 \leftrightarrow 2CH_3NH_2 \tag{6.4}$$

$$(CH_3)_3N + NH_3 \leftrightarrow (CH_3)_2NH + CH_3NH_2 \tag{6.5}$$

$$(CH_3)_3N + CH_3NH_2 \leftrightarrow 2(CH_3)_2NH \tag{6.6}$$

At this conditions, a by-side product – dimethyl ether – is detected, which may also be an intermediate in the synthesis of the corresponding amines:

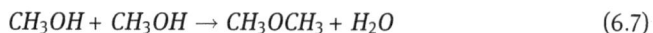

$$CH_3OH + CH_3OH \rightarrow CH_3OCH_3 + H_2O \tag{6.7}$$

$$CH_3OCH_3 + NH_3 \rightarrow CH_3NH_2 + CH_3OH \tag{6.8}$$

By analysing the equilibrium reaction, it can be observed that the mono- and dimethylation are favoured by excess of ammonia (expressed as a N/C ratio), while at equilibrium conditions (N/C = 1) trimethylamine is dominating. The increase in the di- or trimethylamines production can also be obtained by recycling unwanted methylamines, but the low-efficient monomethylamine production can be increased by using dimethyl ether in place of methanol. The ratio of products can also be adjusted to reaction parameters such as temperature of the reaction, residence time, and by selection catalyst. The use of specific dehydration or amination catalysts enables the control of the relative amounts of each amine in a product mixture. The selection of the reaction condition as well type of catalyst is very important for higher alcohols than methanol. In the case of ethanol and alcohols with longer hydrocarbon chain as side reaction, also the dehydration of an alcohol to the corresponding olefin can occur, which negatively influence on the amination efficiency, as well as, can cause a catalyst deactivation.

Most of the technology processes are variations of the Leonard process [22]. At presented flow scheme (Figure 6.2), the methylamine synthesis is carried out in the adiabatic fixed reactor at gas phase at 380–450 °C and 2–5 MPa over the amorphous silica-alumina catalyst. The mole ratio of ammonia to methanol is 4:1, and the reaction leads to produce the equilibrium mixture of methylamine, dimethylamines and trimethylamines, but the shift of equilibrium in favour of the selected methylamines can be achieved by recycling unwanted products (see eq. (6.1)–(6.8)). In the first

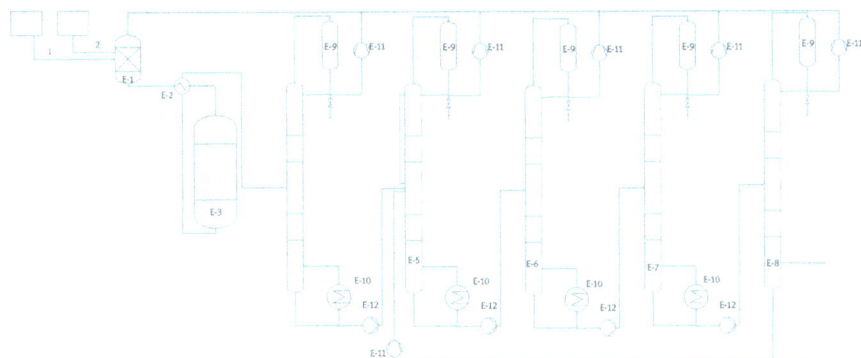

List of equipment: (E-1) mixer, (E-2) heat exchanger, (E-3) reactor, (E-4)-(E-8) distillation columns, E-9 reflux column, E-10 heat exchanger, E-11 pumps for liquid stream , E-12 pumps for gasses.

Figure 6.2: Scheme of installation of vapour phase amination of methanol with ammonia.

stage of the process, the reactants are fed via the mixer (E-1) and the heat exchanger (E-2) into the reactor (E-3), in which the amination occurs. After amination, the eluent is directed to the series of distillation columns and separated. In the first column (E-4), which includes the re-boiler and the condenser, ammonia is separated and re-cycled, while the mixture of methylamine, dimethylamine, trimethylamine and methanol are sent to the extractive distillation column (E-5). In the column at pressures about 1.8 MPa. The separation of methylamine from dimethylamine and trimethylamine is carried out. In the column (E-6) at pressures about 0.33 MPa methanol is separated from dimethylamine, and trimethylamine (methanol is removed at the bottom of the column). In the final distillation column (E-7), operates at pressures about 0.4 MPa, the mixture of heptane and dimethylamine are used as the feed from the absorber. From the bottom of the column the heptane – dimethylamine mixture is received, while from the top trimethylamine. In the last column (E-8) unreacted methanol is separated from water and is recycled back to the reactor (E-3), while the bottom fraction is sent to extractive distillation column (E-5).

Most of used the alcohol amination system operated in identical scheme, however using more or less selective catalyst [23]. In the case of an amorphous silica-alumina catalyst, the methylamines can be produced continuously; however, even at low methanol conversions the processes produce more trimethylamine than mono- and dimethylamine. Moreover, to produce a product mix different from the equilibrium composition, unwanted methylamines recovered downstream must be recycled to the reaction to suppress their formation. Production of methylamine can be achieved when equilibrium is reached, and at high methanol conversion [24]. Another options are the using of high C/N ratio, and lower temperature of the reaction being optimal for monoalkylation. The optimal temperature of the amination is 210–225 °C; however, it is not guarantee success in the synthesis. The temperature of the process should also be adjusted to the performance of the catalyst. Too low temperature results in the insufficient hydrogenation capacity of the catalyst thus the amination rate is slow, while too high temperature lead to the much better hydrogenation capacity of the catalyst but the selectivity of the amination is not so high as expected. At higher temperature, the worse efficiency of monoalkylamine is due to a difficulty in alcohol accessibility [25].

The demand for methylamines is much smaller than that for the dialkyl analogue. Therefore, in most worked installations these products are recirculated in the reaction system and reused. Separation of dimethylamine from methylamine does not create difficulties and can be realised by means of distillation. However, the separation from higher analogue (trimethylamine) is not so easy. Trimethylamine forms a difficult to separate azeotropic mixture with ammonia, and other methylamines, therefore a large-scale distillation is required. This procedure increases the cost of consumed energy of the process. The reduction the production cost of dimethylamine can be realised by limitation in the by-product formation, especially the formation of trimethylamine. This assumption can be achieved by appropriate choice of catalyst. The selectivity to the methylamines is thermodynamically determined for conventional amorphous solid acid catalyst such as alumina or silica alumina. For example at C/N ratio equal to 1 and at $T = 400\,°C$ the equilibrium weight ratio of the methylamines produced, mono- and dimethylamine in 28, trimethylamine in 44%. The increase in yield of dimethylamine to 46% with simultaneals reduction of trimethylamine up to 20% results in about 14% increase of a total fixed capital. In the process developed by Mitsui Toatsu Chemicals, Inc. (Japan) the amination occurs over a silylated mordenite catalyst. This type of catalyst limits production of trimethylamine to level of 7% and favours formation of other oligomer, however the most desired dimethylamine is produced in 63%. This type of catalyst enables also to very high methanol conversion, which is keeping at the level of 90%. Beyond the catalyst, the process is carried out in a disproportionation reactor using the C/N ratio of 0.53, temperature equal to 310 °C, pressure of 1.86 MPa, and maintaining gas hour space velocity of 590 h^{-1}. In this process, trimethylamine is not separated throughout costly distillation, but as an azeotropic mixture with ammonia, and in as mixture with the fresh methanol is recycled to the reactor for further conversion [26–28]. Mitsubishi Rayon has also developed a process that can produce a non-equilibrium product with a high dimethylamine content (e. g. 86%). This selectivity is posed by zeolite-type catalysts (sodium- and hydrogen-sodium mordenite) [25]. Disadvantage of the mordenite using is that its catalytic properties are very susceptible to influence by coke deposition due to its crystalline structure. For this reason, the synthesis temperature (around 400 °C) poses a problem in the aspect of the catalyst life. Solution of this problem is conducting the reaction at a temperature not higher than 340 °C, and using catalyst which has a sufficiently high activity at such a low temperature [29, 30]. Reduction in trimethylamine formation to <10 wt.% has also been achieved by using as catalyst surface treatment of chabazites [31], and a dehydrated crystalline metal aluminosilicate catalyst [18]. In the case of that last the selectivity of the reaction is possible to arrange N-alkylation through the so-called hydrogen borrowing concept, when an alcohol is first dehydrogenated to appropriate carbonyl derivative, which reacts with ammonia or amine to form desired product [32, 33]. Mobil Oil Corporation has also proposed a process for producing primary aliphatic amines, in preference to secondary and tertiary amines [16]. In this process the feed rate of alcohol or ether (with strength or branched alkyl chain) and ammonia being within the ratio from 2 to 4 reacted over a natural or synthetic dehydrated crystalline aluminosilicate having the structure of ZSM-5, ZSM-11 or ZSM-21 at temperature of 350 °C (alcohol) or 400 °C (ether).

6.1.1.1 Case study – kinetic of vapour phase amination of methanol with ammonia

As discussed above the vapour phase amination of methanol with ammonia over amorphous silica/alumina leads to the products which are an equilibrium composition of methylamine (MMA), dimethylamine (DMA) and trimethylamine (TMA), and main reaction eq. (6.1)–(6.4) determine formation of the methylamines.

The kinetics of the above reaction system is described by typical rate balance equations, which read:

$$
\begin{aligned}
r_{NH_3} &= -k_1 C_{NH_3} C_{MeOH} \\
r_{MeOH} &= -k_1 C_{NH_3} C_{MeOH} - k_2 C_{MMA} C_{MeOH} - k_3 C_{DMA} C_{MeOH} \\
r_{MMA} &= k_1 C_{NH_3} C_{MeOH} - k_2 C_{MMA} C_{MeOH} \\
r_{DMA} &= k_2 C_{MMA} C_{MeOH} - k_3 C_{DMA} C_{MeOH} \\
r_{TMA} &= k_3 C_{DMA} C_{MeOH}
\end{aligned}
\tag{6.9}
$$

The corresponding k_i rate constants are defined using typical Arrhenius definition:

$$
k = k_\infty \exp\left(\frac{-\Delta E}{RT}\right)
\tag{6.10}
$$

The three rate constants k_1, k_2 and k_3 are described by Arrhenius type eq. (6.10) with the frequency factors and activation energies given in Table 6.2.

6.1.2 Halogenated hydrocarbon ammonolysis

Ammonolysis of alkyl halides is the most known methods of alkylamines synthesis, however complexity of the process and the formation of corrosive by-products causes that this method is not used on an industrial scale. In the classical procedure ammonolysis is carried out in the presence of an alcohol as the diluent

Table 6.2: Rate coefficients for methylamines synthesis over amorphous silica/alumina [34].

Product	k_i, mol/(kg·s)	ΔE, cal/mol
MMA	$3 \cdot 10^{12}$	40,392
DMA	$6 \cdot 10^{13}$	42,065
TMA	$4 \cdot 10^{10}$	36,090

or a reaction medium and produces the desired alkyl amine as a hydrohalide salt (eq. 6.9–6.11). Then the produced hydrochloride reacts with caustic soda, and as result sodium chloride is formed – highly corrosive and environmental pollutant (eq. 6.12–6.14). Moreover, the released during alkylation operation hydrogen halide reacts at least in part with the ammonia present in the reaction system, and it finally forms insoluble in the reaction medium an ammonium halide:

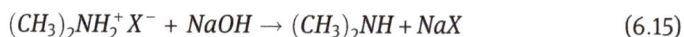

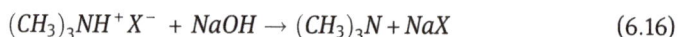

$$CH_3X + NH_3 \rightarrow CH_3NH_3^+ X^- \tag{6.11}$$

$$CH_3X + CH_3NH_2 \rightarrow (CH_3)_2NH_2^+ X^- \tag{6.12}$$

$$CH_3X + (CH_3)_2NH \rightarrow (CH_3)_3NH^+ X^- \tag{6.13}$$

$$CH_3NH_3^+ X^- + NaOH \rightarrow CH_3NH_2 + NaX \tag{6.14}$$

$$(CH_3)_2NH_2^+ X^- + NaOH \rightarrow (CH_3)_2NH + NaX \tag{6.15}$$

$$(CH_3)_3NH^+ X^- + NaOH \rightarrow (CH_3)_3N + NaX \tag{6.16}$$

As a result of the reaction, a second phase is formed, which is a source of serious technological problems, e. g. the process required a removing from the reactor, and then separating mixed liquid phases [35]. The technological problems have not caused a total resignation from the halogenated hydrocarbon ammonolysis. In the case of alkylamines a primary alkyl amine (e. g. methylamine) is employed with the other reactants for the alkylation operation in an amount effective to prevent the formation of insoluble ammonium halide [36]. The synthesised alkyl amine can be recovered from the reaction mixture as liquid phase by using standard procedure such as distillation, and solvent extraction. This invention completely eliminates formation of a mixture of a liquid phase and a solid phase, which makes the classical synthesis less economic.

Ammonolysis of aliphatic halides is still considered for diamines synthesis, wherein after the direct ammonolysis of aliphatic dihalides an appropriate diamine is formed in a predominant amount and under relatively mild conditions. It has been indicated that effecting methodology of dihalide conversion to an appropriate diamine requires the use of elevated temperatures often in excess of 200 °C. Of course, such drastic conditions have been found to result in commercially feasible reaction rates, higher product yield, etc. but also it generates the products derived such as a mixture of various mono- and poly amines and high molecular weight materials which are difficultly separable. Therefore, as in the case of ammonolysis of alkyl halide, the recovery of desired product requires additional treatments, e. g. fractional distillation, extraction. The method presented in US Patent [37] provides an improved process for a direct ammonolysis of aliphatic dihalides, wherein the diamine is selectively formed in a predominant amount under relatively mild conditions. Moreover, the method enables to not only the selective formation of primary diamines, but also minimizes the formation of undesirable by-products. A primary advantage of the presented invention is that the direct ammonolysis of aliphatic dihalides to selectively form the corresponding diamines in a predominant amount without the use of expensive high pressure equipment that an efficient and reliable economically feasible process for diamine formation on a large scale is used. In this process, the aliphatic dihalide reacts with a large excess of anhydrous liquid ammonia at a temperature of up to 50 °C and a pressure of up to 1.4 MPa removing the excess ammonia from the reaction mixture, treating with a base and separating the freed diamine product. The conversion of dihalides is close to 100%. The excess of ammonia is recycled in a continuous process and the metal halide may be recovered for further processing.

6.1.3 Olefin amination

Technology for producing amines from alkenes was first patented by Sinclair Refining Company [38]. This type of process is more simple and lower investment than other ammonolysis. However, it requires higher temperature of the reaction, and higher pressure. A technological development of the process was a consequence a development in catalysts, especially in using of acid catalysts, such as naturally-occurring or synthetic crystalline aluminosilicates, metal supported silica [23, 39–43]. It was noted that using as catalyst a metal supported on a spinel type support, silica and diatomaceous earth polyamines and polyolefins were produced. Noble metal-containing catalysts has also been proposed for producing primary alkyl amines from low molecular weight olefins and ammonia [44], aromatic amines are produced in the presence of ruthenium or iron compound catalysts [45], as well as effective amine formation achieved using a perfluorinated organic acid catalyst [46]. Although many companies have investigated this technology, only BASF has commercialized the process of alkene amination for production of *tert*-butylamine (compound used in the production of accelerators for the rubber, and as a component in pharmaceutical and agricultural industries) [47–49]. The technology is not attractive for industry because the conversion of the reagents is very low, and mostly, where conversion increases, the selectivity of the reaction decreases [50]. Using gallium silicate zeolite as catalyst, the selectivity is relatively high (98%), but quantitative conversion is possible keeping pressure of 30 MPa [51].

The example process for producing *tert*-butylamine by reacting isobutene and ammonia on a silica alumina catalyst is presented on Figure 6.3.

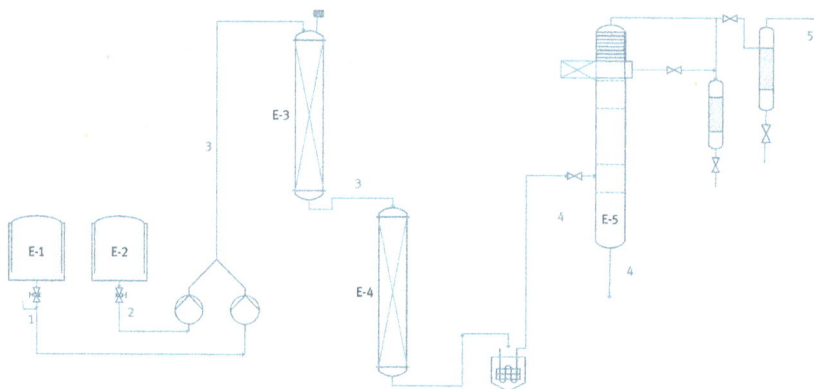

List of equipments:(E-1), (E-2)-vessels   (E-3) -mixer with heating   (E-4) -reactor distillation column with cooling system

Figure 6.3: Scheme of process for *tert*-butylamine production.

In the installation the gaseous ammonia (stream 1) is sent to the mixer (E-1) together with gaseous isobutene (stream 2) at the molar ratio ammonia: isobutene equal to 2. The mixture of the reagents (stream 3) is dosed to the reactor (E-2), in which reaction is carried out over introduced silica catalyst (the pressure in the reactor is set to 15 MPa). The product formed in the reactor (E-2) is directed (stream 4) to the distillation column (E-5), where it is separated from unreacted substrates. The resulting azeotropic mixture of isobutene and ammonia is removed from the column (E-5) as a liquid phase (stream 5) and directed again to the reactor (E-2).

6.1.4 Reductive aldehyde and ketone amination

The reductive amination is next reaction from the group of the most convenient and practical approaches enabling to synthesis of amine with very high selectivity. Very interesting is that a range of amines from ammonia to aromatic amines can be employed.

The mechanism of the reductive amination are dependent on the type of reducing agent and additives; however, generally the proposed mechanism involves a condensation of amine with carbonyl compounds. This stage results in formation of a carbinolamine structure (I), which undergoes dehydration to an appropriate iminium ion (II), and followed by reduction of iminium ion to an appropriate amine product (III) (17):

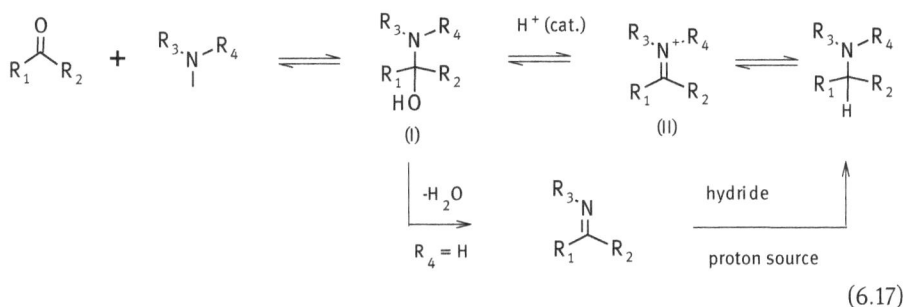

$$(6.17)$$

In general, the reductive amination of carbonyl compounds is performed in the presence of heterogeneous catalysts [52]. This reaction would be of general interest for the synthesis of the industrially most important primary amines, especially containing different alkyl, or alkyl, aromatic groups. However, this process similar to ammonolysis can cause formation of different analogues of the final product (e. g. not only primary, but also secondary, and tertiary amines). Composition of the final mixture depends mainly on reactivity (nucleophilicity) of the initially produced amine, but also condition reaction and type of catalyst. The type of catalyst also influences on the diversification of products, especially possibility to carbonyl compounds reduction [53, 54]. Thus, development of a selective solution for reductive amination of carbonyl, is highly demanded and challenging. Two most commonly used direct reductive amination methods differ in the nature of the reducing agent: homogeneous and heterogeneous catalytic hydrogenation. More promising is to conduct reductive amination heterogeneous platinum, palladium, nickel or ruthenium metal catalysts [55–58], as well as zirconium-based catalysts [59]. This is an economical and effective reductive amination method in large-scale reactions. However, the reaction in the presence of these type of catalysts is not without by-products. Less effective but more selective is amination with using homogenous catalyst such as rhodium(I) complexes with chelating phosphorus ligands [60], cycloocta-1,5-dienyl ligand [52] and also using complexes with Iridium [61–63]. Even though the selectivity of the homogenous catalysts is high, the nucleophilicity of the main product can cause formation of other amine analogues.

The United States Patent [64] has presented the process route of primary amines by conversion of an appropriate aldehyde with ammonia, and catalytic hydrogenation of the resulted imine over the cobalt or nickel containing catalyst. In the technology, presented schematically in Figure 6.4 the synthesise route of butyl-amine is conducted using butanal, ammonia, and hydrogen over the cobalt catalyst (with 10 w% MgO, and 30 w% MnO) incorporated on diatomaceous earth.

In this process butanal (1.1 L/h) from vessel (E-4) is introduced by using pump (E-5) into reactor equipped with a cooling jacket (E-6). Simultaneously ammonia (415 L/h) are conveyed by pump (E-2) via evaporator (E-3) into reactor (E-6) through sieve plate (E-7). The line 3 is used for recalculation of the non-reacted ammonia.

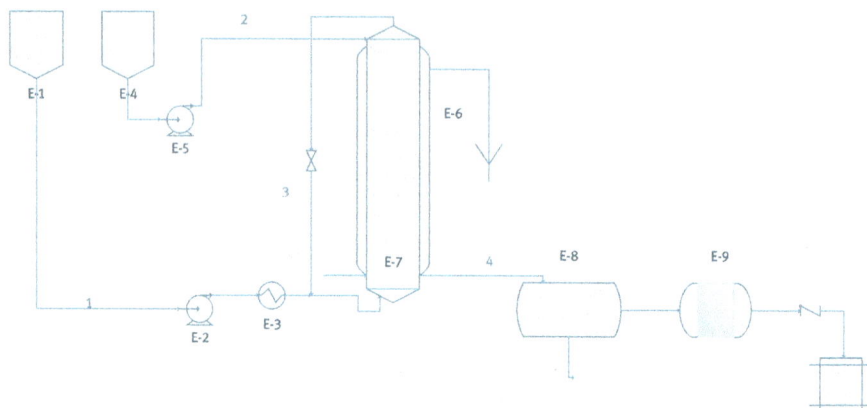

List of equipment: (E-1), (E-4) –vessels, (E-2), (E-5)-pumps, (E-3)-evaporator
(E-6), (E-7) -reactors with sieve plate, (E-8)-settling tank, (E-9)-hydrogenation reactor

Figure 6.4: Scheme of process for butylamine production from butanal.

The reaction is carried out at the temperature of 21 °C. The resulting condensation product (imine) flows into the settling tank (E-8) where generated water is removed from the line. Next, the reaction product is transferred into a hydrogenation reactor (E-9) and converted mainly to butylamine in the presence of the catalyst under a hydrogen pressure of 100 atm., ammonia 715 g/h, and at 120 °C. After separation of ammonia and residual water, butylamine is separated by distillation with purity of 99.4%. As mentioned above the reductive amination of carbonyl compounds is mainly the way of primary amines synthesis, however over last years an effective and selective preparing of secondary amines was also proposed. Example is a synthesis of asymmetric secondary *tert*-butylamines from *tert*-butylamine, appropriate aldehyde and in the presence of a palladium on activated carbon as catalyst [65].

Reductive amination of formaldehyde can be also conducted without the hydrogenation stage, however this procedure leads to formation of cyclic tetramine (1,3,5,7-tetraazatricyclo-[3.3.1.1^{3,7}] decane also known as hexamine or urotropine):

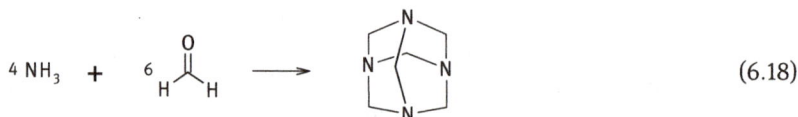

$$4\,NH_3 \;+\; 6 \underset{H}{\overset{O}{\underset{\,}{\parallel}}}\!\!\!\!\!\!\!\!\! H \longrightarrow \quad\text{(structure)} \tag{6.18}$$

Urotropine can be produced using liquid and gases phase method. In the liquid phase method formaldehyde is used as the 37% aqueous solution, while in the gas phase method formaldehyde is firstly synthesised from methanol and directly from the synthesis reacts with ammonia. The selection of method requires the using of suitable product recover procedure, what also influences on economic of the process as well as purity and form of the product [66].

As mentioned above, in the liquid phase production process the formaldehyde as the water solution (37%) reacts with ammonia gas to form hexamine and water according to reaction:

$$6CH_2O\,(liquid) + 4NH_3(gas) \rightarrow (CH_2)_6\,N_4 + 6H_2O \qquad (6.19)$$

Filtered ammonia gas as a mixture with the liquid formaldehyde is sent to the reactor, in which at temperature of 60–65 °C urotropine is formed (conversion approximately 24–27%), and sent to film evaporator to achieve concentration of the product close to 60–65%. The concentrated solution is, then sent to the evaporator port for continuous dehydration and concentration under vacuum condition, when crystal grains are formed. The crystals of is redirected to the centrifuge to separate mother liquor. In the last stage, the solid urotropine is sent to the cyclone separator to receive the final product. Evaporation port system, with relative long crystallization time, enables to produce welcomed by customer's particle size of urotropine. Disadvantage of the process is that the energy consuming is high, and is resulted in high steam consumption in the vaporization, concentration and drying process.

The gas-phase method production process is based on the reaction:

$$6CH_2O(gas) + 4NH_3(gas) \rightarrow (CH_2)_6N_4 + 6H_2O \qquad (6.20)$$

According to the reaction equation, gases formaldehyde reacts with ammonia to form urotropine and water. The reaction also generates a significant amount of heat (76.8 Mcal/mol), which is consumed over the process. The feature of this process is to utilize formaldehyde latent heat and urotropine reaction heat, which works promoting the reaction of the synthesis. However, the reaction temperature must be carefully controlled, because higher temperature can not only affected the process efficiency, but also can cause the by-product formation (trimethylamine forming which reduces the purity of the final product). In order to promote the reaction to proceed towards hexamine forming an excess ammonia must also be assured.

In installation of Jiangsu Kaimao Chemical Technology Co., Ltd illustrated on Figure 6.5. urotropine is produced by using the gas phase method production [67]. In this process high temperature formaldehyde gas (steam 1) converted from methanol oxidation and simultaneously, similar to the liquid phase method, the filtered ammonia gas (steam 2) are directed to the reactor (E-1), where both substrates are bubbled at temperature of 75–80 °C. The formed urotropine (stream 3) is then transmitted to the receiving tank (E-2). The mother liquor (stream 4) is returned to the reactor, while concentrated mixture (stream 5) is sent to a centrifuge spin-drying (E-3) to separate solid phase. Solid urotropine is sent for flash drying (E-4) for further dehydration (to achieve final product in amount more than 99.3%). After that, final product is received. Under vacuum, the water is evaporated and recycled to the process cycle. The Unreacted ammonia is sent to absorption tower for absorption to be, after recovery, recycled back in gasses form. The diluted ammonia solution is transmitted to the ammonia stripper to

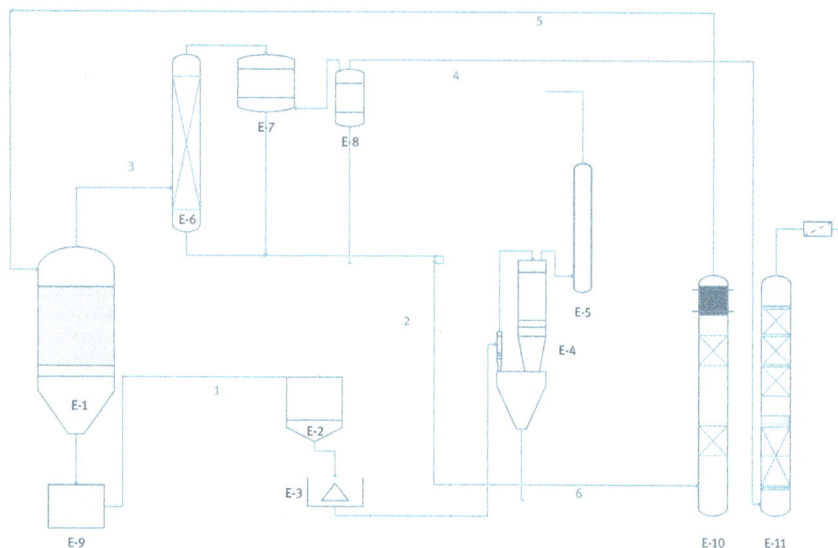

List of equipment: (E-1)-reactor, (E-2)-crystalline receiving groove, (E-3)-centrifuge, (E-4)-drying system, (E-5)-recovery column, (E-6)-ammonia absorption tower, (E-7)-condenser, (E-8), (E-9)-buffer tanks, (E-10)-desorber, (E-11)-absorber off-gas

Figure 6.5: Scheme of process for urotropine production using gases formaldehyde.

recover the ammonia, which is then returned to the reactor to continue the ammonolysis reaction. Diluted methanol is treated in a similar way and returns to the evaporator after recovery in the recovery tower.

6.1.4.1 Case study – kinetic of urotropine synthesis

The kinetic parameters for synthesis urotropine were estimated using continuous stirred tank reactor with temperature control system [68]. Continuous flow kinetic measurements with concentrated and dilute solutions of feed stream, in a single stirred-tank reactor were conducted to determine the rate of formation of hexamine. Authors used two different formulations for chemical reaction rate. The first, simpler but with standard deviation to experimental data at the value of 0.173 is

$$r = k_1 \cdot C_{NH_3} C_{CH_2O}^2 \quad \left[\frac{mol}{dm^3 s} \right] \tag{6.21}$$

The second formulation with exponentials adjusted and standard deviation of 0.150 is as follows:

$$r = k_2 \cdot C_{NH_3}^{0.83} C_{CH_2O}^{1.69} \quad \left[\frac{mol}{dm^3 s} \right] \tag{6.22}$$

Kinetic parameters for both formulations are given in Table 6.3.

Table 6.3: Arrhenius kinetic parameters for rate constants.

Parameter	k_∞, (dm^3/ mol)^2/s	ΔE, cal/ mol
k_1	$1.13 \cdot 10^9$	14,200
k_2	$6.24 \cdot 10^8$	14,200

6.1.5 Hydrogenationof unsaturated amine

Synthesis of cycloaliphatic amine can be realised using different methods, however the most interesting from a commercial point of view are reductive amination of cyclohexanol or phenol [69], hydroamination of alkenes [70] and hydrogenation of aniline catalysed by iridium [71], nickel [72], palladium [73], rhodium [74, 75] and ruthenium [76, 77] compounds. Hydrogenation of aniline to cyclohexylamine is carried out in a liquid phase batch or continuous reactor at temperatures of 100–250 °C, and at pressures of 1–350 bar. Aniline generally is converted to cycloalkylamined approximately in 99%, but selectivity towards cyclohexylamine are not lower than 90%. On the selectivity of cyclohexylamine production mainly influenced catalyst composition, and ammonia dossing into the hydrogenation reactor. An example of continuous aniline hydrogenation, in which cyclohexylamine is recovered in 99.9% is process proposed in US Patent 4,914,239 [78]. In this invention nickel catalyst is used. the catalytic material can also be used as a component of a distillation system, however this methods is much less selective than in classical form. Cyclohexylamine is produced in 65–75% selectivity along with phenylcyclohexylamine and dicyclohexylamine as co-products [79]. The hydrogenation process over a Ru–Pd catalyst incorporated on silica has been proposed as method of dicyclohexylamine synthesis with yield of 80% [80]. Another strategy to synthesize cyclohexylamines is the hydrogenation of nitrobenzene and its derivatives by using ruthenium nanoparticles [81], metal – organic frameworks, which are used as the support for noble metal nanoparticles [82], as well as using bifunctional Pd/MIL-101 catalyst which enables to the selective hydrogenation to aniline or to dicyclohexylamine conducting reaction in dimethylformamide or hexane, respectively [83]. Hydrogenation to cyclohexylamine was also tested at supercritical conditions using known catalysts [84].

Most used manufacture is based on the methods illustrated on Figure 6.6. In this solution [81] a reaction column (E-1) was charged with nickel catalyst supported on diatomaceous earth (e. g. SN-110) in amounts of 0.075 part by weight per part by weight of aniline. The atmosphere in the reactor is purged with hydrogen gas and, then, the hydrogen–ammonia mixture gas is introduced from the bottom of the reactor with the flow 7 cm/s. The internal temperature was increased to 190 °C by external

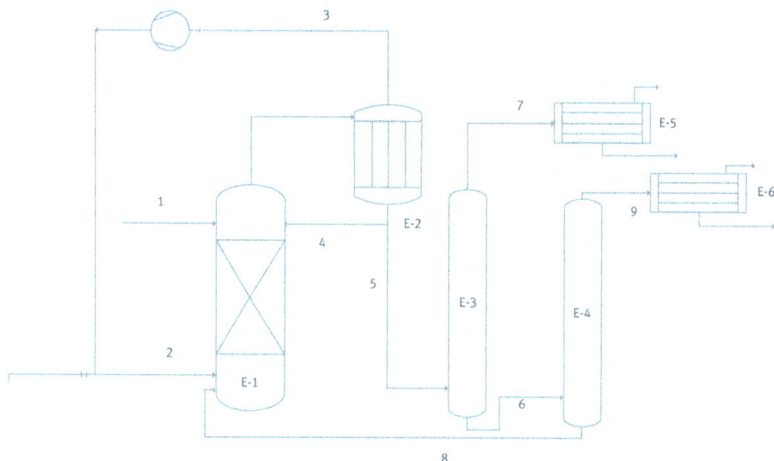

List of equipment: (E-1)- reactor, (E-2) - reboiler, (E-3), (E-4) — fractionating columns, (E-5), (E-6)- condensers

Figure 6.6: Scheme of aniline hydrogenation process.

heating. Thereafter, the reaction pressure and temperature are maintained at 225 °C by controlling the feeding rate of hydrogen-ammonia mixture gas, and the starting material aniline is also continuously supplied to the reactor to keep the liquid level in the column constant. From the reaction column product gas is transported to a condenser (E-2), where is cooled to 60 °C. The cooling procedure enables to separate a vapour phase, consisting mainly hydrogen and ammonia gases, from a liquid phase containing cyclohexylamine, unreacted aniline and byproducts (dicyclohexylamine, N-cyclohexylaniline and hydrocarbons). The liquid phase is sent to a fractionating column (E-3) where at atmospheric pressure a low-boiling fraction of hydrocarbons is recovered from the top of column, but the bottoms are sent to a second fractionating column (E-4), where cyclohexylamine was recovered. The unreacted aniline, dicyclohexylamine and N-cyclohexylaniline are a high-boiling fraction and are recovered as distillation residues from the bottom of the column, and redirected to the reactor. The residue and reaction substrate are introduced from the bottom of the reaction column to maintain a constant level of liquid during the hydrogenation.

6.1.5.1 Case study – kinetic of aniline hydrogenation to cyclohexylamine
In the presented kinetic study, the catalyst used was nickel deposited on clay. The process can be performed in the vapour phase or in the liquid phase:

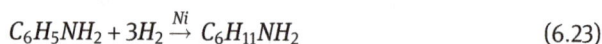

$$C_6H_5NH_2 + 3H_2 \xrightarrow{Ni} C_6H_{11}NH_2 \tag{6.23}$$

The hydrogenation is a first order reaction with respect to hydrogen and zero order with respect to aniline [85]. The rate equation for this reason depends on hydrogen concentration and reads:

$$r = k \cdot C_{H_2} \left[\frac{mol}{cm^3 cat \cdot s} \right] \tag{6.24}$$

The kinetic rate constant k was calculated based on apparent (for given reactor setup) kinetics with the account for the mass transfer limitations and are given in Table 6.4. It is important to note that these values describe the intrinsic (of chemical nature without mass transfer effects) kinetics of the hydrogenation.

Table 6.4: Frequency factor and activation energy for hydrogenation of aniline.

T, °C	k_∞, cm^3/ (cm^3 cat·s)
110	2.78
120	7.08
130	38.62
140	181.23

In a case when the mass transfer is a high limitation factor or the kinetics should be reformulated or the mass transfer explicitly included into the mathematical description of the reaction. The latter typically requires to use appropriate mass transfer description for the whole system. It turned out that the gas-phase hydrogen rate is about 50 times higher than the reaction rate so it is negligible. However, the mass transfer rate for gas-liquid-solid phases is found to be comparable to the rate of reaction and thus it is significant. For the case presented the apparent kinetics, adjusted for specific catalyst setup given in Table 6.5, are described by Arrhenius type parameters.

Table 6.5: Arrhenius kinetic parameters for rate constants.

	k_∞, cm^3/cm^3 cat·s	ΔE, cal/mol
k	$7.902 \cdot 10^5$	12,700

6.1.6 Aromatic amines – hydrogenation of nitro-compounds

The commercial production of aniline can be carried out starting from various material, but the use a hydrogenation of nitrobenzene is the classical and the most frequently used method [86]. Interesting is that on the small scale, aniline can be produced by the chemical reduction of nitrobenzene in a process known as the Bechamp reaction (Fe as catalyst and reductive acid- HCl), however this method currently is more used as a source of iron oxide pigments than aniline. For the production of aniline, hydrogenation with numerous heterogeneous catalysis are used. The mostly catalysts include supported copper, cobalt, palladium and nickel. However, some of them have very high reactivity and require the addition of inhibitors to prevent hydrogenation to alicyclic amine [87]. Promising methods also utilizes the mixture of CO/H_2O as an easily accessible reducing agent [88]. Other route of aniline commercial production involves the amination of chlorobenzene [89], where selectivity is not higher than 91%. Commercially is also used method in which as a starting material phenol is used. It is the amination of hydroxyl group, which is carried out over classical catalyst developed for ammonolysis process (see Section 6.1.1). This method has gained in importance in recent years resulted in the progress in phenol production [90]. The other solution is that used by Du Pont, where aniline is synthesised by benzene amination in the presence of catalyst such as ZrO_2-Ni/NiO. The selectivity of this process is high but benzene conversion does not exceed 13% [91].

In Figure 6.7 scheme of an installation used to a vapour phase hydrogenation of nitrobenzene over the hydrogenation catalyst [92, 93]. The installation consists of three adiabatic reactor (E-2, E-3, and E-4) supplied with a stationary catalyst. The catalytic materials are positioned as single containers having insulated catalyst beds. The catalyst beds are installed on or between gas permeable walls what guarantees very high conversion. Before each catalyst bed, fresh evaporated nitrobenzene (in E-1) is introduced into the circulating gas stream, which consists of hydrogen and nitrogen (freshly added and redirected). This solution eliminates the formation of deposits in the reactor. In the reactor (E-2) the hydrogenation is conducted at temperature of 220 °C, and at pressures approximately 5 bar, in the next

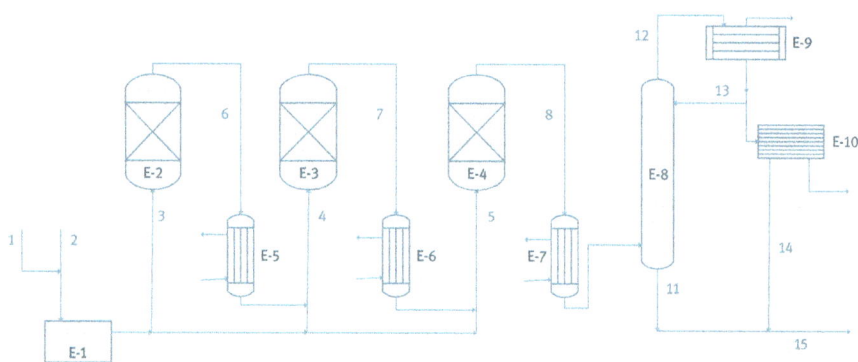

Equipment list: (E-1)-evaporator,(E-2), (E-3), (E-4) –reactors, (E-5), (E-6), (E-7)-steam generator, (E-8)-distillation column, (E-9)-condenser, (E-10)-separating vessel

Figure 6.7: Scheme of nitrobenzene hydrogenation process.

reactors (E-3 and E-4) the temperature increases up to 400 °C. In the heat exchangers (E-5, E-6) located upstream the reactor E-3 and E-4 the circulating gas (stream 6, 7 and 8) is heated to the inlet temperature. After leaving the last heat exchanger (E-7), the product gas is sent to the distillation column (E-8), where aniline – water mixture, and low-boiling by-products are separated from higher boiling components. In further stage the higher boiling fraction is cooled (E-10), what enables to remove the next fraction of the aniline-water. The remaining hydrogen is redirected to the reactor (E-2), after heating up to the inlet temperature.

6.1.7 Amination of olefin oxides

Amination of olefin oxides is a further expensive used methods of synthesis industrially significant components. Ethanolamine can be used to produce ethylenediamine, while diethanolamine to produce morpholine. Other alkanolamines also show very high potential in production of various pharmaceutical compounds. However, this is not the only application of such type of compounds, e. g. diethylethanolamine, and their derivatives are used as compounds of pesticides, corrosion inhibitors, as well as in emulsification, paints and coatings, as additives in polymers, metal fabrication and finishing, as lubricants and stabilizers in drilling muds. Isopropanolamines can also be used as production of detergents, pesticides, pharmaceuticals, and dyes. They are also used to remove acidic constituents from gases.

Ethanolamines are produces by BASF Int., Dow Chemical Company, Euteco S.P. A., Oxiteno, and all of industries synthesise ethanolamines by reaction of ethylene oxide with aqueous ammonia. The formation of the selected compound is determined by the following reactions:

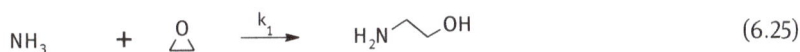

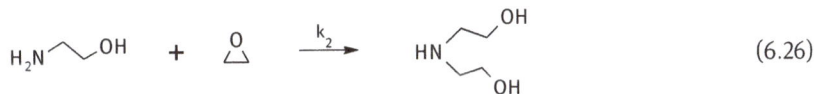

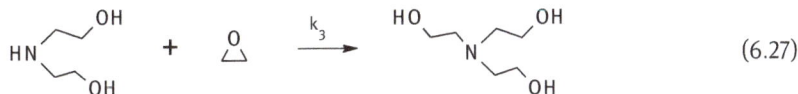

$$NH_3 \quad + \quad \triangle\!\!O \quad \xrightarrow{k_1} \quad H_2N\text{---}OH \qquad (6.25)$$

$$H_2N\text{---}OH \quad + \quad \triangle\!\!O \quad \xrightarrow{k_2} \quad HN(\text{---}OH)_2 \qquad (6.26)$$

$$HN(\text{---}OH)_2 \quad + \quad \triangle\!\!O \quad \xrightarrow{k_3} \quad N(\text{---}OH)_3 \qquad (6.27)$$

Commercially The main process steps is the aqueous route in which the ethylene oxide reacts exothermically with 20–30% aqueous ammonia at 60–150 °C and

30–150 bar in a tubular reactor. The reaction results in formation of ethanolamine, dithanolamine and triethanolamine, as well as other by-products (e. g. polyethanolamine). The amount of formed ethanolamines and the selectivity of the synthesis depend on amount of ammonia added to the reactor. The ammonia is present in significant excess, favours monoethanolamine, whereas a slightly excess favours other analogues and polymeric by-products. The selectivity can also be controlled by presence of water, which works as an accelerator, or by using a fixed-bed catalyst in an anhydrous process.

The aqueous route presented on Figure 6.8 involves reaction of ethylene oxide with excess ammonia at temperature of up to 150 °C and a pressure typically above 20 bar. Given the exothermic nature of the reaction, multiple reactors in series with intermediate cooling and multiple ethylene oxide additions may be used [94]. In this process the ammonia-water mixture in molar ration equal to 0.6 (stream 1) is pumped (E-8) from column (E-1) to be mixed with ethylene oxide – stream 2 (molar ratio ammonia:water:etylen oxide 3:5:10). The mixture is then sent to the isothermal reactor (E-4) maintaining inlet temperature of 30 °C. In the reactor (E-4) ammonia reacts with ethylene oxide and exothermic effect causes an increase of temperature in the reactor up 90 °C. Under these conditions conversion of ethylene oxide to ethanolamines is in about 80%. The rest of the reagents (stream 3) are converted to products in the adiabatic reactor (E-15), in which reaction is finalised within 20 min. at 120 °C. The mixture leaving reactor (E-15) consists mono-, di- and triethanoloamines, unreacted ammonia, water and high-

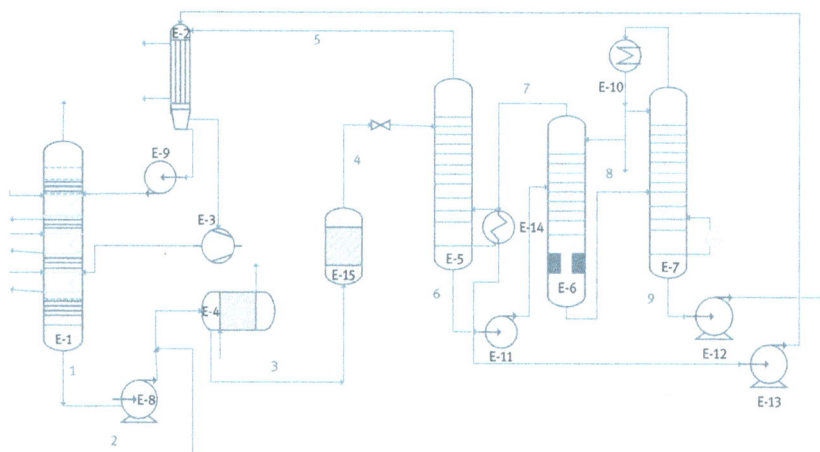

List of equipment: E-1 column   E-2 condenser-absorption apparatus   E-3 compressor   E-4 isothermal reactor   E-5-plate-type column   E-6 evaporator with plate-type column and reboiler   E-8, E-9, E-11, E-12, E-13 pumps   E-10 heat exchanger   E-14 reboiler   E-15 evaporator

Figure 6.8: Scheme of proces for ethanoloamines production.

boiling by-products (stream 4). The reaction mixture is transported to the column (E-5) to desorb ammonia. The desorbed ammonia with water and small amounts of ethanolamines (stream 6) is redirected to (E-1). A liquid flow consisting of the ethanolamines in the feed mixture is directed from the bottom of the column (E-5) the evaporator (E-6), which contained plates column including reboiler. In this column most of water is evaporated from the reacting mixture. The liquid leaving the bottom of the column contains all of ethanolamines produced in the reactor (E-4) and (E-15) and few percent of water. The liquid (stream 8) is then sent to the rectifying column (E-7) with temperature at the bottom of 160 °C, and taken from the column as an anhydrous mixture of ethanolamines (stream 9).

6.1.7.1 Case study – Kinetics of ethanoloamines synthesis from ethylene oxide

The parallel reactions run simultaneously in a reactor, typically of tubular type [95]. The gas phase ethylene oxide (EO) must be added to the concentrated ammonia solution due to the possibility of explosive polymeric reaction risk. Three ethanolamines are the product of the reactive system: monoethanoloamine (MEA), diethanoloamine (DEA) and triethanoloamine (TEA) eq. (6.25–6.27). The kinetics of the above reaction system is described by typical rate balance equations:

$$r_{NH_3} = -k_1 C_{NH_3} C_{EO}$$

$$r_{EO} = -k_1 C_{NH_3} C_{EO} - k_2 C_{MEA} C_{EO} - k_3 C_{DEA} C_{EO}$$

$$r_{MEA} = k_1 C_{NH_3} C_{EO} - k_2 C_{MEA} C_{EO} \tag{6.28}$$

$$r_{DEA} = k_2 C_{MEA} C_{EO} - k_3 C_{DEA} C_{EO}$$

$$r_{TEA} = k_3 C_{DEA} C_{EO}$$

The three rate constants k_1, k_2 and k_3 are described by Arrhenius type eq. (6.10) with the frequency factors and activation energies given in Table 6.6.

Table 6.6: Rate coefficients for ethanoloamines synthesis.

Reaction	k_i, dm³/(mol·min)	ΔE, cal/mol
EO + NH3	$1.581 \cdot 10^7$	19,660
EO + MEA	$4.92 \cdot 10^8$	19,660
EO + DEA	$5.2 \cdot 10^8$	19,660

6.2 Nitro-compounds

Nitrobenzene is the basic raw material for aniline production, and the high demand for aniline determines the development in nitrobenzene production.

Various industrial production processes have been developed to obtain nitrobenzene, but nitration is still an active field of research to improve commercial production. Nitrobenzene is manufactured by direct nitration of benzene using a
mixture of nitric and sulphuric acids, so-called nitrating acid. Since two phases
are formed in reaction mixture and reactant are distributed between them, the
rate of nitration is controlled by transfer between the phases as well as by chemical kinetics [96].

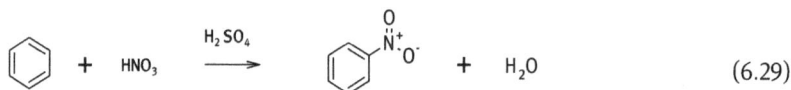

$$
\text{C}_6\text{H}_6 \;+\; \text{HNO}_3 \;\xrightarrow{\text{H}_2\text{SO}_4}\; \text{C}_6\text{H}_5\text{NO}_2 \;+\; \text{H}_2\text{O}
\tag{6.29}
$$

Initially, in commercial installations, a dilute sulfuric acid formed through the
water generation during the nitration process was removed from the system and replaced by concentrated acid. This generated a significant cost factor in the production process. Another problem is the selectivity of the mono-substituted product
synthesis without associated by-products such as dinitrobenzene, which adversely
affects aniline production. The nitration is highly exothermic reaction (nitration
benzene $\Delta H = -117\,\text{kJ/mol}$), and a large excess of sulfuric acid takes up almost quan-
titatively the reaction enthalpy in the form of heat released during the reaction.
Therefore, in the further inventions the heat from the reaction enthalpy stored has
been used to re-concentrate the diluted spent acid. Currently development in the
field of nitrobenzene production concerns minimizing installations and increasing
the selectivity of reactions. An example is a continuous adiabatic nitration process
whereby, mono nitration reaction is favoured, while the dinitro analogue is formed
in slight amount. The innovation is also that the free of dinitro benzene process proceeds at much higher temperatures, thereby markedly increasing both the rate of
reaction and the efficiency [97, 98].

Figure 6.9 shows an example of continuous adiabatic nitration process [99]. A
mixture of 65.7 w% H_2SO_4, 4 w% HNO_3 and 30.3 w% H_2O (stream 1), and benzene
(stream 2) are fed to the adiabatic reactor (E-1) at a molar ratio of benzene to nitric
acid – 1.1:1. The reaction mixture flows through the reactor (E-1) with rate of 173 m^3/h
($T = 135\,°\text{C}$). The stream 4 leaves the reactor (E-1) and is sent to a phase separator (E-2),
from which two streams flow out: spent acid phase (stream 6) and crude nitrobenzene
phase (stream 5). In next part of the line the stream 6 is transferred to the evaporator
(E-3), where the spent acid, at $T = 106\,°\text{C}$ and at $p = 135\,\text{bar}$, is concentrated up to
69 w%. To achieve a higher concentration of acid (70 w%), the stream 7 is fed to the
second evaporator (E-4), in which the temperature and pressure are set at 95 °C and 74
mbar, respectively. The concentrated acid is redirected to the hydrogenation reactor
(E-1). The resulting vapour phases (streams 9 and 12) are fed to the condensers (E-6
and E-8). The non-condensable portions (streams 11 and 14) are discharged as waste
gas stream.

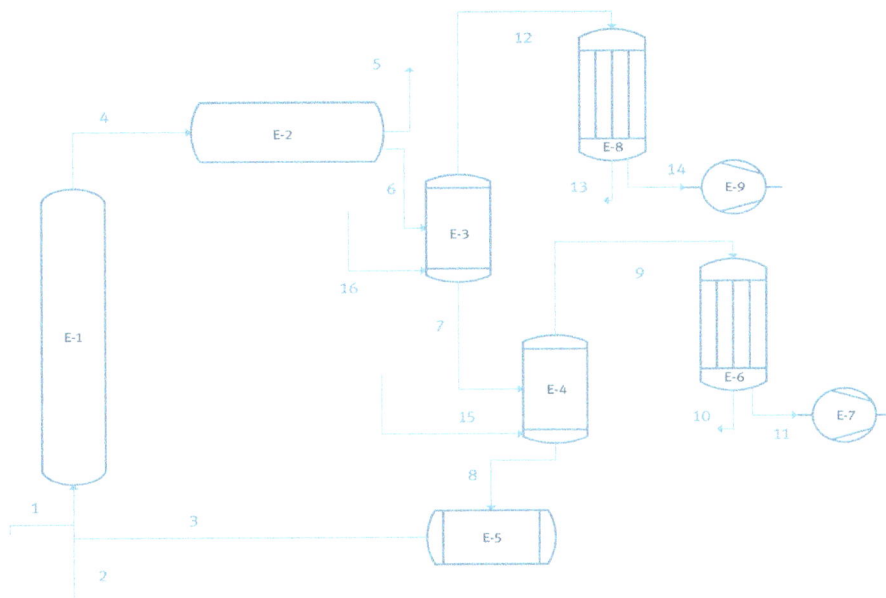

List of equipment: (E-1) - adiabatic reactor, (E-2) - separator, (E-3), (E-4) – evaporator, (E-5) - acid tank, (E-6), (E-8) - condenser, (E-7), (E-9) - pump

Figure 6.9: Scheme of nitrobenzene production process.

6.3 Isocyanates

Total world production of Isocyanates is due to the overall grow of polyurethanes products such as foams, insulation in buildings elastomers, surface coatings, fibres, as well as in the production of paints and varnishes. At least 96% of total world production of Isocyanates is accounted for by two aromatic isocyanates: toluene diisocyanate (TDI) (used in the production of flexible foams) and methylene diphenyl diisocyanate (MDI) (used in the production of rigid foams and other polyurethans products) [100]. For most aromatic isocyanates available on global market, the phosgenation process is the most used in commercial production of the compounds. This type synthesis route is not ideal because it requires the application of extremely toxic phosgene; therefore, non-phosgene routes to isocyanates have also been developed [101]. The most interesting and with commercial potential is the procedure according which amine reacts with methyl carbonate [102], which is also a green solution for the synthesis of methylene diphenyl diisocyanate from aniline [103].

Even though many of non-phosgene procedures have been proposed, commercially TDI and MDI are produced by using classical way, which example is presented in Figure 6.10 [104]. In the first stage aniline, formaldehyde and HCl are dosed to reactor (E-1) and stirred at 40 °C. The reaction mixture is then heated up to

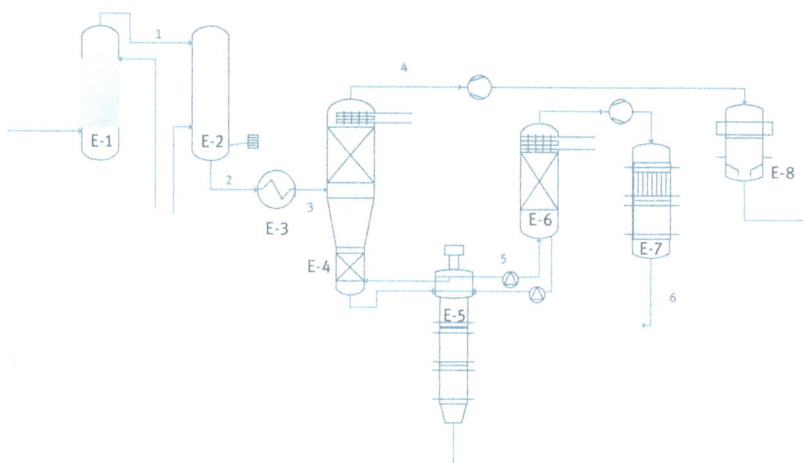

List of equipment: (E-1) – reactor, (E-2) - phosgenation reactor, (E-3) - heat exchanger, (E-4) - distillation column, (E-5) - reboiler, (E-6) - rectifier, (E-7),(E-8) - condenser

Figure 6.10: Schematic 4,4′,-MDI phosgenation process.

65 °C, and after 3 h neutralized with sodium hydroxide. After neutralization, the crude methylene diphenyl diamines (stream 1) are separated, and in the next stage is fed to the phosgenation reactor (E-2), where methylene diphenyl diisocyanates and other polyisocyanate are formed (MDI and PMDI). The crude MDI and PMDI (stream 2) are heated in the heat exchanger (E-3) to a temperature of about 200 °C. The preheated mixture (stream 3) is sent to the distillation column (E-4) to separate the lower boiling components (i.e. 2,2′-MDI and 2,4′-MDI) from the higher boiling components (i. e. 4,4′-MDI and PMDI). The distillation column (E-4) includes both a stripping section and the rectification column. In each of the section temperature and pressure are adjusted to the temperature of compounds to be separated. The lower boiling components (stream 4) are recovered at the stripping section and transported to the condenser (E-8). The mixture of 4,4′-MDI and PMDI are removed from the distillation column (E-4) as bottoms and passed through an evaporative reboiler (E-5). After partial evaporation, 4,4′-MDI (stream 5) is purified in the side rectifier (E-6) and next through a quench condenser (E-7) to obtain very pure 4,4′-methylenediphenyl diisocyanate (stream 6).

References

[1]   Lawrence SA. Amines: synthesis, properties and applications. Cambridge, UK: Cambridge University Press, 2004.
[2]   Ricci A. Amino group chemistry: from synthesis to the life sciences. Weinheim: Wiley-VCH, 2008.
[3]   Roose P, Eller K, Henkes E, Rossbacher R, Amines HH. Aliphatic. In: Ullman's encyclopedia of industrial chemistry. Weinheim: Wiley-VCH Verlag GmbH & Co. KGaA, 2015:1–55.
[4]   https://pharmaceutical.basf.com/en/APIs-Raw-Materials.html. Accessed: 7 Sept, 2015.
[5]   Technavio report, Global aniline market. 2015–19, Published: Sep. 2015.
[6]   Braithwaite J. Kirk-Othmer, encyclopedia of chemical technology, 4th ed. New York: J Wiley & Sons, 1995.
[7]   McKetta JJ, Cunningham WA, editors. Encyclopedia of chemical processing and design. New York: Marcel Dekker, 1977:390.
[8]   Wang H, Xin Li ZJ. Progress in the hydro -amination of alcohols. Topics Chem Mat Eng. 2018;1:191–4.
[9]   Slaugh LH, Schoenthal GW. Amine process using Cu-Sn-Na catalyst, United States Patent, Patent Number: 4,229,374, Date of Patent: Oct. 21, 1980.
[10]   Afanasenko A, Elangovan S, Stuart MC, Bonura G, Frusteri F, Barta K. Efficient nickel-catalysed N -alkylation of amines with alcohols. Cat Sci Technol. 2018. DOI: 10.1039/C8CY01200H.
[11]   Ibáñez J, Kusema BT, Paul S, Pera-Titus M. Ru and Ag promoted Co/Al 2 O 3 catalysts for the gas-phase amination of aliphatic alcohols with ammonia. Cat Sci Technol. 2018. DOI: 10.1039/C8CY01334A.
[12]   Milan M, Kaushik C, Bhaskar P, Bivas CR, Ruthenium SK. (II)-NNN-pincer-complex-catalyzed reactions between various alcohols and amines for sustainable C–N and C–C bond formation. Adv Syn Cat. 2017;3604:722–9.
[13]   Ibáñez J, Araque-Marin M, Paul S, Pera-Titus M. Direct amination of 1-octanol with NH3 over Ag-Co/Al2O3: promoting effect of the H2 pressure on the reaction rate. Chem Eng J. 2018. DOI: 10.1016/j.cej.2018.10.021.
[14]   Dumon AS, Wang T, Ibañez J, Tomer A, Yan Z, Wischert R, et al. Direct n-octanol amination by ammonia on supported Ni and Pd catalysts: activity is enhanced by "spectator" ammonia adsorbates. Catal Sci Technol. 2018;8:611–21.
[15]   Wu Y, Huang Y, Dai X, Shi F. Alcohol amination catalyzed by copper powder as a self-supported catalyst. ChemSusChem. 2018;12:3185–91.
[16]   Kaeding WW. Production of aliphatic amines utilizing a crystalline aluminosilicate catalyst of zsm-5, zsm-11 or zsm-21, United States Patent, Patent Number: 40,82,805 Date of Patent: Apr. 4, 1978.
[17]   Hamilton LA, Pitman NJ. Production of primary and secondary amines, United States Patent, Patent Number: 3,384,667, Date of Patent: May 21, 1968.
[18]   Takeyuki F, Kazumoto O, Kiyonobu N, Michio F. Production of methylamines, European Patent, Patent Number: 07,63,519, Date of Patent: July. 18, 2007.
[19]   Thomas WJ, Portalski P. Thermodynamics in methanol synthesis. Ind Eng Chem. 1958;50:967–70.

[20] Ramioulle J, David A. Hydrocarbon process. Int Ed. 1981;60:113–17.
[21] Van Gysel AB, Musin W. Methylamines. In: Ullmann's encyclopedia of industrial chemistry. Weinheim, Germany: WileyVCHVerlag, 2000.
[22] The Leonard Process Co. Inc. Hydrocarbon Process 58 1979: 194.
[23] Blauwhoff PM, Gosselink JW, Kieffer EP, Sie ST, Stork WH. Zeolites as catalysts in industrial processes. In: Weitkamp J, Puppe L, editors. Catalysis and zeolites. Berlin, Heidelberg: Springer, 1999: 437–538.
[24] Weigert FJ. Preparation of monomethylamine from ammonia and methanol using a zeolitic catalyst, European Patent, Patent Number: 0,025,693 Date of Patent: July 20, 1983
[25] Weigert FJ. Preparation of monomethylamine, United States Patent, Patent Number: 4,254,061, Date of Patent: Mar. 3, 1981.
[26] Kiyoura T, Nakahara S. Process for preparing methylamines avec a mordenite catalyst, European Patent, Patent Number: 0,681,869, Date of Patent: Aug. 9, 2002.
[27] Kiyoura T, Nakahara S. Process for preparing methylamines, United States Patent, Patent Number: 5,959,150, Date of Patent: Sep. 28, 1999.
[28] Kiyoura T, Terada K. Method for preparing methylamines, United States Patent, Patent Number: 5,382,696, Date of Patent: Jun. 19, 1975.
[29] Corbin DR, Schwarz S, Sonnichsen GC. Methylamines synthesis: a review. Cat Today. 1997;37:71–102.
[30] Ashina Y, Fujita T, Fukatsu M, Yagi J. Process for producing dimethylamine in preference to mono- or trimethylamine by vapor-phase catalytic reaction of methanol and ammonia, United States Patent, Patent Number: 4578516, 1986.
[31] Wilhelm FC, Parris GE, Aufdembrink BA, Gaffney TR. Preparation of methylamines using shape selective chabazites, United States Patent, Patent Number: 5,399,769, Date of Patent: Mar. 21, 1995
[32] Bähn S, Imm S, Neubert L, Zhang M, Neumann H, Beller M. The catalytic amination of alcohols. ChemCatChem. 2011;31:1853–64.
[33] Irrgang T, Kempe R. 3D-metal catalyzed N- and C-alkylation reactions via borrowing hydrogen or hydrogen autotransfer. Chem Rev. 2008;119:2524–49.
[34] Staelens N, Reyniers MF, Marin GB. Transalkylation of methylamines: kinetics and industrial simulation Ind. Eng Chem Res. 2004;43:5123–32.
[35] Eller K, Henkes E, Rossbacher R, Höke H. Amines, aliphatic. In: Ullmann's encyclopedia of industrial chemistry. Weinheim, Germany: WileyVCHVerlag, 2005.
[36] Mills KL. Ammonolysis of alkyl halides, United States Patent, Patent Number: 3,399,236, Date of Patent: Aug. 27, 1968.
[37] Steinmetz WE. Ammonolysis of halides, United States Patent, Patent Number: 3,450,765, Date of Patent: June 17, 1969.
[38] Teter JW. Production of organic compounds containing nitrogen, United States Patent, Patent Number: 2,381,470, Date of Patent: Aug. 7, 1945.
[39] Teter JW, Olson LE. Process for reacting ammonia and olefins, United States Patent, Patent Number: 2,623,061, Date of Patent: Dec. 23, 1952.
[40] Teter JW. Catalyst, United States Patent, Patent Number: 2,392,107, Date of Patent: Jan. 1, 1946.
[41] Olin JF, Deger TE. Manufacture of aliphatic amines and acid amides, United States Patent, Patent Number: 2,422,631, Date of Patent: Jene 17, 1947.
[42] Peterson JO, Fales HS. Amines via the amination of olefins, United States Patent, Patent Number: 4,307,250, Date of Patent: Dec. 22, 1981.
[43] Peterson JO, Fales HS. Amines via the amination of olefins, United States Patent, Patent Number: 4,375,002, Date of Patent: Feb. 23, 1983.

[44] McClain DM. Process for the preparation of aliphatic amines, United States Patent, Patent Number: 3,412,158, Date of Patent: Nov. 19, 1968.

[45] Gardner DM. Preparation of amines from olefins using certain transition metal catalysts, US Patent 4454321 United States Patent, Patent Number: 4,536,602 Date of Patent: Aug. 20, 1985.

[46] Deeba M. Amination of olefins using organic acid catalysts, United States Patent, Patent Number: 4,536,602 Date of Patent: Aug. 20, 1985.

[47] Taglieber V, Hoelderich W, Kummer R, Mross WD, Saladin G. Preparation of tert-butylamine from isobutene, United States Patent, Patent Number: 4,929,758, Date of Patent: May 29, 1990.

[48] Taglieber V, Hoelderich W, Kummer R, Mross WD, Saladin G. Production of amines from an olefin and ammonia or a primary or secondary amine, United States Patent, Patent Number: 4,929,759, Date of Patent: May 29, 1990.

[49] Hesse M, Steck W, Lermer H, Fischer R, Schwarzmann M. Verfahren zur herstellung von aminen, European Patent, Patent Number: 431,451, Date of Patent: Aug. 9,1991.

[50] Fales HS, Peterson JO. Catalytic amination process for the production of amines European Patent, 0,039,918, Date of Patent: Feb. 29,1984.

[51] Hesse M, Steck W, Lermer H, Schwarzmann M, Fischer R. A process for the preparation of amines, Germany Patent, Patent Number: 3,940,349, Date of Patent: Dec, 12, 1989.

[52] Gross T, Seayad AM, Ahmad M, Beller M. Synthesis of primary amines: first homogeneously catalyzed reductive amination with ammonia. Org Lett. 2002;4:2055–8.

[53] Baxter EW, Reitz AB. Reductive aminations of carbonyl compounds with borohydride and borane reducing agents. Org React. 2004;1–714.

[54] Miriyala B, Bhattacharyya S, Williamson JS. Chemoselective reductive alkylation of ammonia with carbonyl compounds: synthesis of primary and symmetrical secondary amines. Tetrahedron. 2004;60:1463–71.

[55] Nakamura Y, Kon K, Touchy AS, Shimizu KI, Ueda W. Selective synthesis of primary amines by reductive amination of ketones with ammonia over supported Pt catalysts. ChemCatChem. 2015;7:921–4.

[56] Senthamarai T, Murugesan K, Natte K, Kalevaru NV, Neumann H, Kamer PC, et al. Expedient synthesis of N-methyl- and N-alkylamines by reductive amination using reusable cobalt oxide nanoparticles. ChemCatChem. 2018;10:1235–40.

[57] Gallardo-Donaire J, Ernst M, Trapp O, Schaub T. Direct synthesis of primary amines via ruthenium-catalysed amination of ketones with ammonia and hydrogen. Adv Synth Catal. 2016;358:358–63.

[58] Komanoya T, Kinemura T, Kita Y, Kamata YK, Hara M. Electronic effect of ruthenium nanoparticles on efficient reductive amination of carbonyl compounds. J Am Chem Soc. 2017;139:11493–9.

[59] Zhang H, Tong X, Liu Z, Wan J, Yu L, Zhang Z. The sustainable heterogeneous catalytic reductive amination of lignin models to produce aromatic tertiary amines. Catal Sci Technol. 2018;8:5396.

[60] Tararov VI, Kadyrov R, Börner A, Riermeier TH. On the reductive amination of aldehydes and ketones catalyzed by homogeneous Rh(i) complexes. Chem Comm. 2000;19:1867–8.

[61] Gülcemal D, Gülcemal S, Robertson CM, Xiao J. A new phenoxide chelated IrIII N-heterocyclic carbene complex and its application in reductive amination reactions. Organometal. 2015;34:4394–400.

[62] Lee OY, Law KL, Yang D. Secondary amine formation from reductive amination of carbonyl compounds promoted by lewis acid using the InCl3/Et3SiH system. Org Lett. 2009;11:3302–5.

[63] Ogo S, Uehara K, Abura T, Fukuzumi S. pH-dependent chemoselective synthesis of α-amino acids. Reductive amination of α-keto acids with ammonia catalyzed by acid-stable iridium hydride complexes in water. J Am Chem Soc. 2004;126:3020–1.

[64] Göthel H, Feichtinger H, Tummes H, Falbe J. Process for the production of primary amines United States Patent, Patent Number: 4,041,080, Date of Patent: 9 Aug., 1977

[65] Wigbers CW, Mueller C, Melder JP, Stein B, Meissner H, Haderlein G, et al. Process for preparing unsymmetric secondary tert-butylamines in the liquid phase, United States Patent, Patent Number: 8,445,726, Date of Patent: Apr. 5, 2011

[66] Kovac KA. Energy-efficient hexamine production process. Adv Chem Eng Res. 2013;2:51–4.

[67] Hexamine Production Technology, Jiangsu Kaimao Chemical Technology Co., Ltd. http://enjs-kaimao.com/products_detail/productId=53.html. Accessed: 3 Sept, 2019.

[68] Wood RK, Stevens WF. Reaction kinetics of the formation of hexamethylenetetramine. J Appl Chem. 1964;14:325–30.

[69] Cui X, Junge K, Beller M. Palladium-catalyzed synthesis of alkylated amines from aryl ethers or phenols. ACS Catal. 2016;6(11):7834–8.

[70] Ng EP, Law SP, Rino RM, Juan JC, Farook A. Hydroamination of cyclohexene enhanced by ZnCl2 nanoparticles supported on chiral mesoporous silica. Chem Eng J. 2014;243:99–107.

[71] Rasero-Almansa AM, Corma A, Iglesias M, Sanchez F. Post-functionalized iridium–Zr-MOF as a promising recyclable catalyst for the hydrogenation of aromatics. Green Chem. 2014;16:3522–7.

[72] Mink G, Horváth L. Hydrogenation of aniline to cyclohexylamine on NaOH-promoted or lanthana supported nickel. Reac Kinet Cat Let. 1998;65:59–65.

[73] Immel O, Darsow G, Waldmann H, Petruck GM. Process for producing mixtures containing cyclohexylamines and dicyclohexylamines which may be substituted using a palladium /Al2O3 catalyst, European Patent, Patent number: 0,50,347, Patent date: Feb. 24, 1992.

[74] Sokolski DV, Ualikhanova A, Temirbulatova AE. Aniline hydrogenation in the presence of alumina-supported rhodium under hydrogen pressure. Reac Kinet Cat Let. 1982;20:35–7.

[75] Graham KF, Hindle KT, Jackson SD, Williams DJ, Wuttke S. Stereoselective synthesis of alicylic amines. Top Catal. 2010;53:1121–5.

[76] Cho H-B, Park Y-H. The effects of impeller characteristics in the hydrogenation of aniline on Ru/C catalyst. Korean J Chem Eng. 2003;20:262–7.

[77] Tomkins P, Müller TE. Enhanced selectivity in the hydrogenation of anilines to cyclo-aliphatic primary amines over lithium-modified Ru/CNT catalysts. ChemCatChem. 2018;10:1438–45.

[78] Naramoto I, Kiyuma T. Method for production of cyclohexylamines, United States Patent, Patent Number: 4,914,239, Date of Patent: Apr. 3, 1990.

[79] Hearn D, Nemphos SP. Process for the production of cyclohexyl amine, United States Patent, Patent Number: 5,599,997, Date of Patent: Feb. 4, 1997.

[80] Darsow G, Immel O, Petruck G, Waldmann H, Petruch GM. European Patent, Patent Number: 501,265, Date of Patent: Feb. 17, 1992.

[81] Axet MR, Conejero S, Gerber IC. Ligand effects on the selective hydrogenation of nitrobenzene to cyclohexylamine using ruthenium nanoparticles as catalysts. ACS Appl Nano Mater. 2018;1:5885–94.

[82] Liu H, Chang L, Bai C, Chen L, Luque R, Li Y. Controllable encapsulation of "clean" metal clusters within mofs through kinetic modulation: towards advanced heterogeneous nanocatalysts. Angew Chem Int Ed. 2016;55:5019–23.

[83] Chen X, Shen K, Ding D, Chen J, Fan T, Rongfang W, et al. Solvent-driven selectivity control to either anilines or dicyclohexylamines in hydrogenation of nitroarenes over a bifunctional Pd/MIL-101 catalyst. ACS Catal. 2018;8:10641–8.

[84] Chatterjee M, Sato M, Kawanami H, Ishizaka T, Yokoyama T, Suzuki T. Hydrogenation of aniline to cyclohexylamine in supercritical carbon dioxide: significance of phase behaviour. Appl Catal A Gen. 2011;396:186–93.

[85] Venneti MH, Govindarao KV, Ramana M. Liquid phase hydrogenation of aniline in a trickle bed reactor. J Appl Chem Biotechnol. 1975;25:169–81.

[86] Vogt PF, Gerulis JJ, Amines A. Ullmann's encyclopedia of industrial chemistry. Weinheim, Germany: WileyVCHVerlag, 2000.

[87] Downing RS, Kunkeler PJ, van Bekkum H. Catalytic syntheses of aromatic amines. Catal Today. 1997;37:121–36.

[88] Krogul-Sobczak A, Cedrowski J, Kasperska P, Litwinienko G. Reduction of nitrobenzene to aniline by CO/H_2O in the presence of palladium nanoparticles. Cat. 2009;9:404–15.

[89] Green RA, Hartwig JF. Nickel-catalyzed amination of aryl chlorides with ammonia or ammonium salts. Angew Chem Int Ed Engl. 2015;54:3768–72.

[90] Katada N, Iijima S, Igi H, Niwa M. Synthesis of aniline from phenol and ammonia over zeolite beta, Editor(s): Hakze Chon, Son-Ki Ihm, Young Sun Uh, Stud. Surf Sci Cat. 1997;105:1227–34.

[91] Squire EN. Synthesis of aromatic amines by reaction of aromatic compounds with ammonia, United States Patent, Patent Number: 3,919,155, Date of Patent: Nov. 11, 1975.

[92] Langer R, Buysch HJ, Pentling U, Wagner P. Process for the production of aromatic amines, United States Patent, Patent Number: 5,877,350, Date of Patent: Mar. 2, 1999.

[93] Langer R, Buysch HJ, Pentling U. A process and catalyst for the production of aromatic amines by gas phase hydrogenation, Germany Patent, Patent Number: 19,521,670, Date of Patent: June 14, 1995.

[94] Cocuzza G, Torreggiani G. Process for the preparation and the recovery of ethanolamines, United States Patent, Patent Number: 4,169,856, Date of Patent: Oct. 2, 1979.

[95] Zahedi G, Amraei S, Biglari M. Simulation and optimization of ethanolamine production plant. Korean J Chem Eng. 2009;26:1504–11.

[96] Quadros PA, Oliveira NM, Baptista CM. Continuous adiabatic industrial benzene nitration withmixed acid at a pilot plant scale. Chem Engin J. 2005;108:1–11.

[97] Alexanderson V, Trecek JB, Vanderwaart CM. Adiabatic process for nitration of nitratable aromatic compounds, United States Patent, Patent Number: 4,021,498, Date of Patent: Dec. 9, 1975.

[98] Alexanderson V, Trecek JB, Vanderwaart CM. Continuous adiabatic process for the mononitration of benzene, United States Patent, Patent Number: 4,091,042, Date of Patent: May. 23, 1978.

[99] Pennemann B, Munnig J, Dietrich W, Rausch AK. Method and system for producing nitrobenzene, United States Patent, Patent Number: 9,139,508 Date of Patent: Sep. 22, 2015.

[100] ACC Center for the Polyurethanes Industry 2014 End-Use Market Survey, ACC's Economic & Statistics Department, Plastics and Polymer Composites in Light Vehicles, Polyurethane Foams, Chemical Economics Handbook, 2014.

[101] Kreye O, Mutlu H, Meier MA. Sustainable routes to polyurethane precursors. Green Chem. 2013;15:1431–56.

[102] Litwinowicz M, Kijeński J. Carbamoylation of primary, secondary and aromatic amines by dimethyl carbonate in a flow system over solid catalysts. Sustain Chem Process. 2015;3:1–7.

[103] Li F, Miao J, Wang Y, Zhao X. Synthesis of methyl N-phenyl carbamate from aniline and dimethyl carbonate over supported zirconia catalyst. Ind Eng Chem Res. 2006;45:4892–7.

[104] Merenov AS, Jansma IG, Gillis PA. Process for the production of methylene diphenyl diisocyanate isomer mixtures with high 2,4'-methylene diphenyl diisocyanate purity, United States Patent, Patent Number: 9,080,005, Date of Patent: Jul. 14, 2015.

Katarzyna Staszak

7 Halogened hydrocarbons – current trends

Abstract: Current trends in industrial synthesis of selected halogened hydrocarbons (HHCs) were presented in detail. Special emphasis was put on the most popular compounds, such as chloromethanes (CH_3Cl, CH_2Cl_2, $CHCl_3$, CCl_4), chloroethene (vinyl chloride $H_2C = CHCl$), chlorobenzene, 1-4-dichlorbenzene and epichlorohydrin. The possibility methods of modification technologies based on literature information were also reported. Presented simple schema, with production conditions, should be helpful in understanding the issues discussed.

Keywords: halogened hydrocarbons, halocarbons, halides, chlorination

7.1 Introduction

Halocarbon compounds are a large class of synthetic and natural compounds that contain one or more halogens (fluorine, chlorine, bromine, or iodine) combined with carbon. Halogenated hydrocarbons (HHCs) include aliphatic, alicyclic, aromatic, polyaromatic, and heterocyclic hydrocarbons. The typical classification of such compounds depends of the type of chain as alkyl, vinyl, aryl, or acyl halides, wherein the last group is also called acid halides (Figure 7.1). It should be noted that class of halocarbon depends on the type of the carbon to which the halogen atom is bonded. For example, benzyl chloride ((chloromethyl)benzene, with the formula $C_6H_5CH_2Cl$), does not belong to aryl chloride group but to alkyl chloride one due to the directly bond between chloride atom and methyl group, or 3-chloroprop-1-ene (known as allyl chloride, with the formula $CH_2 = CHCH_2Cl$) is a type of alkyl chloride, despite of the double band in the chain.

Natural halogen-containing organic compounds are found in small amounts in plant and animal organisms, mainly marine [1–3], as chlorine, bromine, iodine, and fluorine derivatives. One of the examples is tyrian purple (6,6′-dibromoindigo; (2E)-6-bromo-2-(6-bromo-3-oxo-1H-indol-2-ylidene)-1H-indol-3-on), obtained from several species of predatory sea snails in the family Muricidae, and rock snails Murex [4]. Such type of compounds could be found also in human body, i.e. thyroid hormones composed of iodine: (S)-thyroxine and (S)-triiodothyronine [5].

Due to their diverse properties, organic halogen compounds are widely used in the industry. The first association that comes is, of course, the polymer industry including polyvinyl chloride (PVC) – world's third-most widely produced synthetic

This article has previously been published in the journal Physical Sciences Reviews. Please cite as: Staszak, K. Halogened hydrocarbons – current trends *Physical Sciences Reviews* [Online] 2020, 5. DOI: 10.1515/psr-2019-0032.

https://doi.org/10.1515/9783110656367-007

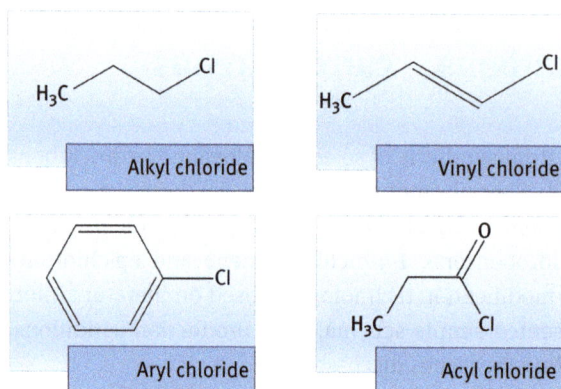

Figure 7.1: Examples of types of organohalogen compounds, based on organochlorine compounds.

plastic polymer, after polyethylene and polypropylene, produced by polymerization of the vinyl chloride monomer. Well known are also trihalomethanes (THMs) – chemical compounds in which three of the four hydrogen atoms of methane (CH_4) are replaced by halogen atoms. They are or sometimes were (due to their toxicity, carcinogenic, and greenhouse effects [6]) used in the chemical industry as solvents or refrigerants. The most popular are as follows: (i) trichloromethane, known as chloroform or R-20 (common solvent in organic chemistry, used as precursor to polytetrafluoroethylene, PTFE, and various refrigerants), (ii) trifluoromethane, known as Freon 23, R-23 or HFC-23, and (iii) chlorodifluoromethane, known as R-22, HCFC-22 (both used as refrigerants, while fluoroform is also used in the semiconductor industry in plasma etching [7]). Additionally, chlorinated hydrocarbons have been widely used as pesticides, mainly insecticides, including DDT (dichlorodiphenyltrichloroethane, 1-chloro-4-[2,2,2-trichloro-1-(4-chlorophenyl)ethyl]benzene), and its metabolite DDD (dichlorodiphenyldichloroethane, 1-chloro-4-[2,2-dichloro-1-(4-chlorophenyl) ethyl]benzene), lindane (*gamma*-hexachlorocyclohexane, (1*r*,2*R*,3*S*,4*r*,5*R*,6*S*)-1,2,3,4,5,6-hexachlorocyclohexane), chlordane (1,2,4,5,6,7,8,8-octachloro-3a,4,7,7a-tetrahydro-4,7-methanoindane), aldrin (1,2,3,4,10,10-hexachloro-1,4,4a,5,8,8a-hexahydro-1,4:5,8-dimethanonaphthalene), and dieldrin ((1a*R*,2*R*,2a*S*,3*S*,6*R*,6a*R*,7*S*,7a*S*)-3,4,5,6,9,9-hexachloro-1a,2,2a,3,6,6a,7,7a-octahydro-2,7:3,6-dimethanonaphtho[2,3-b] oxirene) [8]. There are also few examples of chlorocarbon compounds which have been used as herbicides, especially 2,4-D ((2,4-dichlorophenoxy)acetic acid) and 2,4,5-T ((2,4,5-trichlorophenoxy)acetic acid). Most of these products are banned or significantly restricted due to their harmfulness. They are persistent in the environment, and accumulate in organisms, sometimes causing toxicity. Another example of a compounds that have been produced and used (as dielectric and coolant fluids in electrical transformers, carbonless copy paper and in heat transfer fluids) on a large scale and then banned because of its high harmfulness is PCBs (polychlorinated biphenyls, with the formula $C_{12}H_{10-x}Cl_x$) [9].

There are several methods of preparation of halogenated hydrocarbons. The most important are presented in Table 7.1.

Depending on the type of organic halogen compounds, different ways of synthesis (see Table 7.1), as well as different technologies, are proposed in the literature and industry, which will be discussed, in detail, in the following sections. Particular attention was given to the compounds with the widest industrial applications.

7.2 Alkyl halides

Alkyl halides, known as haloalkanes, are a group of organic chemical compounds derived from alkanes in which one or more hydrogen atoms have been replaced by a halogen atom (see Figure 7.1). Similar to alcohols, alkyl halides could be divided into different classes depending on how the halogen atom is positioned on the chain of carbon atoms: i. primary alkyl halides (RCH_2X), ii. secondary alkyl halides (R_2CHX) and iii. tertiary alkyl halides (R_3CX).

7.2.1 Chloromethanes

The simplest example of haloalkanes is chloromethane (known also as methyl chloride, Refrigerant-40, R-40 or HCC 40). In industrial practice, two methods of its synthesis are applied: chlorination of CH_4 or hydrochlorination of methanol. Global chloromethane market was esteemed at 1.8 billion US$ in 2017 and is expected to reach 2.5 billion US$ in 2026, at a CAGR of 4.21 % during a forecast period [16].

Due to ta cascade reaction in chlorination process, the other products are also obtained: dichloromethane, chloroform, and carbon tetrachloride. From the practical point of view, generally synthesis of all compounds is realized in the industry process simultaneously. Dichloromethane (known also as Refrigerant-30, Freon-30, R-30, DCM) is a popular solvent in organic chemistry. Carbon tetrachloride (known also as Freon-10, Refrigerant-10, Halon-104) was often used as solvent and in fire extinguishers, as a precursor to refrigerants and as a cleaning agent, in the last century. Nowadays, due to its toxicity the application is limited.

7.2.1.1 Chlorination of methane
The production of chloromethane from CH_4, presented schematically in Figure 7.2, is based generally on the natural gas as a source of CH_4 (stream no. 1). Natural gas is a mixture of hydrocarbon gases, mainly CH_4 (up to 99 %), and ethane, propane, butane, pentane. Raw natural gas also contains water vapor, hydrogen sulfide, carbon dioxide, nitrogen, helium, and other impurities [17]. Because the higher hydrocarbons (C_{2+}) could disturb in the chlorination process (i.e. by the formation of

Table 7.1: Synthesis methods for HHCs.

Substrates/products	Description
S: Alkanes P: Mixture of compounds mono- or multihalogenated at various positions	–Type of halogenation: radical halogenation [10]; –Substitution reaction initiated by ultraviolet radiation or sunlight; –Reactivity, as well as thermal effect, depends of the type of halogens: F>Cl>Br>I; –Important! The thermal effect of the fluorination reaction is particularly large. This process is relatively rarely used in industry because it requires the use of special technology; –Exemplary reactions: $CH_4 + Cl_2 \rightarrow CH_3Cl + HCl$; $CH_3Cl + Cl_2 \rightarrow CH_2Cl_2 + HCl$; $CH_2Cl_2 + Cl_2 \rightarrow CHCl_3 + HCl$; $CHCl_3 + Cl_2 \rightarrow CCl_4 + HCl$
S: Alkenes and alkynes P: mono-haloalkane or alkene, or dihalide in the reaction with hydrogen halides or halides, respectively	Reaction with hydrogen halides: –addition reactions, with Markovnikov's rules, –reaction rates increase in the order HF<HCl<HBr<HI and as the alkene gets more complicated [11], –exemplary reaction: $CH_3\text{-}CH = CH\text{-}CH_3 + HCl \rightarrow CH_3\text{-}CH_2\text{-}CHCl\text{-}CH_3$, $HC \equiv CH + HCl \rightarrow CH_2 = CHCl$ Reaction with halides: –addition of halogens is stereospecific and produces vicinial dihalides with anti addition [12], –mainly Br and Cl as halogens, in thermodynamical terms I is too slow, and F is too vigorous and explosive, –exemplary reaction: $R_2C = CR_2 + Cl_2 \rightarrow R_2CCl\text{-}CR_2Cl$.
S: Alcohols P: Mono-haloalkane or alkene, depends of type of alcohols	–The order of reactivity of alcohols is 3°>2°>1° methyl [13]; –The order of reactivity of the hydrogen halides is HI>HBr>HCl (HF is generally unreactive); –The reaction is acid catalyzed; –Exemplary reaction: $R\text{-}OH + HX \rightarrow RX + H_2O$.
S: Benzene or aniline P: Mixture of mono- and multi aryl halides or mono aryl halides in reaction with benzene and aniline, respectively	Friedel–Crafts reaction: –directly halogenation to the aromatic ring: –other benzene derivatives i.e. di-, tri-, chlorobenzene with different positions of the halogene atoms attached to the benzene ring could be formatted. Sandmeyer reaction [14]: –diazonium salt reacts with copper salts as reagents or catalysts:

Table 7.1 (continued)

Substrates/products	Description
S: Carboxylic acids or esters P: Acyl halogens	Directly reaction of carboxylic acids or esters [15] with source of halogens, i.e. thionyl chloride (SOCl$_2$), phosphorus trichloride (PCl$_3$), phosphorus pentachloride (PCl$_5$):

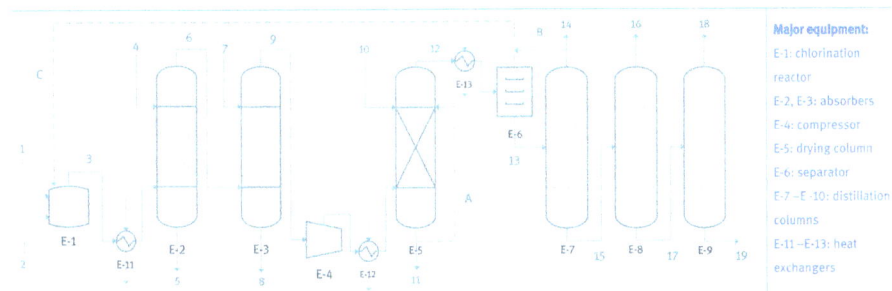

Figure 7.2: Schema of production of chloromethanes based on chlorination of methane.

other alkyl chlorides, difficult to separate from the reaction mixture) usually in the first step the hydrocracking of natural gas is proposed. For example, presented in natural gas ethane reacts with chlorine and obtained 1,1-dichloroethane is difficult to separate from trichloromethane, because their boiling points are similar: 57.2 °C and 61.2 °C, respectively. Moreover, dichloromethane forms azeotropic mixtures with presented in natural gas pentane.

Chlorine is the second feedstock in this technology (stream no. 2). Depending on the expected final product, recycled CH$_4$ from the separator E-6 (steam C), and/ or chloromethane (if the production is focused on higher chlorine-based CH$_4$ derivatives, stream B) are introduced into the reactor E-1. The reaction needs high temperature (350–500 °C), and therefore the reaction mixture (stream no. 3) have to be cooled in a heat exchanger (E-11). It should be noted that the composition of methyl chlorides depends on the chosen reaction's temperature as well as of the initial concentrations of reagents. If the production is focused on obtaining tetrachloromethane slight CH_4 deficiency in relation to chlorine is used (0.8:1), while for the production of chloromethane four/five times excess of CH_4 is applied. The hydrochloric acid formed in the reaction is removed on two adsorption columns (E-2 and E-3). The first of them (E-2) is sprayed with water – here the recovery of 31 % HCl is obtained (stream no. 5). The remaining acid, as well as chlorine (stream no. 6) is removed on a second column (E-3) by washing by sodium hydroxide (stream no. 7). These residues are removed from the installation in stream no. 8. The mixture of unreacted CH_4 and methyl chlorides (stream no. 9) are compressed (compressor E-4), cooled (heat exchanger E-12) and dried (drying column E-5). The drying process involves spraying the reaction mixture by 98 % sulfuric acid (stream no. 10). Depending on the water content in the reaction mixture, H_2SO_4 may be recycled (stream A) until the concentration reached 70–78 %. Discharged acid is removed from the installation in stream no. 11. The dried mixture of gases (stream no. 12) is cooled in heat exchanger E-13, and CH_4 is separated in separator E-6. CH_4, and optional chloromethane, are returned to the reactor E-1 (stream C), as was described above. The liquefied condensate (stream no. 13) is separated by distillation under pressure into its pure components: monochloromethane (stream no. 14) in distillation column E-7, dichloromethane (stream no. 16) in distillation column E-8, and trichloromethane (stream no. 18) and tetrachloromethane (stream no. 19) in distillation column E-9. If the production is focused on the higher substituted of chloromethanes the stream no. 15 is composed by the mixture of ca 70 wt % dichloromethane, 27 wt % trichloromethane, and 3 wt. % tetrachloromethane. Depending of the purity of raw materials the product yields are in the level 70–85 % based on CH_4, and 95–99 % based on chlorine [18].

It should be noted that there is a problem with selectivity of the chlorination reaction in this technology. Usually, in addition to monosubstituted chloromethane, the di-, tri- and even tetra-chloromethanes are formed. This is a result of few-steps reaction of chlorination (see the first row in Table 7.1). To avoid this problem the much higher concentration of CH_4 in comparison to chloride should be used. This problem does not occur in the second approach of chloromethane synthesis, when the methanol is used as a substrate (see Section 7.2.1.2).

There are several propositions described in the literature of modification of methyl chloride production from CH_4. Such solutions are not applied in the industrial actually, but it is worth to mention about it here. Podkolzin et al. [19] proposed selective production of methyl chloride by a reaction of CH_4, hydrogenchloride, and oxygen over lanthanum-based catalysts (LaOCl, LaCl$_3$, and La phases with an intermediate extent of chlorination) according to the reaction:

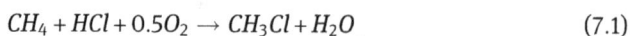

$$CH_4 + HCl + 0.5O_2 \rightarrow CH_3Cl + H_2O \qquad (7.1)$$

Nowadays, such oxidative chlorination reaction is adapted in industrial to vinyl chloride production. The results showed that at low CH_4 concertation the catalytic reaction was 100 % selective toward chloromethane. At higher concentrations, despite main product CH_3Cl, the co-products were also obtained: dichlormethane, carbon mono- and dioxides and trace amounts of trichloromethane. There was no tetrachloride in the reaction mixture. The potential installation scheme will be similar to Figure 7.2. The main differences are as follows: (i) in stream no. 2 should be hydrochloric acid instead of chlorine, (ii) additionally stream with source of oxygen (air) should be added and the air should be introduced to the reactor E-1, and (iii) because of the lack of tetrachloromethane in the reaction mixture the last distillation column (E-9) is unnecessary and could be removed.

Similarly, Paunović and Pérez-Ramírez [20] proposed the application of catalysts to improve the selectivity of the classical non-catalytic radical-mediated halogenation (chlorination and bromination) of CH_4. Several potential catalysts were tested including carriers: quartz, SiO_2, SiC, α-Al_2O_3, γ-Al_2O_3 and carbon, noble metals: Pt, Pd and Ru, metal oxides: Fe_2O_3 and CeO_2, chlorides: $PdCl_2$ and $CuCl_2$ and oxyfluorides: $TaOF_3$ supported on SiO_2, γ-Al_2O_3, carbon or H-ZSM-5 carriers, sulfated systems: S-ZrO_2, S-ZrO_2-SBA-15, S-TiO_2, S-Nb_2O_5, S-Ta_2O_5 and Nafion, as well as zeolites: 3A, H-USY, H-MOR, H-SAPO-34, H-BETA and H-ZSM-5. The obtained results showed that zeolites and Pt/carbon weren't selective to halomethanes and promoted undesirable polyhalogenation. For other catalysts tested selectivity was similar to the non-catalytic reactions, wherein higher in bromination ($S_{CH_3Br} = 80$–95 % versus $S_{CH_3Cl} = 52$–90 % at $X_{CH_4} = 5$–18 %). Unfortunately, application of catalysts caused the decomposition of halomethanes for several materials studied in the chlorination process and for all catalysts in bromination process.

As mentioned in the first paragraph of this chapter, the purification of natural gas by removing components such as ethane from the CH_4 feed stream is necessary in this solution. Unfortunately, this process is expensive but desirable to obtain of high-quality chlorinated CH_4. Also the using of high purity of CH_4 in the production, although it would be beneficial from the point of view of purity of the products obtained, is too expensive in industrial practice. To solve this problem in United States Patent Application [21], the novelty process for the production of chlorinated CH_4, with avoidance of expensive CH_4 purification, thus directly from natural gas, was proposed. The processes involve a dehydrochlorination and/or chlorination step and allow to converts almost complete (up to 100 %) of the higher chlorinated alkanes. These more highly chlorinated alkanes can be easily removed from the process stream. The schema of chlorination of natural gas, based on the patent, is presented in Figure 7.3.

To the chlorination reactor E-1 two mainly streams are introduced: stream no. 1 containing natural gas or optionally CH_4, and stream no. 2 with chlorine. The reaction occurs in temperature in the range from 425 °C to 475 °C, pressure in the range 5–12 bar. Additionally, recycled streams A-C could be also entered to the reactor.

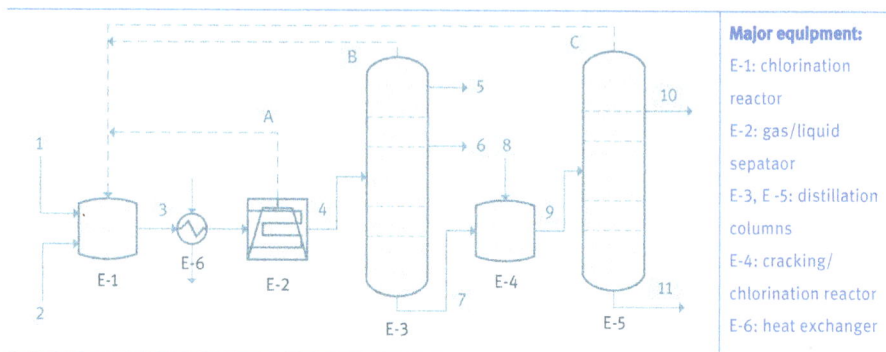

Figure 7.3: Schema of production of chloromethanes based on chlorination of natural gas.

After reaction in stream no. 3 several compounds are presented: chlorinated CH_4, chlorinated ethanes, chlorinated ethylenes, chlorinated higher hydrocarbons, HCl and unreacted CH_4. Usually, in such condition of chlorination reaction, complete chlorine conversion occurs, but small amounts of unreacted chlorine may also be present in stream no. 3. The reaction mixture is cooled in heat exchanger E-6 and enters to the gas/liquid separator E-2. Part of the unreacted CH_4 and hydrogen chloride, as well as optionally chlorinated CH_4, are recycle as gaseous stream A to the chlorination reactor E-1. The condensed liquid from gas/liquid separator E-2 (stream no. 4), containing chlorinated CH_4 products, chlorinated ethanes, ethylenes and chlorinated higher hydrocarbons and HCl, are separated in distillation columns. In Figure 7.3, these columns are presented schematically as one distillation column E-3. Number of columns depends of the composition of feed stream (stream no. 4). Distillation columns allow to separate HCl (stream B), methyl chloride (stream no. 5) and dichloromethane (stream no. 6). Depends of the kind of final product only HCl or its mixture with CH_3Cl or CH_3Cl and CH_2Cl_2 are recycled to the chlorination reactor E-1 in stream B. Residue from the distillation processes in stream no. 6 is introduced to catalytic cracking/chlorination reactor E-4. In this stream the following compounds are presented: trichloromethane and carbon tetrachloride, as well as chlorinated ethanes, ethylenes and chlorination products of higher hydrocarbons. Additionally, to the reactor E-4 the next portion of chloride is added in stream no. 8. In catalytic cracking/chlorination reactor E-4 chlorinated ethane and chlorinated higher alkanes are dehydrochlorinated (cracked) and chlorinated to obtain heavier components, such as 1,1,2-trichloroethane or 1,1,2,2-tetrachloroethane. A catalysts type – Y molecular sieve, $FeCl_2$, $FeCl_3$, $AlCl_3$ are proposed in this solution. The product in stream no. 9 is introduce to next series distillation columns, schematically shown as one column E-5. The number of columns must be chosen to ensure effective separation of products. Bottoms stream no. 9 contain carbon tetrachloride and chlorinated ethanes and ethylenes, as well as higher compound, such as chlorinated propanes and butanes produced in the process. Overhead stream no. 10 include tri-chloromethane, HCl and unreacted chlorine. Hydrogen chloride and chlorine may be, optionally, recycled in stream C to chlorination reactor E-1.

7.2.1.1.1 Cases study

One of the most important elements of chemical reaction modeling and reactors design is proper description of chemical kinetic. The rate at which the components convert from substrates to products is crucial to estimate the geometry constraints of the reactor unit. In any given process condition, particularly taking into account product demand and substrate flowrates, the volume required to obtain desired conversion is associated with reactor residence time. The proper residence time, in other words time of chemical reaction run, can be obtained by providing unit of appropriate volumetric size. The kinetics of chemical reaction allows also to study the reaction in different geometries and in different acceptable process conditions.

In the case of gas-phase thermal chlorination of CH_4, the following results can be taken into account. The process of thermal chlorination of CH_4 [22] was studied in the temperature range of 400–420 °C. The reaction was investigated in the pilot reactor at an excess pressure of no higher than 1.5 atm. The pilot reactor with internal recycle was a tubular unit of diameter 0.8 m and length 3 m. The following reactions were taken into account [23]:

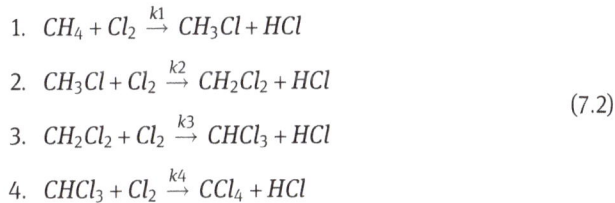

$$\begin{aligned}
&1. \quad CH_4 + Cl_2 \xrightarrow{k1} CH_3Cl + HCl \\
&2. \quad CH_3Cl + Cl_2 \xrightarrow{k2} CH_2Cl_2 + HCl \\
&3. \quad CH_2Cl_2 + Cl_2 \xrightarrow{k3} CHCl_3 + HCl \\
&4. \quad CHCl_3 + Cl_2 \xrightarrow{k4} CCl_4 + HCl
\end{aligned} \tag{7.2}$$

The molar ratio of reagents used was $CH_4/Cl_2 = 5$. The kinetic parameters were estimated for two reaction mechanism that are running simultaneously, for cross termination and quadratic termination. Both mechanisms fit the Arrhenius equation type and the ratio between them was related to substrates concentrations ratio:

$$\frac{r_q}{r_c} = 0.12\frac{C_{CH_4}}{C_{Cl_2}} \tag{7.3}$$

The rates for both mechanism termination are given by

$$\begin{aligned}
r_{q_i} &= k_{q_i}[Cl_2]^{1.5} \\
r_{c_i} &= k_{c_i}[R]^{0.5}[Cl_2]
\end{aligned} \tag{7.4}$$

where R corresponds to CH_4 (i = 1) and i + 1 chlorosubstituted substrates. The kinetic data are given in Table 7.2 and Table 7.3.

Table 7.2: Frequency factors and activation energies for methane chlorination reactions according to cross termination mechanism.

Reaction chlorination step i	Frequency factor, mol/dm^{3}s	Activation energy, J/mol
1	$2.31 \cdot 10^{11}$	129,000
2	$5.37 \cdot 10^{11}$	130,900
3	$5.14 \cdot 10^{11}$	133,700
4	$1.56 \cdot 10^{11}$	136,100

Table 7.3: Frequency factors and activation energies for methane chlorination reactions according to quadratic termination mechanism.

Reaction chlorination step i	Frequency factor, mol/dm^{3}s	Activation energy, J/mol
1	$1.32 \cdot 10^{12}$	128,900
2	$2.09 \cdot 10^{12}$	131,900
3	$1.57 \cdot 10^{12}$	136,200
4	$2.94 \cdot 10^{11}$	138,900

7.2.1.2 Hydrochlorination of methanol

The second industrial proposition of chloromethane production is process based on the hydrochlorination (esterification) of methanol using hydrogen chloride (see reaction (6) [24]). In the current time, this method becomes more important in comparison to the CH_4 chlorination method described above. The advantages of this method in comparison to the CH_4 chlorination, which appeal for its application, are as follows: (i) management of waste acid instead of its production, (ii) high selectivity of the reaction – only one product: chloromethane, (iii) easy to transport and storage, as well as low-cost substrate – methanol. Further reaction of chloromethane with chlorine can produce dichloromethane, trichloromethane and tetrachloromethane, according to the technique described in Section 7.2.1.1.

In this technology, as presented schematically in Figure 7.4, methanol (stream no. 1) and hydrochloric acid (stream no. 2) as a feedstock react in reactor E-1 with the following conditions: 280–350 °C and ca. 0.5 MPa (5 bar). In methanol hydrochlorination usually the alumina catalyst is used [25, 26]. The reaction mixture (stream no. 3) contains unreacted methanol and hydrochloric acid, as well as main product chloromethane and co-product water. Moreover, the dimethyl ether is presented in the system due to the dehydration of methanol reaction:

$$2\, CH_3OH \rightarrow (CH_3)_2O + H_2O \tag{7.5}$$

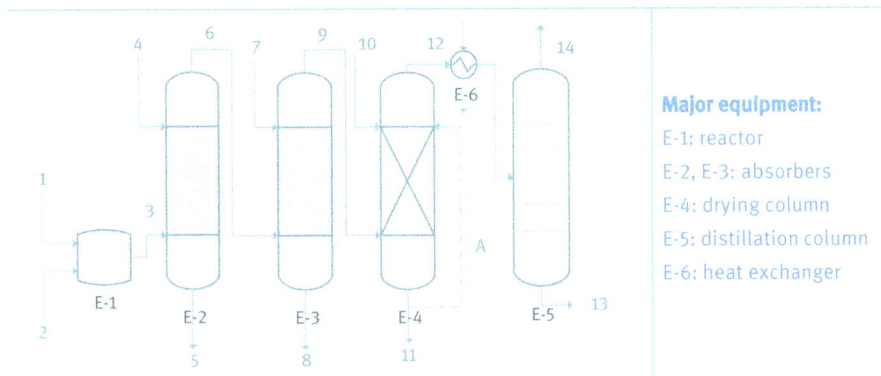

Figure 7.4: Schema of production of chloromethanes based on hydrochlorination of methanol.

In two adsorption columns (E-2 and E-3), these unnecessary products are removed. The first quench tower (E-2) is sprayed with water – here the recovery of 21 % HCl is obtained (stream no. 5), as well as removal of by-product water and small amounts of rested methanol. The remaining acid (stream no. 6) is removed on a second column (E-3) by washing by sodium hydroxide (stream no. 7). These residues are removed from the installation in stream no. 8. In the next step the mixture (stream no. 9) is dried in drying column E-4. The drying process involves spraying the reaction mixture by 96 % sulfuric acid (stream no. 10). Depending on the water content in the reaction mixture, H_2SO_4 may be recycled (stream A) until the concentration reached 80 %. In this process also dimethyl ether is removed – dimethyl ether reacts with sulfuric acid to form onium salts and methyl sulfate. Dry mixture containing mainly chloromethane (stream no. 12) is separated by distillation (distillation column E-5). Pure chloromethane (stream no. 14) could be treated as a product or substrate to chlorination process to obtain other chloromethanes. The residue are removed from installation in stream no. 13. The overall yield of the process, calculated on the basis of methanol, is ca. 99 %, and thus much higher in comparison to the chlorination of CH_4 process [18, 27].

Generally, in this process the excess of hydrogen chloride relative to methanol is applied. As a consequence unreacted hydrogen chloride and dimethyl ether (see reaction (5)) as a by-product are presented in the system. To avoid the necessity to remove these reagents, thus necessity to use absorption and dryer columns (see Figure 7.3), the following solution was proposed in USA patent [28]. In this patent instead of one hydrochlorination reactor, two reactors were recommended. Moreover, the reaction occurred by contacting hydrogen chloride with at least a stoichiometric amount of methanol. Feed solution of methanol was divided into at least two streams. To the first hydrochlorination reactor the portion containing about 60 to 95 % of the methanol was added and reacted with hydrogen chloride at a temperature in the range of about 115 to 170 °C. The reaction mixture and remaining amount of methanol were introduced to the second reactor. This reaction occurred in slightly lower temperature in the range of about 100 to 160 °C.

7.2.1.2.1 Cases study

The hydrochlorination of methanol in liquid phase is industrial method of consuming spent hydrochloric acid [29]. Hydrochlorination of methanol was performed in a continuous stirred tank reactor with volume 45 mL. The stirrer was working at 1500 rpm, which guaranteed the ideal mixing assumption true. The kinetics of the reaction was investigated between 40–65 °C and with water concentrations of 0.5 to 13 mol/dm^3:

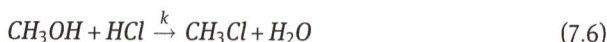

$$CH_3OH + HCl \xrightarrow{k} CH_3Cl + H_2O \tag{7.6}$$

The kinetic rate equation used reads:

$$r = k \cdot C_{CH_3OH} C_{HCl} \quad \left[\frac{mol}{dm^3 s}\right] \tag{7.7}$$

The rate constant measured obey the Arrhenius temperature dependence law:

$$k = k_\infty \exp\left(\frac{\Delta E}{RT}\right) \tag{7.8}$$

The measured values of frequency factor k_∞ and activation energy ΔE are as follows: 7.23·10^9 dm^3/mol·s and 98,550 J/mol for equal water and hydrogen chloride concentrations.

For different ratios of the water to the hydrogen chloride concentration the γ coefficient was introduced into the kinetic equation, which modifies frequency factor and activation energy term in kinetic rate constant:

$$k_\infty = 1.89 \cdot 10^{10} \gamma^{1.38} \quad \left[\frac{dm^3}{mol \cdot s}\right] \tag{7.9}$$

and

$$\Delta E = 101850 + 4800 \ln(\gamma) \quad \left[\frac{J}{mol}\right] \tag{7.10}$$

where

$$\gamma = \frac{C_{H_2O}}{C_{HCl}} \tag{7.11}$$

7.2.1.3 Mixed system

As mentioned in Section 7.2.1.2, chloromethane obtained in hydrochlorination of methanol could be used as a substrate, instead of CH$_4$, for production of higher substitutes chloromethanes derivatives. The connection of both processes allows to use the hydro-chloric acid generated in chlorination processes to the hydrochlorination of methanol. As a consequence, inevitable accumulation of hydrochloric acid can be decreased.

The schema of production of methyl chlorides by such mixed system is presented in Figure 7.5. In this technology, the chlorine (stream no. 1) and chloromethane (stream no. 2 and returned from the process streams no. 15 and 22) are introduced into chlorination reactor E-1. The reaction occurs with the following conditions: pressure of 0.8–1.5 MPa (8–15 bar) at elevated temperature (350–400 °C). The reaction mixture (stream no. 3) is initially separated in the quench system E-2, cooled in heat exchanger E-11 and separated in condensation unit E-3. In the next step chloromethanes (stream no. 5) are dried (E-4) and separated by distillation into pure components: dichlorome-thane (stream no. 18) in distillation column E-9, and trichloromethane (stream no. 20) and tetrachloromethane (stream no. 21) in distillation column E-10. The unreacted sub-strate – chloromethane is separated in the first step of multistage distillation in column E-8, and returned in stream no. 22 to the chlorination reactor E-1. Optionally, the resi-due from dryer E-4 is returned in stream B to the quench system. After condensation gaseous hydrogen chloride, including slight contents of chloromethane (stream no. 6), is introduced to the hydrochlorination reactor E-5 and reacted with methanol (stream no. 7), according to reaction (6). The chloromethane, as a main product of reaction, and unreacted substrates (stream no. 8) are separated in absorber E-6 by water wash-ing (stream no. 9). In this step the unreacted hydrochloric acid is removed in stream no. 10 with concentration about 20 %. Similar to the hydrochlorination of methanol installation, in the next stage (stream no. 11) the residues are removal from the process using drying (E-7) by 96 % sulfuric acid (stream no. 12). Depending on the water con-tent in the reaction mixture, H_2SO_4 may be recycled (stream A) until the concentration reached 80 % (stream no. 13). In this process also dimethyl ether (product of reaction (5)) is removed – dimethyl ether reacts with sulfuric acid to form onium salts and methyl sulfate (stream no. 14). Chloromethane is returned in stream no. 15 into the chlorination reactor E-1.

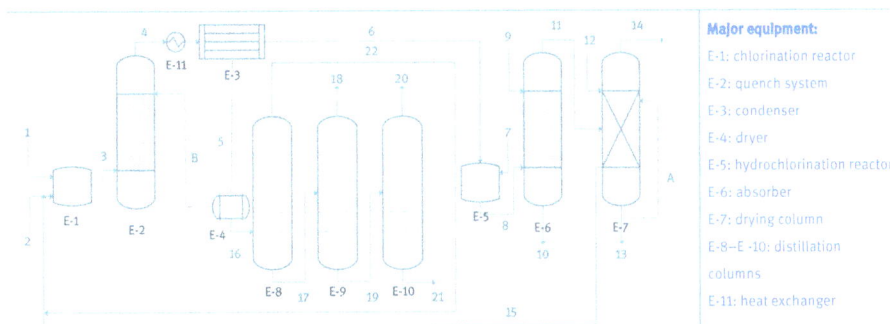

Figure 7.5: Schema of production of chloromethanes based on chlorination of chloromethane in mixed system.

The chloromethane plants are usually focused on the production of mainly three products: chloromethane, hydrochloric acid and carbon tetrachloride. For example phoenix equipment corporation offers the plant with the capacity 26 000 ton per years (TPY) of CH_3Cl, 26 000 TPY of 30 % HCl, and 3 000 TPY CCl_4 [30].

7.2.2 Chloroalkanes with longer alkyl chain

In the industrial practice the chloroalkanes with longer alkyl chain are also produced. The procedure of their production depends on the carbon numbers in the alkyl chain. In the case of hydrochlorination of alcohols main problem is insolubility of longer-chain alcohols in reaction environments – water. In laboratory scale the substitution reaction of chloride to alcohol molecule is usually performed using chlorination agents such as $SOCl_2$, PCl_3, PCl_5, or $COCl_2$ [31]. Unfortunately, application of these reagents in industrial scale is inappropriate from economical, safety and environmental aspects. For longer-chain alkyl chlorides, the direct hydrochlorination process often requires the presence of a Lewis acid or phase-transfer catalyst, long reaction times and elevated temperature regimes [32].

The most popular chloroalkanes from this group are chloroethane and 1- and 2-chloropropane. Chloroethane is commonly used in producing tetraethyllead, a gasoline additive, and as ethylating agent. Similar to the chloromethane production the chloroethane could be obtained in the reaction of ethane (in Figure 7.2, stream no. 1, instead of CH_4) and chlorine, as well as in the reaction of ethanol (in Figure 7.3, stream no. 1, instead of methanol) and hydrogen chloride.

However, nowadays the post popular method of chloroethane production is hydrochlorination of ethylene (in Figure 7.4, stream no. 1, instead of methanol), according to the reaction:

$$C_2H_4 + HCl \rightarrow C_2H_5Cl \tag{7.12}$$

1-chloropropane, as well as 2-chloropropane is used usually as a solvent in organic synthesis. These chloropropanes are produced in the reaction of appropriate propanol (propan-1-ol or propan-2-ol, respectively) with phosphorus trichloride in the presence of a zinc chloride catalyst. Moreover, 2-chloropropane is obtained in the reaction of HCl and propylene, similar to reaction (7.12) for chloroethane.

7.2.2.1 Alkane (C_2-C_4) chlorination with atomic chlorine

7.2.2.1.1 Cases study
Kinetics of reaction of atomic chlorine with C1-C4 alkanes (CH_4, ethane, propane, n-butane and isobutene) was measured in the temperature of 25 °C [33]. The presence

of molecular chlorine in the gaseous mixture is unavoidable but it leads to restoration
of atomic chlorine according to quite fast reaction with alkanes ($R = C_2-C_4$):

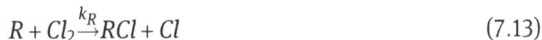

$$R + Cl_2 \xrightarrow{k_R} RCl + Cl \tag{7.13}$$

The kinetics for the reaction is first order with respect to atomic chlorine
concentration:

$$r = k_R[Cl] \tag{7.14}$$

For the above reaction, the kinetic rate constant is as follows [34]:

$$k_R = (1.2 - 3.6) \cdot 10^{10} \frac{dm^3}{mol \cdot s} \tag{7.15}$$

These difficulties could be decreased using concentrations of molecular chlorine
below 10^{10} molecules/cm^3 ($1.6 \cdot 10^{-11}$ mol/dm^3). The kinetics rate constants for chlorination of alkanes with atomic chlorine according to reaction:

$$RH + Cl \xrightarrow{k} RCl + HCl \quad (R = C_1...C_4) \tag{7.16}$$

Table 7.4 presents the values of kinetics rate constant corresponding to first four
homologies of alkanes chlorinated by atomic chlorine.

Table 7.4: Kinetics rate constants for the alkane chlorination with atomic chlorine.

Alkane	CH_4	C_2H_6	C_3H_8	$n\text{-}C_4H_{10}$	$i\text{-}C_4H_{10}$	CH_3Cl	CH_2Cl_2	$CHCl_3$
k, $dm^3/$ $mol \cdot s$	$5.66 \cdot 10^7$	$3.33 \cdot 10^{10}$	$7.41 \cdot 10^{10}$	$1.27 \cdot 10^{11}$	$8.43 \cdot 10^{10}$	$2.65 \cdot 10^8$	$1.93 \cdot 10^8$	$6.63 \cdot 10^7$

7.3 Vinyl halides

From the perspective of applications, the most important compound from the vinyl
halides class is chloroethene, known as vinyl chloride. It is widely used in industrial
as a precursor to PVC. Chloroethene belongs to the top twenty largest petrochemicals
(petroleum-derived chemicals) in world production [35]. Because chloroethene is most
often used, it will be discussed in detail in the next section. Another examples of commercial product is 1,1-difluoroethene, known as vinylidene fluoride. It is used, as precursor, in the production of fluoropolymers such as polyvinylidene fluoride (PVDF).

7.3.1 Chloroethene

Historically the production of chloroethene based on the reaction of hydrogen chloride and acetylene. Actually, the most often used method in industrial practice is

based on ethylene conversion to ethylene dichloride, which is thermally cracked to produce chloroethene and by product: hydrogen chloride.

7.3.1.1 Chlorination of ethylene

The most popular method for production of chloroethene is based on three steps: 1. Direct chlorination ethylene to 1,2-ethylenedichloride (see reaction 1 in eq. (7.19); 2. Recovery of HCl from step 3 and oxychlorination of ethylene to 1,2-ethylenedichloride (see reaction 2 in eq. (7.19); 3. Thermal cracking (pyrolysis) of 1,2-ethylenedichloride, from step 1 and 2, to chloroethene. Overall equations for this process may be presented as follows [36]:

$$C_2H_4 + 0.5\ Cl_2 + 0.25\ O_2 \rightarrow C_2H_3Cl + 0.5\ H_2O \tag{7.17}$$

The schema of the process contains these three steps, as well as the procedures of separation is presented in Figure 7.6. According to reactions (7.17) and (7.19), the stream of ethylene (stream no. 1) is divided into half. One part of ethylene with chlorine (stream no. 2) is introduced to the chlorine chlorination reactor E-1, while the other half of ethylene is transported to the oxychlorination reactor E-3. Reactor E-3 is fed also by oxygen (stream no. 3), and HCl (stream no. 4). Some of the hydrogen chloride is recycled from the cracking process (steam B). Moreover, to the reactor E-3 the unreacted substrates are recycled in stream A.

In the chlorination reaction usually the Lewis-acid type catalyst is used, mainly $FeCl_3$ (concentration: 0.1 to 0.5 wt%). In addition to the main reaction – 1,2-ethylenedichloride production, the others reactions take place in the reactor

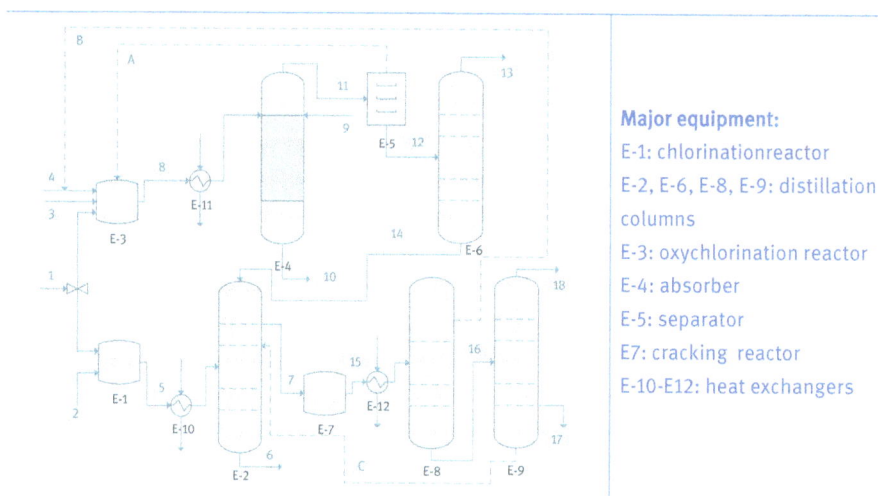

Major equipment:
E-1: chlorinationreactor
E-2, E-6, E-8, E-9: distillation
columns
E-3: oxychlorination reactor
E-4: absorber
E-5: separator
E7: cracking reactor
E-10-E12: heat exchangers

Figure 7.6: Schema of production of chloroethene based on chlorination and oxychlorination of ethylene.

E1. The most significant by product is the 1,1,2-trichlorethane, which is obtained according to the chlorination reactions of ethylene and 1,2-ethylenedichloride:

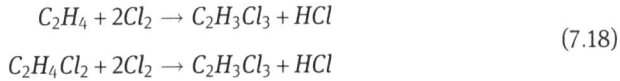

$$C_2H_4 + 2Cl_2 \rightarrow C_2H_3Cl_3 + HCl$$
$$C_2H_4Cl_2 + 2Cl_2 \rightarrow C_2H_3Cl_3 + HCl$$

(7.18)

In the industrial practice chlorination reaction may by conducted in two ways, as a low-temperature chlorination or high-temperature chlorination. The operation conditions in the first solution (temperature between 50 and 70 °C, and thus below the mixture boiling point) allow to achieve higher selectivity, over 99 %. Whereas in high-temperature chlorination process the reactor works at the boiling point of 1,2-ethylenedichloride (at 1.5 to 5 bar and 90 to 150 °C). In this solution reactor may be integrated as a reboiler of a distillation column and the purification of products may be achieve in one process. On the contrary, the high-temperature chlorination process indicates lower selectivity in comparison to the low-temperature chlorination process. Depends of the solution selected the reaction mixture (stream no. 5) is cooled in heat exchanger (E-10) and then separated in distillation column E-2, at low-temperature chlorination process, as it is presented in Figure 7.6. For second solution the reaction mixture is separated directly in the reactive distillation column, not presented in the schema. In stream no. 6 the by products, such as 1,1,2-trichlorethane, are removed from the process, while 1,2-ethylenedichloride (as a sum of product form chlorination, oxychlorination, and cracking), in stream no. 7 is entered to the cracking reactor E-7.

Oxychlorination of ethylene to 1,2-ethylenedichloride occurs in reactor E-2 at temperatures around 200 °C and pressures of 1.5–5 bar. This reaction, similar to the chlorination reaction, is a catalytic one. In industrial mainly $CuCl_2$ impregnated on alumina is applied as catalyst. The conversion of ethylene is on the level 93–97 %, with selectivity for 1,2-ethylenedichloride on the level 91–96 %. As mentioned above, the selectivity of chlorination reaction is much higher – 99 %; thus, in this process amount of impurities (by products) is much higher. The main by products are as follows: 1,1,2-trichloroethane, trichloroacetaldehyde, trichloroethylene, 1,1-and 1,2-dichloroethylenes, ethyl chloride, chloromethanes (chloromethane, dichlormethane and trichloromethane), as well as polychlorinated high-boiling components. This is the reason that reaction mixture (stream no. 8) should be purified in few operations steps. After cooling in heat exchangers E-11, the products of oxychlorination are washing by water or alkaline solution, mainly NaOH (stream no. 9) in absorption column E-10. It is important to remove especially trichloroacetaldehyde in this process, because of its high tendency to polymerization. The residue reaction mixture (stream no. 11) is again cooled in separator E-5, and unreacted gases in stream A are recycled to the oxychlorination reactor E-3. The water presented in liquid phase (stream no. 12) is separated in distillation column E-6 in stream no. 13. The mixture containing main product of oxychlorination of ethylene – 1,2-ethylenedichloride, as well as by products, such as

1,1,2-trichlorethane (stream no. 14) is separated in distillation column E-2 together with reaction mixture from chlorination process (stream no. 5) and recycled stream from cracking process (stream C).

Purified 1,2-ethylenedichloride (stream no. 7) is decomposed into chloroethene and HCl cracking reactor, according to reaction (3) in eq. (7.19). The reaction occurs at high temperature from 480 to 550 °C and pressures from 3 to 30 bar. In the cracking a lot of by products are formed, such as chloromethanes, butadiene, 2-chlorobuta-1,3-diene, trichloroethylene. Moreover, level of 1,2-ethylenedichloride conversion in this process is in the range 50–60 %. Thus, the separation operations should be applied after reaction. After cooling, reaction mixture in stream no. 15 is introduced to the distillation column E-8, where formed in cracking reaction HCl is recycled (stream B) to the oxychlorination reactor E-3. In next step (stream no. 16), in distillation column E-9 unreacted 2-ethylenedichloride is separated and recycled in stream C to the distillation column E-2. The low-boiling compounds such as chloromethane are also separated (stream no. 17) in the same column, and could be used in other processes. The main product chloroethene is obtained in stream no. 18.

7.3.1.1.1 Cases study

The major operating steps used in the ethylene based balanced process for the synthesis of chloroethene (vinyl chloride) are presented as follows [37]:

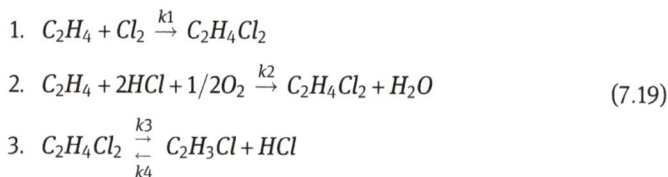

$$1. \quad C_2H_4 + Cl_2 \xrightarrow{k1} C_2H_4Cl_2$$

$$2. \quad C_2H_4 + 2HCl + 1/2O_2 \xrightarrow{k2} C_2H_4Cl_2 + H_2O \tag{7.19}$$

$$3. \quad C_2H_4Cl_2 \underset{k4}{\overset{k3}{\rightleftarrows}} C_2H_3Cl + HCl$$

Additionally the several by products are produced which are not covered here. Direct chlorination of ethylene to 1,2-ethylenedichloride (reaction 1, eq. (7.19)) runs through molecular and radical addition and substitution reactions. Its kinetic formulation reads:

$$r_1 = k_1[C_2H_4][Cl_2]$$

$$k_1 = 0.132 \quad \left[\frac{m^3}{mol \cdot s}\right] \tag{7.20}$$

The kinetics for oxychlorination of ethylene to 1,2-ethylenedichloride (reaction 2, eq. (7.19)) also runs through molecular and radical addition and substitution reactions. The kinetics of this reaction reads:

$$r_2 = k_2 p_{C_2H_4} p_{Cl_2}^{0.5} \quad \left[\frac{mol}{dm_{cat}^3 \cdot h}\right] \tag{7.21}$$

The rate constant is given by Arrhenius equation:

$$k_2 = k_{2\infty} \exp\left(\frac{-\Delta E}{RT}\right) \tag{7.22}$$

where

$$k_{2\infty} = 10^{4.2} \quad \left[\frac{mol}{dm_{cat}^3 \cdot h \cdot kPa^{1.5}}\right]$$

$$\Delta E = 40.1 \quad \left[\frac{kJ}{mol}\right] \tag{7.23}$$

The above kinetics is defined by the use of partial pressures of components.

The kinetics for the pyrolysis of 1,2-ethylenedichloride to chloroethene (reaction 3, eq. (7.19)) is described as equilibrium reaction. The kinetic equation reads:

$$r_3 = k_3[C_2H_4Cl_2]$$

$$r_4 = k_4[C_2H_3Cl][HCl] \tag{7.24}$$

The rate constants for the above kinetic description of equilibrium reaction read:

$$k_3 = k_{3\infty} \exp\left(\frac{-\Delta E_3}{RT}\right) \quad [s^{-1}]$$

$$k_4 = k_{4\infty} \exp\left(\frac{-\Delta E_4}{RT}\right) \quad \left[\frac{m^3}{kmol \cdot s}\right] \tag{7.25}$$

The Arrhenius data values are as follows:

$$k_{3\infty} = 10^{13.6} \quad [s^{-1}]$$

$$\Delta E_3 = 234461 \quad \left[\frac{J}{mol}\right]$$

$$k_{4\infty} = 0.3 \cdot 10^9 \quad \left[\frac{m^3}{kmol \cdot s}\right] \tag{7.26}$$

$$\Delta E_4 = 184219 \quad \left[\frac{J}{mol}\right]$$

The above three reaction describe the key steps in the process of synthesis of chloroethene.

Moreover, Orejas [38] presented more detailed equations for the first step of production of chloroethene, and thus for chlorination of ethylene. Author considered two possible reactions in this process:

$$C_2H_4 + Cl_2 \rightarrow C_2H_4Cl_2$$
$$C_2H_4 + 2Cl_2 \rightarrow C_2H_3Cl_3 + HCl \tag{7.27}$$

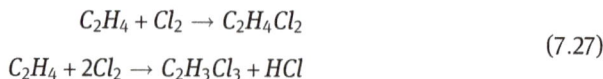

The kinetic equation reads:

$$r_1 = k_1[C_2H_4][Cl_2]$$
$$r_2 = k_2[C_2H_4][Cl_2]^2 \tag{7.28}$$

The rate constants for the above kinetic description of equilibrium reaction read:

$$k_1 = 11493.0\exp\left(\frac{-2156.58}{T}\right)\left[\frac{m^3}{kmol \cdot s}\right]$$
$$k_2 = 8.157 \cdot 10^9\exp\left(\frac{-7282.21}{T}\right)\left[\frac{m^6}{kmol^2 \cdot s}\right] \tag{7.29}$$

For low-temperature chlorination process, in temperature 60 °C, rate constants are equal to 17.7 and 2.71, in the above units and are in good agreement with the experimental data present in work [39] – 13.2 and 2.39, respectively.

7.3.1.2 Reaction of chloromethanes

Although the method of obtaining chloroethene from ethylene is the most popular, alternative solutions are still being search. Main reason for this is relative high cost of ethylene, which is 45 % weight of chloroethene. In 2007 the novelty method of producing from chloromethane and dichloromethane were development and described in United State patent [40]. The schema of the process proposed is presented in Figure 7.7.

The substrates – chloromethane gas in stream no. 1 and dichloromethane vapor in stream no. 2 are introduced to the reactor E-1 and reacted in vapor phase in the presence of catalyst, according to reaction (30). As catalyst Al_2O_3 gel, γ- Al_2O_3, $ZnCl_2$

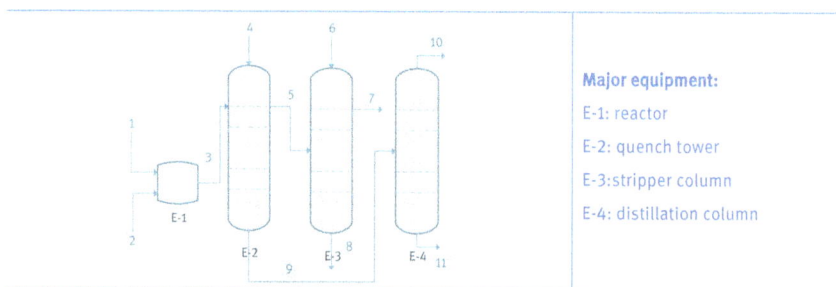

Major equipment:
E-1: reactor
E-2: quench tower
E-3:stripper column
E-4: distillation column

Figure 7.7: Schema of production of chloroethene based on reaction chloromethane and dichloromethane.

on active alumina, silica-alumina, zeolite, and silicon-aluminum-phosphorus oxide are proposed. The operation conditions are as follow: temperature in the range from 300 °C to 500 °C, pressure from about 1 to 10 atmospheres:

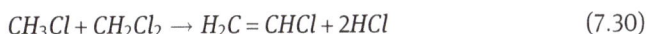

$$CH_3Cl + CH_2Cl_2 \rightarrow H_2C = CHCl + 2HCl \tag{7.30}$$

Presence of catalyst in this process is important from the two reasons. The first one is to ensure high selectivity of the main reaction. For example, dichlorome-thane could react with itself to form 1,2-dichloroethylene. The second reason re-sults from the reaction mechanism (see (7.31)). In this reaction small amount of water is necessary, and appropriate catalyst contains an effective amount of water for reacting step:

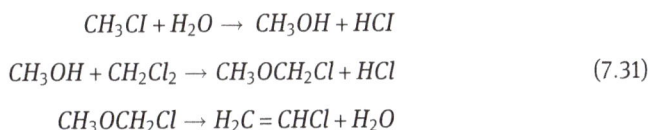

$$CH_3Cl + H_2O \rightarrow CH_3OH + HCl$$

$$CH_3OH + CH_2Cl_2 \rightarrow CH_3OCH_2Cl + HCl \tag{7.31}$$

$$CH_3OCH_2Cl \rightarrow H_2C = CHCl + H_2O$$

The mixture of gases from the reactor E-1 are cooled in quench tower E-2 by di-chloromethane (stream no. 4), which absorbs the chloroethene and heavy ends (by-products, for example 1,2-dichloroethylene). The non-condensed gases from quench tower are transported in stream no. 5 to stripper column E-8, which is spread by water in stream no. 6. The adsorbed hydrogen chloride is removed in stream no. 8, while gases in stream no. 7. Whereas the residue liquid from quench tower is pumped to distillation column E-4. In distillation process the separation of chloroethene (stream no. 10) and heavy ends (stream no. 11) is possible.

Similar solution – the reaction of appropriate chloroalkanes was also proposed for production of vinyl chlorides with longer alkyl chain. For example in US patent [41] process for the production of chlorinated propenes based on 1,2-dichloropropane (by product in the production of chlorohydrin) or with mixture with 1,2,3-trichloro-propane. In the chlorination and/or dehydrochlorination processes the by product – anhydrous HCl is also obtained. As a catalysts the following materials may be used $AlCl_3$, I_2, $FeCl_3$, sulphur, iron, antimony pentachloride, boron trichloride, lanthanum halides, metal triflates, while as chlorination agents – Cl_2 or SO_2Cl_2. The conversion of 1,2-dichloropropane is greater than 50 %.

7.4 Aryl halides

Aryl halides are also known as haloarene or halogenoarene. They have a halogen (one or more) directly bonded to a carbon of an aromatic ring. The most popular ex-ample of this group compound is chlorobenzene (C_6H_5Cl). However, there also a lot examples of natural aryl halides with more complicated structure, such as Vancomycin (used in medicine as antibiotic), Thyroxine, also called 3,5,3',5'-

tetraiodothyronine (natural hormone). Aryl halides are prepared by two major methods: halogenation of the aromatic ring and reactions involving diazonium salts. In industrial practice also oxidative halogenation is proposed.

7.4.1 Chlorinated benzenes

The most often utilize aryl halide is chlorobenzene. It is used as a solvent in organic reaction, as substrate for nitrobenzene production, as well in the production of commodities such as herbicides, dyestuffs, and rubber.

It should be noted that more chlorinated derivatives may formed during the benzene chlorination process. For example in industrial practice the common technology is production of 1,4-dichlorobenzene or 1,2-dichlorbenzene. The first one is used in space deodorant for toilets and refuse containers, and as a fumigant for control of moths, molds, and mildews, as well as in the production of polyphenylene sulfide (PPS) resin. 1,2-dichlorbenzene is proposed as a precursor to 3,4-dichloroaniline herbicides and in production of dyes. The production of chlorinated benzenes is rather no complicated. The amount of the each chlorinated benzenes depends mainly of the using catalyst [42]. For example proportion of 1,4-dichlorbenzene in the dichlorobenzene fraction is equal to 88 % using L-type zeolite, 77 % for $FeCl_3$ catalyst, 50 % for $MnCl_2$ catalyst. The schema of production 1,4-dichlorbenzene provided by Nippon Light Metal Co., Ltd with Tsukishima Kikai's CCCC process [43] is presented in Figure 7.8.

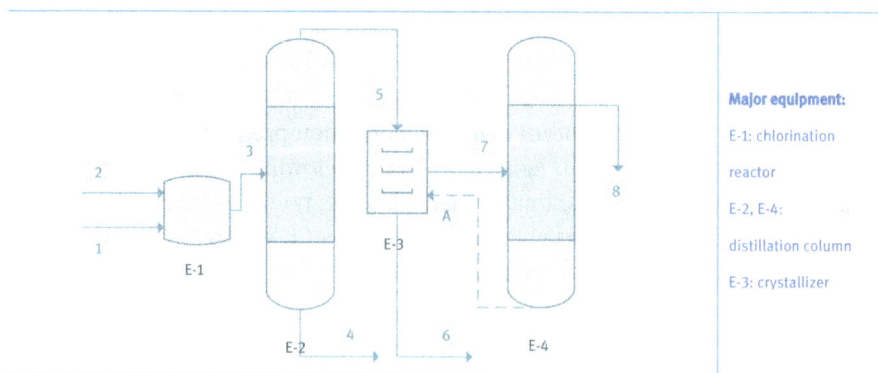

Figure 7.8: Schema of production of 1,4-dichlorbenzene.

The feedstock solution contains benzene (stream no. 1) and chlorine (stream no. 2). The catalyst reaction occurs in chlorination reactor E-1. The reaction mixture (stream no. 3) is introduced to the distillation column E-2, where 1,4-dichlorbenzene, and high-bowling impurities are removed in stream no. 6. In the next step (stream no. 5) the crystallization process, in crystallizer E-3, help to separated unreacted benzene and co-product 1,2-dichlorbenzene and chlorobenzene (stream no. 6). Such separation is possible due to the relative high melting point for 1,4-dichlorbenzene (53.5 °C), while other isomeric, and chlorobenzene melt bellow room temperature. Optionally, the final purification to obtain 99.99 % 1,4-dichlorbenzene (stream no. 8) occurs in distillation column E-4.

In patent application [44] the modification of the process is proposed in context of reaction conditions. The chlorination reaction of benzene and/or chlorobenzene with chlorine, in the presence of a Lewis acid catalyst and a phenothiazine analogue compound was recommended.

7.4.2 Oxidative chlorination of benzene

7.4.2.1 Cases study
Low temperature halogenation permits formation of chloro- and bromotoluenes, mono-chloro-p-xylene, 1-bromonaphthalene, and other species in reasonably high yields without the requirements of toxic molecular halogens use [45]. Moreover, the reaction of benzene chlorination in which there are no undesired side products like dichlorobenzenes or trichlorobenzenes is favorable in many cases. The main reaction reads:

$$C_6H_6 + Cl_2 \xrightarrow{k} C_6H_5Cl + HCl \tag{7.32}$$

The chlorination is conducted in the oxidative system, which is achieved in the two reactions:

$$Na_2O_2 + 2HCl \rightarrow 2NaCl + H_2O_2 \tag{7.33}$$

together with:

$$H_2O_2 + 2HCl \rightarrow Cl_2 + 2H_2O \tag{7.34}$$

The reaction kinetics is described by the rate equation that reads:

$$r = k[C_6H_6][Cl_2] \tag{7.35}$$

The rate of the main reaction (7.32) is highly sensitive to the ratio of the initial H_2O and Na_2O_2 concentrations q:

$$q = \frac{[H_2O]}{[Na_2O_2]} \tag{7.36}$$

Table 7.5 presents the values of activation energies and frequency factors for different q values and different temperatures:

Table 7.5: Kinetic data for the oxidative chlorination of benzene to monochlorobenzene.

q	T, K	$k \cdot 10^4$, kg/mol·s	ΔE, kJ/mol
12.8	298	1.06	38
12.8	303	1.52	38
12.8	313	2.33	38
19.3	298	3.81	12.1
19.3	303	4.29	12.1
19.3	313	4.9	12.1
31.3	298	1.18	70.2
31.3	303	2.95	70.2
31.3	313	5.29	70.2

7.5 Acyl halides

Acyl halides are the group of compounds derived from an oxoacid by replacing a hydroxyl group with a halide group. The most popular group is acyl chloride with compounds such as acetyl chloride or benzoyl chloride. However, these compounds are not as widely used as another oxygen-containing halogen compound – epichlorohydrin. Therefore, only this compound will be discussed, in detail, in the next section.

7.5.1 Epichlorohydrin

Epichlorohydrin is widely used in the production of glycerol, plastics, epoxy glues and resins, and elastomers. Traditionally, epichlorohydrin is produced from propylene which would undergo an allylic chlorination and hydrochlorination to obtain a 1,3-dichloropropanol and 2,3-dichloropropanol mixture (see reaction A in eq. (7.37)) [46]. The main drawback of this method is very low yield of chlorination reaction. Only 25 % chlorine atoms react with propylene; thus, 75 % of the chlorines may be treated as waste [47]. Thus, as alternative, the glycerol (main by product in biodiesel production) is proposed as a substrate in reaction [48, 49], as is presented in reaction B in eq. (7.37). The glycerol based-method allows to eliminate chlorine and reduce chlorine including by products. On the contrary, the reaction indicates low conversion of glycerol to epichlorohydrin, as well as low selectivity for epichlorohydrin [50].

Obtained 1,2-dichloropropanol, independently of the synthesis method, reacts with sodium hydroxide to obtain epichlorohydrin (the reaction, similar to eq. (7.38)):

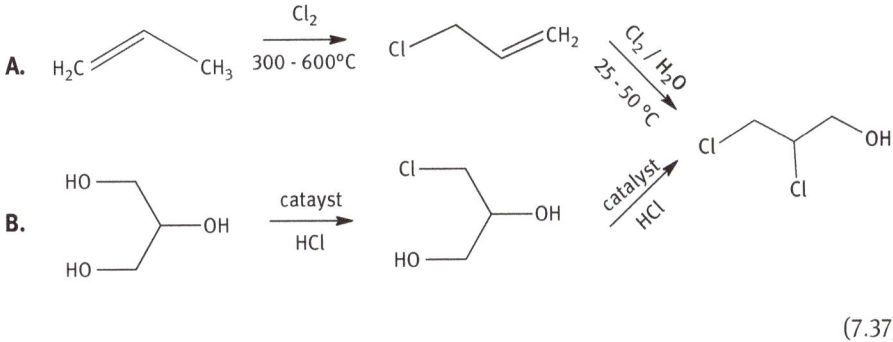

$$(7.37)$$

The schema of production of epichlorohydrin is presented in Figure 7.9. The composition of feed solutions (stream no. 1) for allyl chloride synthesis in reactor E-1 depends of the applied method. According to eq. (7.37), it may be propylene and chloride (reaction A) or glycerol and HCl (reaction B). The stream no. 2 containing reaction mixture is separated in purification column (E-2), to remove unreacted substrates, as well as by products such as other chloride derivatives (stream no. 3). In the next step dehydrochlorination reaction occurs in reactor E-3, in which the dichloropropanol (1,3 or 1,2) are converted to epichlorohydrin in presence of an alkali, with the formation of water and of the corresponding chloride salt, according to reaction (38). Usually as the alkali sodium hydroxide is used (stream no. 5). The rate of conversion is high, faster for the 1,3 isomer in comparison to the 1,2 isomer, and increases with the temperature or the alkalinity. Moreover, some other product may be formed in this reaction such as monochlorohydrin, glycidol or glycerol. Reaction mixture (stream no. 6) is separated in distillation column E-4. The heavy by products as glycidol, glycerol and poly-glycerol, as well as formed in the reaction inorganic salt, are removed in stream no. 7. The product epichlorohydrin is received in stream no. 8.

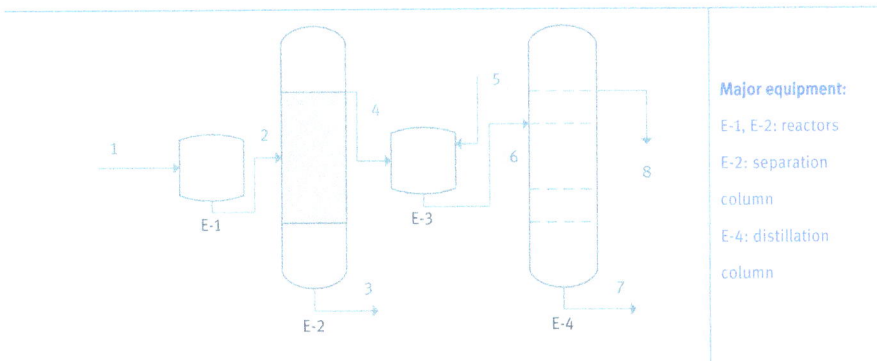

Figure 7.9: Schema of production of epichlorohydrin.

7.5.1.1 Cases study

The main route for the synthesis of epichlorohydrin (EPCH) is the chlorohydrination of allyl chloride obtained from high-temperature chlorination of propylene or hydrochlorination of glycerol. Alas, the method creates a considerable quantity of chlorinated by products and expends lots of energy because of high process temperature. Epichlorohydrin can be made by dehydrochlorination of 1,3-dichloropropanol (DCP) by sodium hydroxide [51]:

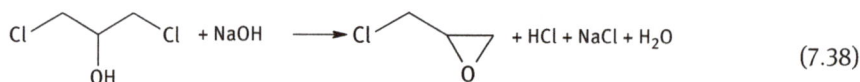

$$Cl\diagup\diagdown\underset{OH}{\diagup}\diagdown Cl \ + NaOH \longrightarrow Cl\diagup\diagdown\underset{O}{\diagup}\diagdown \ + HCl + NaCl + H_2O \tag{7.38}$$

The side hydrolysis reaction of epichlorohydrin to glycerol may also occur at temperatures above 70 °C:

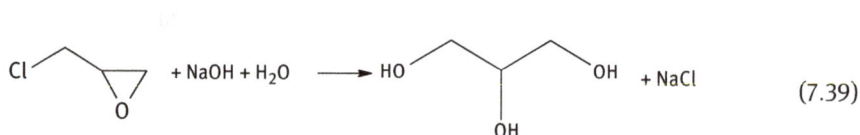

$$Cl\diagup\diagdown\underset{O}{\diagdown}\diagup \ + NaOH + H_2O \longrightarrow HO\diagup\diagdown\underset{OH}{\diagup}\diagdown OH \ + NaCl \tag{7.39}$$

The reaction kinetics is first order and is given by

$$r = k[DCP] \tag{7.40}$$

in which rate constant k obeys Arrhenius rate law:

$$k = k_\infty \exp\left(\frac{-\Delta E}{RT}\right) \tag{7.41}$$

where

$$k_\infty = 1.62 \cdot 10^7 \quad [s^{-1}]$$
$$\Delta E = 38.8 \quad \left[\frac{kJ}{mol}\right] \tag{7.42}$$

The effect of molar ratio of DCP/NaOH was studied and the best molar ratio regarding conversion was found to be 1/5 and the stoichiometric ratio for reaction yield.

The undesired hydrolysis reaction of epichlorochydrine (EPCH) may also be taken into account. The kinetics of this reaction is reported in the work of Ma et al. [52] and is also given by typical Arrhenius temperature dependence for second-order equation:

$$r = k[EPCH][NaOH] \tag{7.43}$$

The rate constant is defined by the same Arrhenius type equation and the frequency factor and activation energy are given by

$$k_\infty = 5.66 \cdot 10^{10} \quad \left[\frac{dm^3}{mol \cdot min}\right]$$

$$\Delta E = 70.790 \quad \left[\frac{kJ}{mol}\right]$$

(7.44)

Increasing temperature is good for improving dehydrochlorination reaction rate, but it will also enhance the hydrolysis reaction rate. For a low contact time, the problem is minimized, because for a low contact time, the probability of hydrolysis reaction is significantly reduced, which means that the yields of EPCH could hardly be affected by the hydrolysis reaction [53].

Acknowledgements: This work was supported by the Polish Ministry of Science and Higher Education [grant number 0912/SBAD/2010].

References

[1] Gribble GW. Naturally occurring organohalogen compounds – A comprehensive update. Vienna, Austria: Springer, 2010978-3-211-99322-4.

[2] Field JA. Natural production of organohalide compounds in the environment. In: Adrian L, Löffler F, editors. Organohalide-respiring bacteria. Berlin, Heidelberg: Springer, 2016.

[3] Bradley PM. History and ecology of chloroethene biodegradation: a review. Bioremediat J. 2003;7:81–109.

[4] Cooksey CJ. Tyrian purple: 6,6'-dibromoindigo and related compound. Molecules. 2001;6:736–69.

[5] Manna D, Mondal S, Mugesh G. Halogen bonding controls the regioselectivity of the deiodination of thyroid hormones and their sulfate analogues. Chemistry. 2015;21:2409–16.

[6] Valdivia-Garcia M, Weir P, Frogbrook Z, Graham DW, Climatic WD. Geographic and operational determinants of trihalomethanes (THMs) in drinking water systems. Sci Rep. 2016;6:35027.

[7] Zhao M, Yang X, Tsui GC, Miao Q. Trifluoromethylation of anthraquinones for n-type organic semiconductors in field effect transistors. J Org Chem. 2019. DOI: 10.1021/acs.joc.9b01263 (in press).

[8] Malik JK, Aggarwal M, Kalpana S, Gupta RC. Chapter 36 – Chlorinated hydrocarbons and pyrethrins/pyrethroids. In: Gupta RC, editor. Reproductive and developmental toxicology, 2nd ed. San Diego, United States: Academic Press, 2017:633–55.

[9] Rosner D, Markowitz G. Persistent pollutants: a brief history of the discovery of the widespread toxicity of chlorinated hydrocarbons. Environ Res. 2013;120:126–33.

[10] Davis R, Durrant JL, Rowland CC. Free radical halogenation of alkanes initiated by transition metal complexes. J Organomet Chem. 1986;316:147–62.

[11] Lenoir D, Chiappe C. What is the nature of the first-formed intermediates in the electrophilic halogenation of alkenes, alkynes, and allenes? Chem Eur J. 2013;9:1036–44.

[12] Chiappe C, Capraro D, Conte V, Pieraccini D. Stereoselective halogenations of alkenes and alkynes in ionic liquids. Org Lett. 2001;37:1061–3.

[13] Classon B, Liu Z, Samuelsson B. New halogenation reagent systems useful for the mild one-step conversion of alcohols into iodides or bromides. J Org Chem. 1988;53:6126–30.

[14] Hanson P, Jones JR, Gilbert BC, Timms AW. Sandmeyer reactions. Part 1. A comparative study of the transfer of halide and water ligands from complexes of Cu^{II} to aryl radicals. J Chem Soc Perkin Trans. 1991;2:1009–17.

[15] Greenberg JA, Sammakia T. The conversion of tert-butyl esters to acid chlorides using thionyl chloride. J Org Chem. 2017;82:3245–51.

[16] Global chloromethane market – Industry analysis and forecast (2018-2026) – by type, application, and region. Maximize Market Research Pvt. Ltd Report ID 27558. Published Date Feb 2019.

[17] Korpyś M, Wójcik J, Synowiec P. Methods for sweetening natural and shale gas. Chemik Sci-Technique-Market. 2014;68:213–15.

[18] Rossberg M, Lendle W, Pfleiderer G, Tögel A, Torkelson TR, Beutel KK Chloromethanes – chapter 9. In: Elvers Barbara, et al., editors. Ullman's encyclopedia of industrial chemistry. Weinheim: Wiley-VCH Verlag GmbH & Co. KGaA, 2012:16–42.

[19] Podkolzin SG, Stangland EE, Jones ME, Peringer E, Lercher JA. Methyl chloride production from methane over lanthanum-based catalysts. J Am Chem Soc. 2007;129:2569–76.

[20] Paunović V, Pérez-Ramírez J. Catalytic halogenation of methane: a dream reaction with practical scope? Catal Sci Technol. 2019;9:4515–30.

[21] Myers JD, Grandbois ML. Process for the production of chlorinated methanes. United States Patent Application Publication, Pub. No.: US 2018/0258016 A1. Pub. Date: Sep. 13, 2018.

[22] Rozanov VN, Treger YA. Kinetics of the gas-phase thermal chlorination of methane. Kinet Catal. 2010;51:635–43.

[23] Ouellette RJ, Rawn D. Alkanes and cycloalkanes, in Principles of organic chemistry. Amsterdam, Netherlands: Elsevier Inc, 2015.

[24] Ouellette RJ, Rawn D. Alcohols: reactions and synthesis, in organic chemistry, 2nd ed. London, United Kingdom: Elsevier Inc, 2018.

[25] Ji Y, Zhang F, Yu F, Zhang J, Zhang J. Methyl chloride synthesis over metal chlorides-modified mesoporous alumina catalyst. Catalysts. 2018;8:99.

[26] McInroy AR, Lundie DT, Winfield JM, Dudman CC, Jones P, Lennon D. Improved atom efficiency via an appreciation of the surface activity of alumina catalysts: methyl chloride synthesis. Appl Catal B. 2007;70:606–10.

[27] Methyl chloride handbook. OxyChem, USA. Technical Information 11/2014. Web site: www.OxyChem.com.

[28] Crow RD, Roberts NP. Process for manufacturing methyl chloride. United States Patent, Patent Number: 6,111,153, Date of Patent: Aug. 29, 2000.

[29] Makhin MN, Zanaveskin LN, Dmitriev GS. Kinetics of the liquid-phase hydrochlorination of methanol. Kinet Catal. 2014;55:163–6.

[30] Phoenix Equipment Corporation, www.phxequip.com.

[31] Reichart B, Tekautz G, Kappe CO. Continuous flow synthesis of n-alkyl chlorides in a high-temperature microreactor environment. Org Process Res Dev. 2012;17:152–7.

[32] Carey FA, Sundberg RJ. Advanced organic chemistry, part B: reactions and synthesis. New York: Springer, 2007:217–23.

[33] Beichert P, Wingen L, Lee J, Vogt R, Ezell MJ, Ragains M, et al. Rate constants for the reactions of chlorine atoms with some simple alkanes at 298 K: measurement of a self-consistent set using both absolute and relative rate methods. J Phys Chem. 1995;99:13156–62.

[34] Timonen RS, Gutman D. Kinetics of the reactions of methyl, ethyl, isopropyl, and tert-butyl radicals with molecular chlorine. Phys Chem. 1986;90:2987–91.

[35] Dreher EL, Beutel KK, Myers JD, Lübbe T, Krieger S, Pottenger LH. Chloroethanes and chloroethylenes. Ullmann's encyclopedia of industrial chemistry. Weinheim, Germany: Wiley-VCH Verlag GmbH & Co. KGaA, 2014.

[36] Dimian AC, Bildea CS. In: Elvers Barbara, et al., editors. Chemical process design: computer-aided case studies. Weinheim, Germany: Wiley-VCH Verlag GmbH & Co. KGaA, 2008.

[37] Lakshmanan A, Rooney WC, Biegler LT. A case study for reactor network synthesis: the vinyl chloride process. Comput Chem Eng. 1999;23:479–95.

[38] Orejas JA. Model evaluation for an industrial process of direct chlorination of ethylene in a bubble-column reactor with external recirculation loop. Chem Eng Sci. 2001;56:513–22.

[39] Wachi S, Morikawa H. Liquid-phase chlorination of ethylene and 1,2-dichloroethane. J Chem Eng Japan. 1986;19:437–43.

[40] Stauffer J. Method for producing vinyl chloride monomer. United States Patent, Patent No.: US 7253328 B2. Date of Patent: Aug. 7, 2007.

[41] Tirtowidjojo MM, Kruper Jr WJ, Fish BB, Laitar DS. Process for the production of chlorinated propenes. United States Patent, Patent No.: US 8907149 B2. Data of patent Dec. 9, 2014.

[42] Rossberg M, Lendle W, Pfleiderer G, Tögel A, Dreher EL, Langer E, et al. Chlorinated hydrocarbons. In: Elvers Barbara, editor. Ullmann's encyclopedia of industrial chemistry. Weinheim, Germany: WileyVCHVerlag, 2002.

[43] https://www.tsk-g.co.jp/en/tech/industry/pdcb.html.

[44] Okamoto T. Process for producing p-dichlorobenzene. European Patent Application, EP 2 351 724 A1. Date of publication: 03. 08.2011.

[45] Rudakova NI, Erykalov YG. Kinetics of low-temperature oxidative chlorination of benzene in aqueous acetic acid. Russ J Gen Chem. 2005;75:748–50.

[46] Pembere AM, Yang M, Luo Z. Small gold clusters catalyzing the conversion of glycerol to epichlorohydrin. Phys Chem Chem Phys. 2002;19:25840–5.

[47] Almena A, Martin M. Technoeconomic analysis of the production of epichlorohydrin from glycerol. Ind Eng Chem Res. 2016;55:3226–38.

[48] Krafft P, Gilbeau P, Gosselin B. Process for producing epichlorohydrin. United States Patent USOO9663427B2. Published data May 30, 2017.

[49] Simola F, Iosco M. Continuous process for producing epichlorohydrin from glycerol. Word Patent WO2014/049625A1. Publication data 03. 04.2014.

[50] Wang S-J, Wong DS, Lim IJ, Chen YT, Huang C-C. Design and control of a novel plant-wide process for epichlorohydrin synthesis by reacting allyl chloride with hydrogen peroxide. Ind Eng Chem Res. 2018;57:20–5.

[51] Herliati R, Robiah Y, Lubena YW. Kinetics of epichlorohydrin syntesis using dichloropropanol and sodium hydroxide. In: The 3rd International Seminar on Chemstry, Bandung Indonesia, 2014.

[52] Ma L, Zhu JW, Yuan XQ, Yue Q. Synthesis of epichlorohydrin from dichloropropanols: kinetic aspects of the process. Chem Eng Res Des. 2007;85:1580–5.

[53] Carrá S, Santacesaria E, Morbidelli M, Cavalli L. Synthesis of propylene oxide from propylene chlorohydrins - I: kinetic aspects of the process. Chem Eng Scie. 1979;34:1123–32.

Daria Wieczorek and Dobrawa Kwaśniewska

8 Novel trends in technology of surfactants

Abstract: Surface active agents are amphipathic molecules that consist of a non-polar hydrophobic portion, usually a straight or branched hydrocarbon or fluoro-carbon chain containing 8–18 carbon atoms (tail), which is attached to a polar or ionic portion (hydrophilic head). The classification of surfactants refers to the behavior of head in solutions. Therefore, the hydrophilic portion can be nonionic, zwitterionic or ionic and accompanied by counter ions in the last case. The production technology of surfactants depends on the final structure of surfactant. The manuscripts describe the methods of obtaining the most important compounds such as soap, linear alkylbenzene sulfonates, methyl ester sulfonate, alcohol sulfates, alcohol ethoxysulfates, sulfosuccinates. Not only the anionic surfactant production method is presented but also cationic and amphoteric synthesis routes are shown. In the nonionic group of surfactants, several classes can be distinguished: alcohol ethoxylates, alkyl phenol ethoxylates, fatty acid ethoxylates, sorbitan ester ethoxylates, monoalkaolamide ethoxylates, fatty amine ethoxylates and ethylene oxide–propylene oxide copolymers. Another important class of nonionics is the multihydroxy products such as glycol esters, glycerol (and polyglycerol) esters, glucosides (and polyglucosides) and sucrose esters. Amine oxides and sulphinyl surfactants represent nonionics with a small head group. They are all produced and used widely throughout the world in a multitude of industries.

Keywords: anionic surfactants, surfactant production technology

Surface active agents are amphipathic molecules which consist of a non-polar hydrophobic portion, usually a straight or branched hydrocarbon or fluorocarbon chain containing 8–18 carbon atoms (tail), which is attached to a polar or ionic portion (hydrophilic head). The classification of surfactants refers to the behavior of head in solutions. Therefore, the hydrophilic portion can be nonionic, zwitterionic or ionic and accompanied by counter ions in the last case. In all groups of surfactants the hydrocarbon chain weak interaction with the water molecules in an aqueous environment is observed. Whereas the polar or ionic head group interacts strongly with water molecules via dipole or ion–dipole interactions, depending of the nature of head [1].

In the 1970s, the world production of surfactants amounted to 18 million tons, over the next 20 years, this value increased to 25 million tons, another 10 years

This article has previously been published in the journal Physical Sciences Reviews. Please cite as: Wieczorek, D., Kwaśniewska, D. Novel trends in technology of surfactants *Physical Sciences Reviews* [Online] 2020, 5. DOI: 10.1515/psr-2019-0036.

https://doi.org/10.1515/9783110656367-008

brought another increase to the value of 40 million tons. Comparing these date with the production volume observed in the 1940s (1.6 million tons), it can be clearly stated that the surfactants market is constantly growing [2]. The increase in the production volume is accompanied by the increase in the value of the global market. In 2017, the value of the global market of surfactants was evaluated at over 43 million USD and it is estimated that by 2025 this market will be growing at a rate of 5.4% per annum. If these estimates worked well in 2025, the value of the surfactants market will exceed 66 million USD. Analyzing the global market of surfactants, it can be stated that it is highly diversified depending on the type, application and region. The type division is obvious and includes anionic, cationic, amphoteric, nonionic and other surfactants. Taking into account the application, the market for surfactants is divided into household detergents, personal care products, cleaning products, food processing industry, petroleum chemicals, agricultural chemistry, textiles, plastics, paints and coatings, adhesives. In the household cleaning sector, currently 46% of surfactants produced in the world are consumed, while in the production of personal care products, it absorbs 14% of the global production [https://www.alliedmarketresearch.com/surfactant-market; https://ihsmar kit.com/products/chemical-surfactants-scup.html]

Currently, many surfactants that belong to different groups and have different applications are obtained. For the synthesis of surfactants obtained in the largest amounts, substrates that can be grouped into three categories:

- minerals
- fossil resources
- biomass-derived materials.

As materials derived from biomass, vegetable oils, tallow and maize are to be understood. Often used is coconut oil which is recovered by extraction from the dried fruit of the coconut palm, by means of extraction yields also palm kernel oil. In turn, palm oil is obtained in the pulp of the outer flesh of tropical palm tree.

Numerous surfactants are obtained from substrates resulting from the processing of crude oil or natural gas (Figure 8.1).

8.1 Anionic surfactants

8.1.1 Soap

Undoubtedly, the oldest anionic surfactants are soaps. Depending on the literature, the origins of soap history are dated differently. The authors argue that soap was already known 5000 years ago, others assume that soap history is at least 2300 years. During the Roman Empire, soaps were obtained from animal fats and plant ashes. However, in the nineteenth century did the saponification

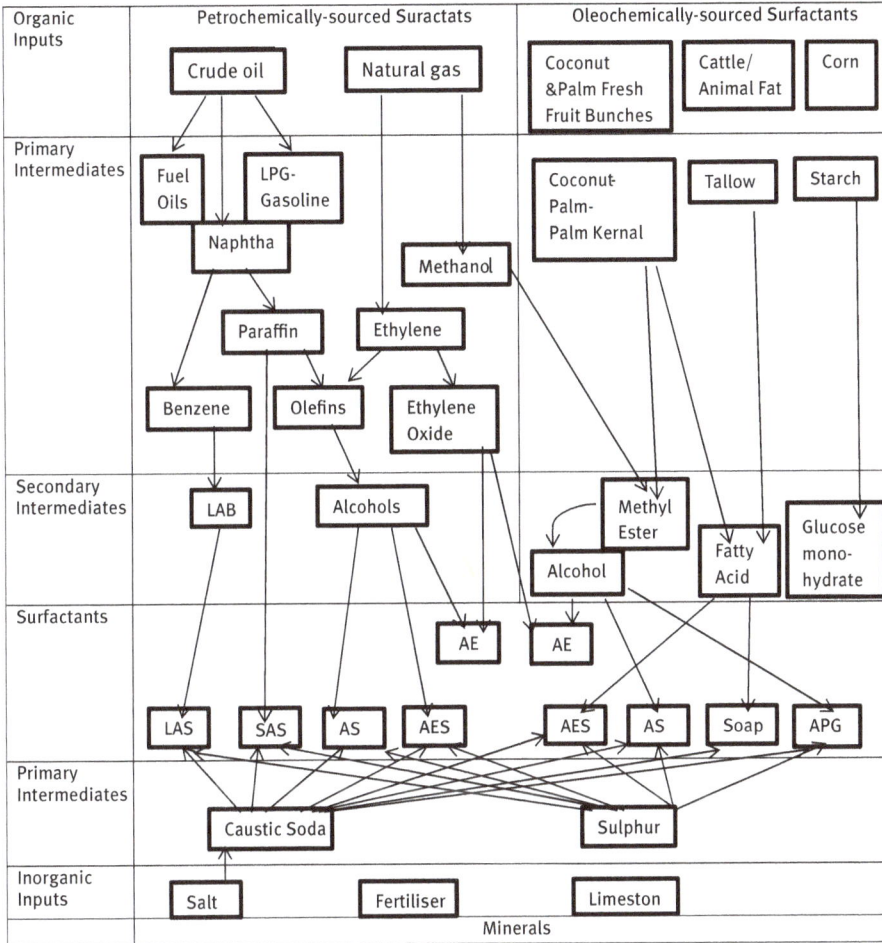

Figure 8.1: Sourcing diagram for surfactants [3].

process be explained in chemical terms. At present, fatty acid salts with alkaline metals are treated as soaps, but also fatty acid salts with ammonia and trietha-nolamine. For the commercial production of soap, fats of various origins, both animal (tallow, lard, marine oil) and vegetable (palm oil, palm kernel oil, coco-nut oil) fats are used. Substrates used in the production of soap have a key im-pact on the quality of the final product, therefore, before the saponification process, the fats are subjected to a series of processes to purify fat from impuri-ties. These impurities can be: waxes, sterols, fat-soluble vitamins, products of ox-idation processes and many others.

Large-scale soap can be obtained in three ways:
- saponification process of neutral oils (triglicerides)

- saponification process of fatty acids obtained from fats and oils
- saponification process of the fatty acid methyl esters derived from fats and oils.

It is estimated that the most important industrial process of obtaining soaps is saponification of neutral oils and fatty acids. Both processes can take place in either batch or continuous mode [4, 5].

8.1.2 Linear alkylbenezene sulfonates (LAS)

It is estimated that inventing alkylbenzene sulfonates began the creation of the detergent industry. Initially, these compounds were obtained in the reaction the Friedel–Crafts alkylation of benzene with propylene tetramer, at the same time, the tetramer was a mixture of 12-carbon olefins. The next step was the sulfonation of alkylbenzene with oleum or sulfur trioxide and neutralization with sodium hydroxide or soda ash. Reactions carried out in this way made it possible to obtain branched derivatives. These compounds proved to be detergents with very good functional properties, however, due to slow biodegradation in the 1960s, linear alkylbenezene sulfonates started to be used. Over the years, there has been a significant increase in the amount of linear alkylbenezene sulfonates produced. Available data indicate that production in the 1980s was one million tons, ten years later produced 1.8 million tons, currently about 3 million tons [6, 7]. These data clearly indicate how big despite the passage of years and the appearance of other surfactants is the production of linear alkylbenzene sulfonates. It is not surprising that so far developed several methods of LAS obtaining. The starting stage for commercial preparation of LAS is to obtain a thereof precursor LAB (linear alkylbenzene). In large-scale industrial processes LAB is obtained by alkylating benzene with linear monoolefins or alkyl halides, such as chloro paraffins, using HF or $AlCl_3$ as an alkylation catalyst. Historically, the first commercially used process of LAB preparation was alkylation with the use of $AlCl_3$, while in the 1960s, the technology with the use of HF was widely used. In the 1990s, a heterogeneous catalyst was started to be used in fixed bed reactors, this technology was adopted as evidenced by the volume of LAB production obtained by this method. Literature data indicate that in 2005, 75% of LAB has been obtained using the HF method, 20% in fixed-bed reactor, while only 5% as a result of the alkylation reaction in the presence of $AlCl_3$ [8].

8.1.3 Methyl ester sulfonate

The increase in ecological awareness and the increase in prices on the petrochemical market have proven that methyl ester sulfonate is currently of great interest. In the production of methyl ester sulfonate, recycled raw materials such as palm or coconut oil are currently used. Typically, methyl ester sulfonate on an industrial scale is obtained in falling-film reactor. This process is carried out in stages with the production of intermediate products. These steps describe the reactions (Figure 8.2) [9].

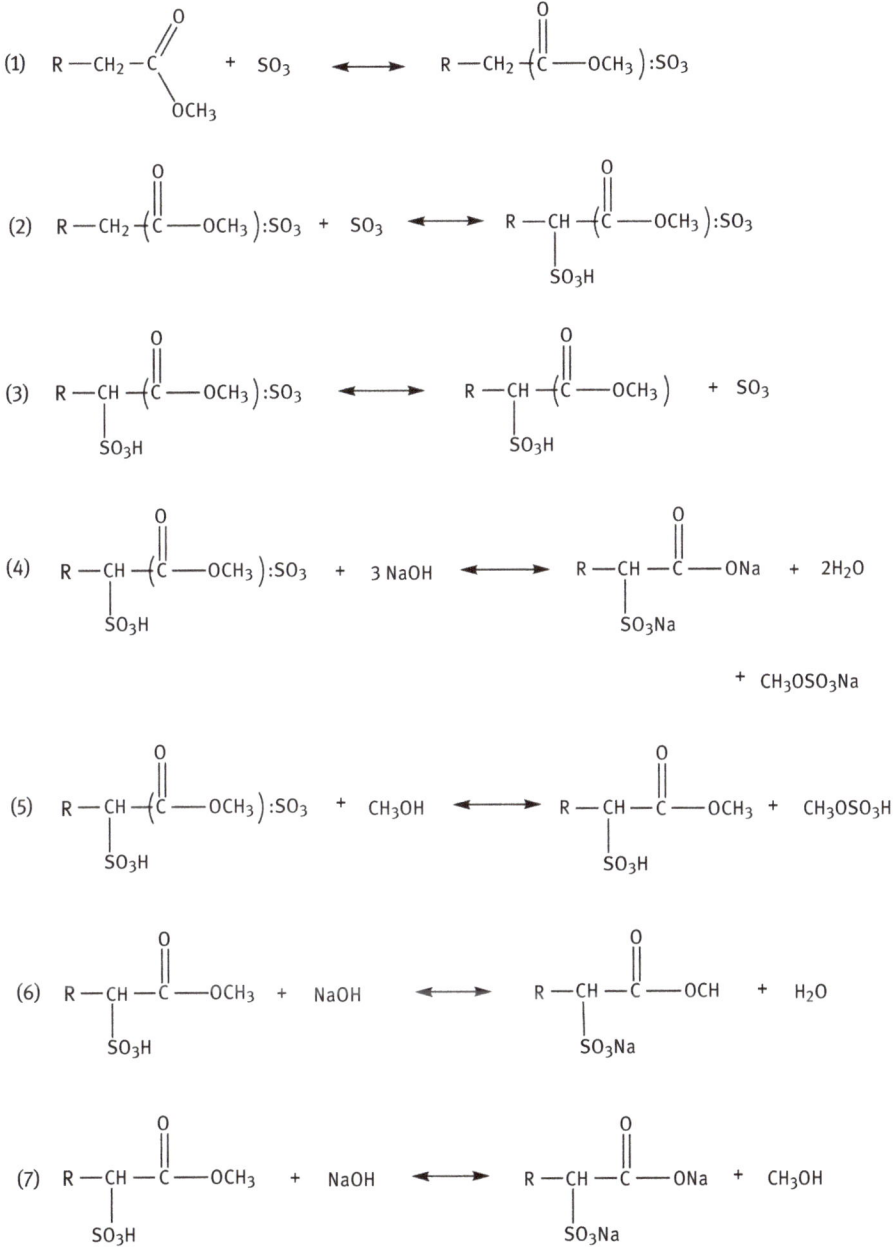

(1) $R-CH_2-C\overset{O}{\underset{OCH_3}{\diagdown}}$ + SO_3 $\longleftrightarrow$ $R-CH_2-(C-OCH_3):SO_3$

(2) $R-CH_2-(C-OCH_3):SO_3$ + SO_3 $\longleftrightarrow$ $R-CH-(C-OCH_3):SO_3$, SO_3H

(3) $R-CH-(C-OCH_3):SO_3$, SO_3H $\longleftrightarrow$ $R-CH-(C-OCH_3)$, SO_3H + SO_3

(4) $R-CH-(C-OCH_3):SO_3$, SO_3H + $3\,NaOH$ $\longleftrightarrow$ $R-CH-C-ONa$, SO_3Na + $2H_2O$

 + CH_3OSO_3Na

(5) $R-CH-(C-OCH_3):SO_3$, SO_3H + CH_3OH $\longleftrightarrow$ $R-CH-C-OCH_3$, SO_3H + CH_3OSO_3H

(6) $R-CH-C-OCH_3$, SO_3H + $NaOH$ $\longleftrightarrow$ $R-CH-C-OCH$, SO_3Na + H_2O

(7) $R-CH-C-OCH_3$, SO_3H + $NaOH$ $\longleftrightarrow$ $R-CH-C-ONa$, SO_3Na + CH_3OH

Figure 8.2: Synthesis scheme of methyl sulfonate ester [9].

The first stage involves the adsorption of sulfur trioxide to the methyl ester and the formation of an intermediate product called an adduct or anhydride. The next stage of the process involves the sulfonation reaction, which leads to the formation of an intermediate which is rearranged leading to the formation of methyl ester sulfonic acid. The last expected step of the process is neutralization, as shown in the diagram. The most important stages of the methyl ester sulfonate preparation process are also accompanied by additional reactions which are presented in the diagram (Figure 8.3) [9].

8.1.4 Alcohol sulfates

Natural alcohol sulfates in washing products were used already in the 1930s. Currently, coconut oil, palm kernel and tallow fat are used for the production of alcohol sulfates. These fats differ in composition of triglycerides, coconut and palm kernel oils dominate derivatives with 12–14 carbon chains, whereas in tallow fat more often occur triglycerides with 16–18 carbon chains. In order to obtain alcohol from fat, a two-step reaction is necessary. The first step involves the reaction of triglyceride with methanol in the presence of a base catalyst. This reaction makes it possible to obtain high-yield fatty methyl ester, in addition to the formation of glycerin. Due to the difference in the density of both products, it is easy to separate them, in addition, impurities and catalyst remain in the glycerin phase. The stage preceding the conversion to alcohol is the distillation of esters and the division into fractions. The stage of the discussed process is the hydrogenation of methyl esters, this reaction is carried out under specific pressure and temperature conditions, namely 3000 psi and the temperature exceeding 200 °C. This process leads to the synthesis of alcohols with a high yield of 95%.

Alcohols for the production of surfactants are also often obtained from petroleum derivatives, there are known applications of ethylene, propylene and also kerosene. Alcohols obtained from propylene due to the branched chain, poor biodegradation are hardly used for the synthesis of surfactants. Ethylene and kerosene are used extensively, and long-chain olefins are obtained from them as a substrate for the synthesis of alcohols. Ethylene is subjected to an oligomerization process at a temperature of 80-120 °C and a pressure of 1000-2000 psig, transition metal complexes are used as the catalyst. In this process, 1,4-butanediol can be used as a solvent, moreover, if the above-mentioned conditions with the stored process do not lead to the formation of high-molecular weight olefins.

One of the processes that can be used to convert olefins into alcohols is hydroformylation. In this process an aldehyde is obtained which can eventually be converted to alcohol by hydrogenation. The hydroformylation process is carried out in the

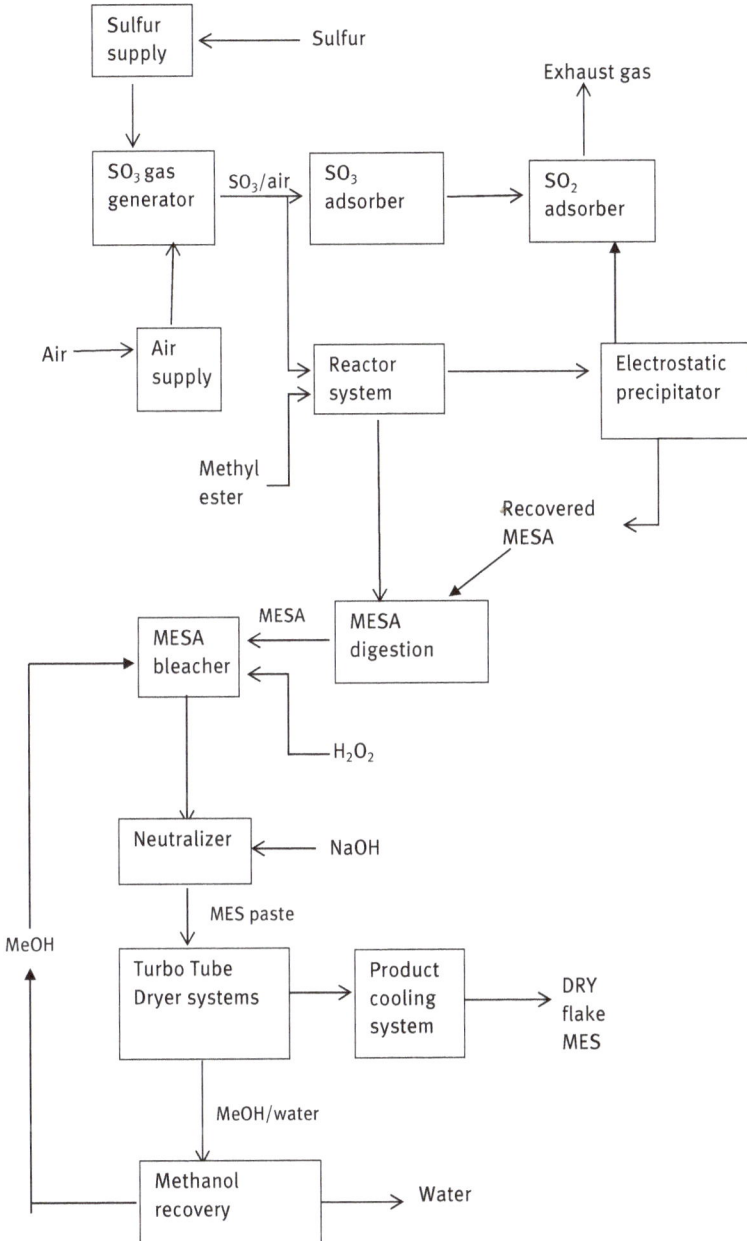

Figure 8.3: Overview of ME sulfonation plant [9].

presence of a catalyst, usually used cobalt carbonyl or cobalt carbonyl/tert-phosphine complexes, the final effect of the reaction is linear aldehyde and 2-alkyl branched aldehydes.

The resulting aldehyde is subjected to hydrogenation, which leads to the formation of alcohol. Above all, the synthetic method presented is used by producers, but the conditions of the reaction as well as the catalyst, cocatalyst or substrate used are different for manufacturers.

Regardless of the origin of the alcohols, natural or synthetic, in order to obtain alcohol sulfates, it is necessary to carry out the sulfation reaction. As the sulfating agent, the following are usually used: sulfur trioxide, sulfuric acid, oleum, chlorosulfuric acid, amidosulfonic acid, sulfur trioxide complex. The alcohol sulfation mechanism presupposes the initial formation of a metastable pyrosulfate which breaks down into alkyl hydrogen pyrosulfate, which in turn reacts with the alcohol molecule to form a mixture of products containing alcohol sulfate. Currently, the most commonly used sulfation factor is sulfur trioxide. The reactors operating all over the world use a continuous process, often work on the principle of continuous falling film SO_3 process [9].

8.1.5 Alcohol ethoxysulfates

Alcohol ethoxysulfates are widely used surfactants, and it is estimated that their annual consumption in Europe is 276,000 tons. The general principle of the production of ethoxylated alcohol sulfates based on the sulfation process the primary alcohol ethoxylates using sulfur trioxide or chlorosulfonic acid, the indispensable stage is the neutralization with the base. The alcohol ethoxylates used as substrats have on average 2 to 3 molecules of ethylene oxide. Alcohol etoxysulfates on an industrial scale are produced using "'multitube falling-film'" reactor. This process involves several stages, in the first one pyrosulfate are formed:

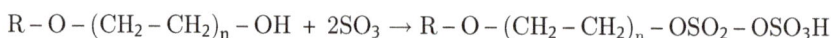

$$R-O-(CH_2-CH_2)_n-OH + 2SO_3 \rightarrow R-O-(CH_2-CH_2)_n-OSO_2-OSO_3H$$

In the further stage, the pyrosulfate decomposes to the sulfuric acid half-ester in the presence of alcohol:

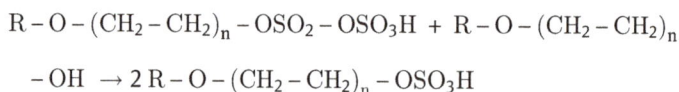

$$R-O-(CH_2-CH_2)_n-OSO_2-OSO_3H + R-O-(CH_2-CH_2)_n$$
$$-OH \rightarrow 2R-O-(CH_2-CH_2)_n-OSO_3H$$

The process ends with neutralization using aqueous alkaline solutions, e. g. NaOH, KOH or NH_4OH [4]:

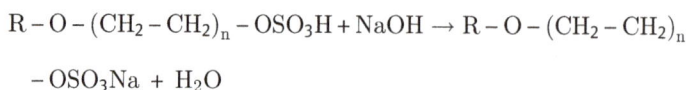

$$R-O-(CH_2-CH_2)_n-OSO_3H + NaOH \rightarrow R-O-(CH_2-CH_2)_n$$
$$-OSO_3Na + H_2O$$

8.1.6 Sulfosuccinates

Sulfosuccinates are anionic surfactants produced in a two-step reaction. The first step involves the reaction of maleic anhydride with an alcohol or ethoxylated alcohol. The next step is the sulfated process with sodium sulfite. Both reactions are exothermic processes, therefore, during the course of the reaction temperature should be strictly controlled, because too high can lead to discoloration of the product. These reactions are also carried out under a nitrogen atmosphere to avoid yellowing of sulfosuccinate [10]. The diagram of the production process is shown in Figure 8.4.

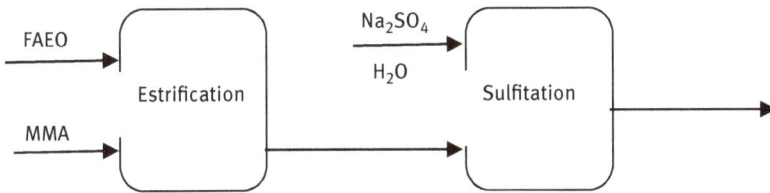

Figure 8.4: Scheme of the sulfosuccinates production process (FAEO – Fatty alcohol ethoxylate; MAA – Maleic acid anhydride) [9].

8.2 Amphoteric surfactants

Another group of surface active agents are zwitterionic often called amphoteric surfactants. *Zwitterionic* means that their molecules are at the same time both anionic and cationic. These structures are characterized by the presence of positive and negative charges in hydrophilic portion of the molecule which is internally neutralized. Amphoteric surfactants are the substances that show a varying charge including positive, zwitterionic as well as negative. The main feature of amphoteric surfactants is their dependency on the pH solution in which they are were dissolved. As they exhibit a varying charge, which can be positive, zwitterionic or negative, their behavior in solutions is differentiated. Thus, in acid pH solutions, the particle gains a positive charge and behaves like a cationic surfactant. In alkaline pH solutions, the molecule becomes negatively charged and behaves like an anionic one. Whilst there can be defined a specific pH value at which part of the molecule with both ionic groups presents equal ionization. That characteristic point is called the isoelectric point of the molecule [40, 11]. Class of zwitterionic surfactants also will also consider those surfactants that exhibit a zwitterionic form irrespectively of changing pH value of the solution. Thus zwitterionic surfactants can be classified into two main groups pH-sensitive and pH-insensitive [11].

pH-sensitive group of zwitterionic surfactants show the properties of cationics at low pHs and of anionics at high pHs. In the pH values close to their isoelectric points they can exist mainly as zwitterionics. In zwitterionic form they exhibit minimum solubility in water, foaming, wetting, and detergency [12].

Representatives of pH-sensitive zwitterionic surfactants are:

- β-N-Alkylaminopropionic acids, $RN^+H_2CH_2CH_2COO^-$, which are used as disinfectants, pigment dispersion aids, inhibitors of corrosion, cosmetics, alkaline cleaning products which contain high alkali and electrolyte,
- N-Alkyl-β-iminodipropionic acids, which are used in fabric softeners,
- Imidazoline carboxylates, which are used in cosmetic and toilet preparations, textile softener.
- N-Alkylbetaines, $RN^+(CH_3)_2CH_2COO^-$ and amidoamines and amidobetaines. These substances are zwitterionic at pHs, which is at and higher than their isoelectric points (neutral and alkaline pHs). While they stay in cationic form below their isoelectric points (acid pHs). As a result of above mentioned they show no anionic properties. Their applications in industry are bonded with cosmetics and personal care products such as liquid soaps, shampoos, facial cleaners. They exhibit not only the mildness on the skin but also the compatibility with cationic, anionic, and nonionic surfactants.

Probably, the most significant class of amphoteric surfactants are just amidobetaines, in particular very popular compound Cocamidopropyl Betaine (CAPB) [13]. This substance can be found in the personal care products and household chemicals [14].

Cocamidopropyl betaine (commercial available in concentration of 30% active) is obtained in a two-step batch process (Figure 8.5). In the first step, coconut oil or fatty acids, which are hydrolyzed from coconut oil (C_{12}–C_{18}) are reacted in aqueous solution and at the temperature about 160 °C with dimethylaminopropylamine (**I**). It is worth mentioned that the coconut oil has a mixed composition of fatty acid, which varies slightly, due to the fact that it is a natural product. The major fatty acid is lauric acid containing 12 carbon atoms in the alkyl chain. The second step of CAPB synthesis is if the reaction of the resultant dimethylaminopropyl cocoamide (amidoamine – **II**) with sodium monochloroacetate (**III**). The reaction should be conducted under alkaline conditions. The final product (CAPB) is obtained as an aqueous solution of surfactant, usually containing 30% of active substance [15].

pH-insensitive group of zwitterionic surfactants are zwitterionics at all pHs. The representatives of this group are sulfobetaines, $RN^+(CH_3)_2(CH_2)_xSO^-$. They are compounds, which structure is similar to alkylbetaines, however, in the head of the molecule carboxylic acid is replaced by a sulfonate group. Compounds classifies to the sulfobetaines can be divided into several subgroups. They differ from each other with the length of the spacer between the quaternary ammonium center and the sulfonate group. The presence of a hydroxyl group in the spacer also divides

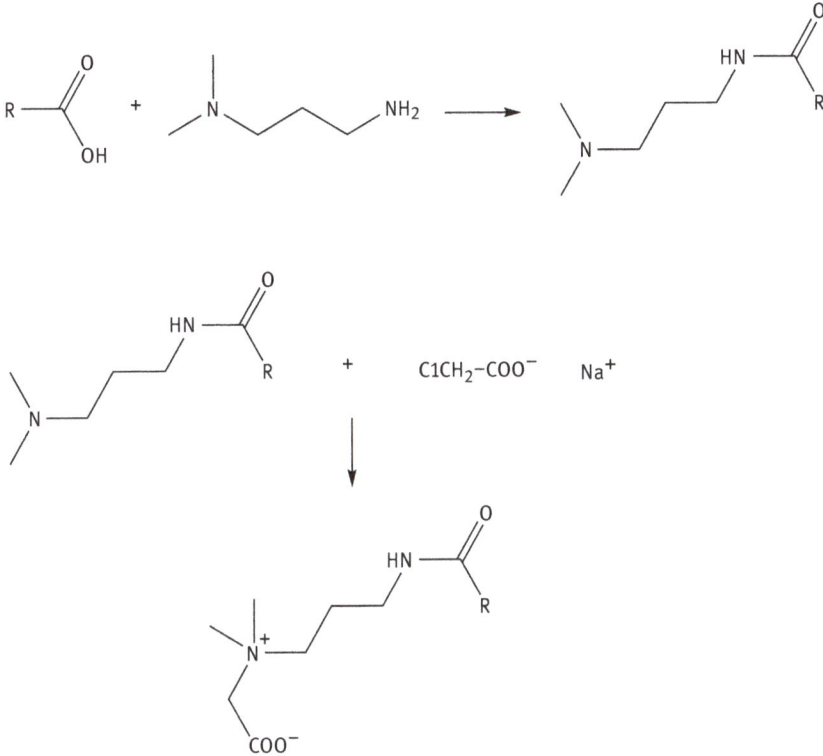

Figure 8.5: Reaction scheme of alkyl amido betaines.
Source: [HERA].

the sulfobetaine surfactants. [[31]]. Since 1990s the is a great interest in gemini surfactants. A characteristic feature of gemini surfactants is that they possess in one molecule two hydrophilic headgroups and hydrophobic tails. The sulfobetaines are obtained in two step reaction (Figure 8.6). In first step tertiary amines are obtained, while in second synthesis of surfactants is conducted. Sulfobetaines can be obtained in the reaction of tertiary amines with sultones or sulfonic salts. This reaction is called quaternization reaction due to the fact that quaternary nitrogen atom is created. In this reaction simple sulfobetaines, homo- and hetero-gemini sulfobetaines can be obtained [16].

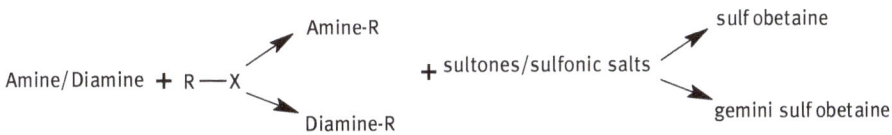

Figure 8.6: Synthesis rout of sulfobetaines.

Zwitterionic surfactants exhibit very good surface properties. They show low values of both surface tension and critical micelle concentration. Zwitterionic surfactants can be used as wetting agents. The characteristic feature of these compounds is they can form high, moderate or low foam [17]. They are chemically stable while they placed both in acids and alkalis. They are compatible with other types of surfactants. Zwitterionic surfactants exhibit less irritating potential to skin and eyes than other surfactants. They can form May hydrophobic film on the negatively or positively charged surfaces, because they are adsorbed onto. Due to their structure they also show antimicrobial and antifungal properties [11, 18].

8.3 Cationic surfactants

Cationic surfactants in comparison with anionic and non-ionic have a small share in the global production of surfactants. For the first time, surfactants based on the quaternary nitrogen atom have been produced on a large scale in the 1940s, using a tallow as the raw material for their production. Cationic surfactants, possessing a positively charged atom in their molecule, gain specific properties and thus many applications. However, it is estimated that the production of fabric softeners absorbs the largest part of the world production of cationic surfactants. Since the 1990s, in Europe, mainly alkanolamine esters have been used in domestic softeners.

8.3.1 Alkanolamine esters (esterquats)

The most important substrates for the production of esterquats are alkanolamines: triethanolamine and methyldiethanolamine. Alkanolamines esterified with fatty acids at 200 °C, the generated water is removed by stripping with an inert gas or under vacuum. It is also advisable to use an acid catalyst. The described process usually leads to a mixture of compounds. The following Figure 8.7 shows an example of the course of the reaction using triethanolamine as a substrate [9, 10].

This reaction leads to the main product triethanolamine diester. To transform the diester into esterquat, the quaternization process should be carried out. As a quaternizing factor, dimethyl sulfate can be used. The reactions are carried out at a temperature of from 40 to 60 by gradually adding dimethyl sulfate, it is also necessary to lower the viscosity by using a solvent such as isopropanol [9, 10].

8.3.2 Cationic surfactants based on alkylamines

Surfactants based on alkylamines are also significant group of cationic surfactants. The oldest commercially used surfactant belonging to this group of compounds is

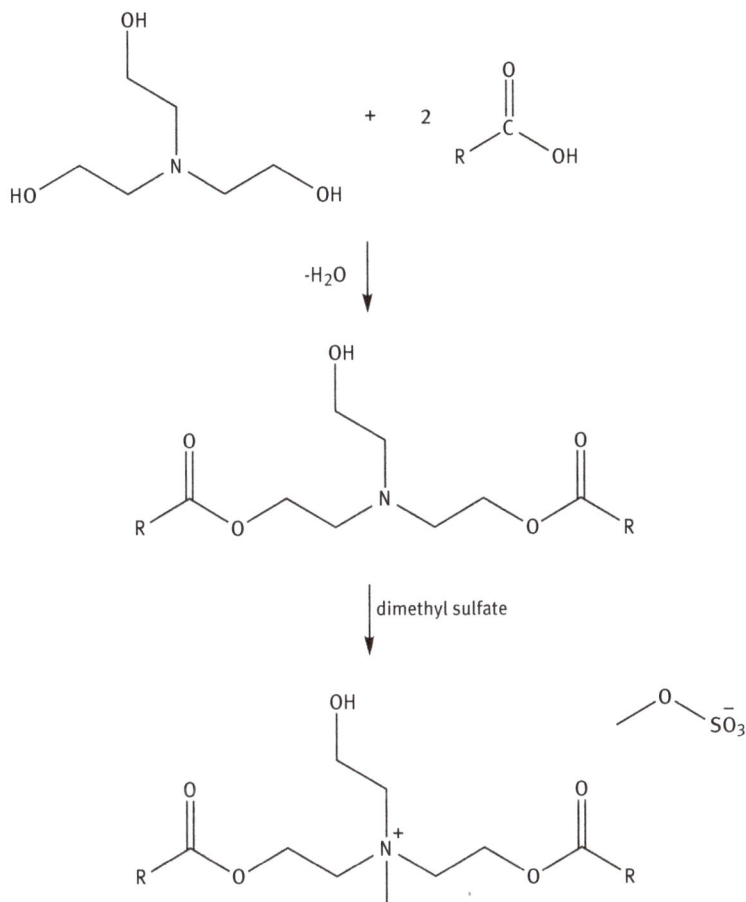

Figure 8.7: The esterquat synthesis scheme [9].

dimethyl ammonium chloride distearyl (DSDMAC), for the first time it was produced in 1949. The substrates for the preparation of this group of cationic surfactants are primary or secondary alkylamines. In industrial production, these amines are produced from fatty acids by a series of reactions. This process begins the reaction of fatty acids with ammonia at a temperature of over 150 °C, it leads to obtaining ammonium soap. The amide thus obtained is converted to the nitrile by catalyzed dehydration. The next step is to conduct the hydrogenation, this process initiates in the presence of a catalyst, the reaction temperature in which the reaction is carried out is wide and can range from 50 °C to 200 °C. To obtain cationic surfactants with a more branched structure, it is necessary to use secondary or tertiary amines as substrates. This amines in industrial processes can be obtained in several ways e. g. in the reaction of catalytic conversion of fatty alcohol to dimethylalkylamine.

Another method involves converting the fatty alcohol into alkyl chloride, which then reacts with dimethylamine. In addition, secondary amines are also obtained by reductive alkylation of primary amines using a nickel catalyst. The final stage in the preparation of cationic surfactants is the quaternization reaction. Usually, the amine is reacted with a suitable alkylating agent, e. g. dimethyl sulfate, benzyl chloride, ethyl chloride, methyl chloride, alkyl chloride. Due to the possibility of corrosion, this reaction is carried out in glass or stainless steel reactors [9, 10].

8.4 Nonionic surfactants

In the global market of surfactants nonionic surfactants are important group of compounds.

The term "non-ionic surfactant" usually is connected with derivatives of ethylene oxide and/or propylene oxide with a long chain alcohol, which contains an active hydrogen atom. Nonionic surfactants, which structure is based on ethylene oxide (ethoxylated surfactants) are considered to be the most common. Over two million tons of commercial nonionic surfactants worldwide produced by over 150 different producers contain at least 50% alkoxylated alcohols. The decrease in ethoxylated nonylphenol production is observed and now equal to 20% of the market, while 15% of the market is alkoxylated fatty acids and another 10% of the market is fatty acid amides and sugar esters [19].

In the nonionic group of surfactants several classes can be pointed: alcohol ethoxylates, alkyl phenol ethoxylates, fatty acid ethoxylates, sorbitan ester ethoxylates, monoalkaolamide ethoxylates, fatty amine ethoxylates and ethylene oxide–propylene oxide copolymers (sometimes referred to as polymeric surfactants). To the nonionic surfactants can be also classified multihydroxy substances such as glycol esters, glycerol (with polyglycerol) esters, glucosides (with polyglucosides) and sucrose esters. In the nonionic group of surfactants can be found amine oxides and sulphinyl surfactants, which are examples of substances with a small head group [1]. All above mentioned groups of nonionic surfactants are produced widely in the world and find application in a multitude of industries.

8.4.1 Polyoxyethylene surfactants

As mentioned before an interesting group of nonionic surfactants is the polyoxyethylene (POE) family. Many nonionic surfactants, especially the derivatives of polyoxyethylene exhibit an inverse temperature–solubility relationship, it means that an temperature increase results in decrease their solubility in water.

The most common synthesis route of the POE surfactants is the process in which ethylene oxide reacts with a hydrophobic substance possessing in the molecule at least one reactive hydrogen, as in the case of alcohols, acids, amides, mercaptans, and related compounds [20]. However this brief description does not present that the reaction route is much more complicated (Figure 8.8). In general there are three main steps to the production of an ethoxylate or in general alkoxylate. The process begins in the pre-treatment vessel, where an amount of accurately charged initiator or catalyst (usually NaOH or KOH) is put together with the some of the feedstock which is to be ethoxylated. The mixture is warmed up to 120–130 °C and the dehydration process is continued until the water level is less than 200 ppm. The rest of the feedstock material, which is to be ethoxylated and the initiator are placed in the reactor, which is warmed to the temperature of 130–140 °C and dried under vacuum. The heating is switched off right after the reactants are dried. Subsequently the vacuum conditions are broken with the addition of nitrogen and ethylene oxide, which are added to achieve the desired degree of ethoxylation. The temperature is set at the level of about 150 °C and the addition of the oxide until is controlled just to meet the specification. An

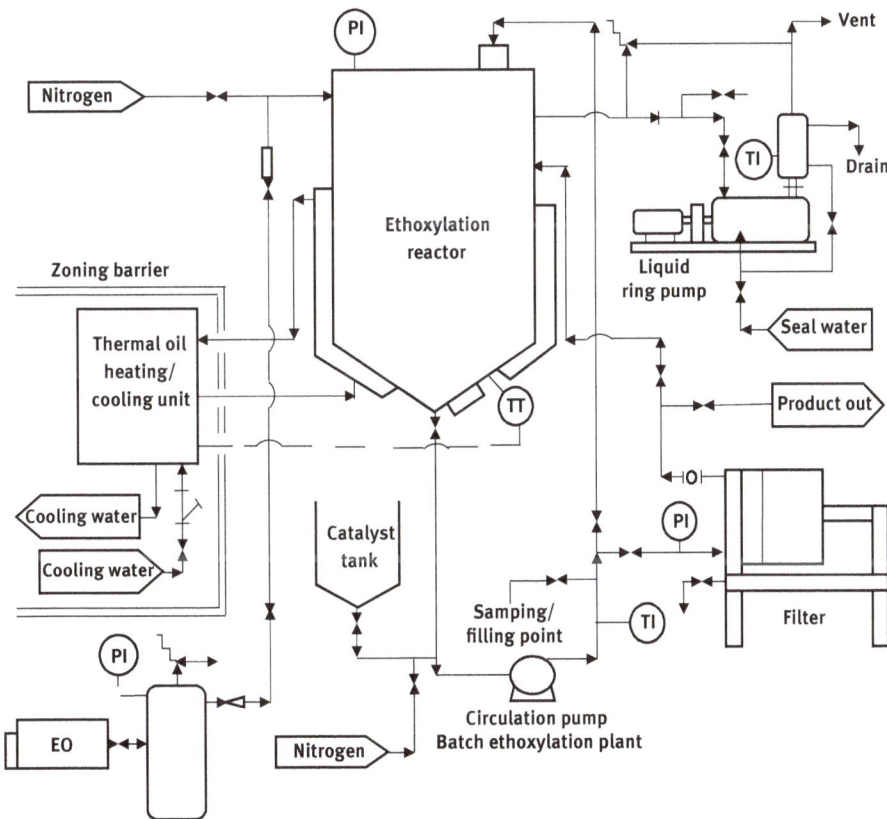

Figure 8.8: Batch ethoxylation unit [19].

ethoxylate reactions temperature usually ranges between 120 °C and 180 °C. In the aim to reduce free ethylene oxide to 1 ppm level the product is sparged with nitrogen. Next, it is cooled and neutralized. Subsequently the product is filtered to remove catalyst. In the post-treatment vessel takes place product neutralization, filtration, bleaching, if all required. While the product is still in the reactor the catalyst for the next batch to be made can be placed in the pre-treatment vessel. Thus the system can be considered as batch continuous [19]. Schematic of batch/continuous process is presented in Figure 8.9.

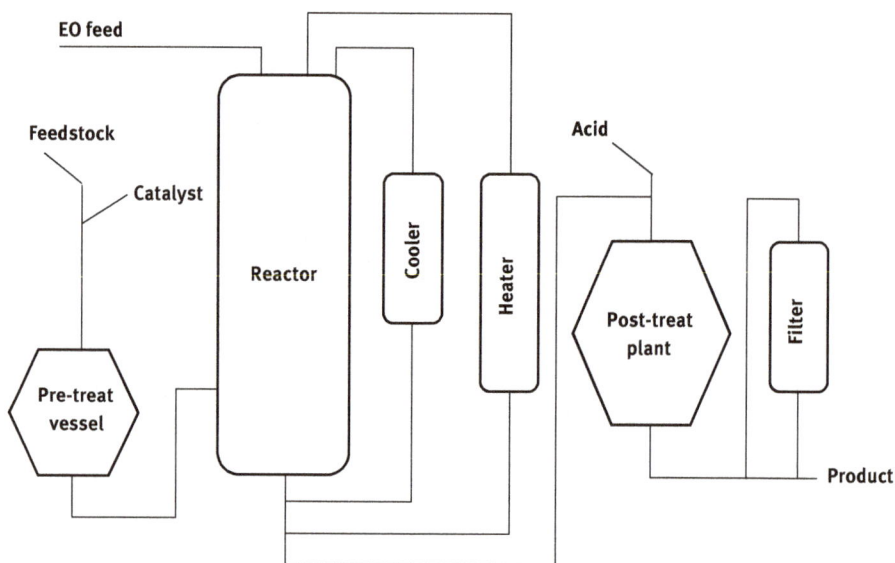

Figure 8.9: Schematic of batch/continuous process [19].

The general formula of polyoxyethylene-based surfactants is $RX(CH_2CH_2O)_nH$ where R is a typical for surfactant hydrophobic tail – alkyl chain, but may also be a tail which is hydrophobic polyether such as polyoxypropylene, and X is O, N, or another functionality capable of linking the POE chain to the hydrophobe. The n is the average number of ethoxylate (OE) units introduced in the hydrophilic group. The average number of the OE groups is bonded with the ethoxylation process, during which the mixture of compounds is obtained [20]. The degree of ethoxylation affects the final properties of POE.

The group of nonionic surfactants contains a great variety of compounds thus the classification is not standardized. One of suggested classification divides nonionic surfactants into: alcohols, ethers, alkanolamides, esters, amine oxides [21].

Alcohols are not generally considered as surfactants; however, they can be used as co-emulsifiers and co-surfactants. Alcohols are also substrates in synthesis of other surfactants, for example ethoxylated alcohols, which are assumed to be the most widely used nonionic surfactants so far [21].

Alcohol alkoxylates, another group of nonionic surfactants, can be obtained in the synthesis with the use of either oleochemical or petrochemical alcohols. Consequently, the hydrophobic part of the surfactants molecule can be highly linear when the substrate (alcohol) is derived from oleochemical and some petrochemical sources, or it can be highly branched when the alcohol come from other petrochemical sources [12]. This class of nonionic surfactants consist of ethoxylated $R(OC_2H_4)nOH$ and propoxylated alcohols $R(OC_3H_6)nOH$. Ethoxylated alcohols are obtained in the reaction of alcohols with ethylene oxide, while propoxylated alcohols are obtained with propylene oxide. Usually during in the production of alcohol alkoxylates the result of the synthesis is a mixture of various substances which differ in both the carbon chain length and the number of alkoxymers. Fatty alcohols have no direct natural source. All of them have to be synthesized. Five major processes of fatty alcohols synthesis can be distinguished:

1. Catalytic hydrogenation of fatty acids from natural fats and oils
2. The OXO process from alpha olefins
3. Ziegler process to give linear even numbered chains
4. From *n*-paraffins to give essentially linear even and odd numbered chains
5. From the Shell "SHOP" process to give linear even and odd numbered chains

Above mentioned processes lead to the production of alcohols which possess hydrophobic tail from C6 to C20 [19].

A decrease in the EO units in the alcohol ethoxylated molecule leads to reduction of the surface tension of the solutions. Also it can observed that the viscosity of solution of nonionic surfactant increases progressively with a growth in its concentration, until it reaches a critical concentration, when the viscosity increases much more rapidly, finally appears a gel-like structure and the structure of hexagonal type liquid crystalline is formed. The value of this critical concertation depends on the tail structure and EO chain length. There can be considered other case, when the viscosity reaches a maximum value and after maximum a decline is observed. It can be explained by the formation of other structures such as lamellar phases [1].

Alcohol ethoxylates exhibit excellent detergency. They can remove oily soil and can be used in especially laundry liquids products. They exhibit excellent emulsifying properties and are suspending agents, thus they found numerous industrial applications [12].

The production process of alcohol ethoxylates is still modified by the introduction of some newer catalysts, which make the reactions quicker and decrease the consumption of catalyst.

8.4.2 Alkyl phenol ethoxylates

Alkyl phenol ethoxylates, $RC_6H_4(OC_2H_4)_nOH$, are prepared in the synthesis route where ethylene oxide react with the appropriate alkyl phenol. The most common such obtained nonionic surfactants are those build on nonylphenol. The family of nonylphenol are liquid at room temperature up to about the 12-mol ethoxylate and do not require heated storage. These surfactants are cheap to produce and are used for reduction of oil–water interfacial tension and are distinguished in removing oily soils. The major drawback is biodegradability and potential toxicity which is bonded with toxicity of the by-product of degradation is nonyl phenol. Despite these many industrial applications can be mentioned such as: water-insoluble types of alkyl phenol ethoxylates can be used as W=O emulsifying agents, agents which control the foaming properties, cosolvents; Water-soluble types of alkyl phenol ethoxylates can be also used as O=W emulsifying agents for paints, in the emulsions for agricultural purpose, in numerous cosmetic and industrial emulsions. Alkyl phenol ethoxylates with high (EO) content (>15 mol EO) found application as detergents and emulsifiers, where the strong electrolyte systems is applied and as agents which entrain the foam in concrete. Alkyl phenol ethoxylates are also used in liquid detergents and as dyeing retarders for cellulose. They are excellent dispersing agents for carbon [1, 12, 19].

8.4.3 Fatty acid ethoxylates

Fatty Acid Ethoxylates $RCOO\text{-}(CH_2CH_2O)_nH$ are produced in the synthesis where ethylene oxide react with a fatty acid or a polyglycol. When a polyglycol is applied, a mixture of mono- and di-esters $(RCOO\text{-}(CH_2CH_2O)_n\text{-}OCOR)$ is produced. Ethoxylation process can be carried out in the same plants and manner as described before in the case of alcohol ethoxylates. At first, the initial amount of ethylene oxide reacts with the acid in the aim to produce ethylene glycol monoester, $RCOO(CH_2CH_2O)H$. Then the reaction is more rapid, due to the fact that further ethylene oxide units are introduced. The product of the reaction is polyethoxylated $RCOO(CH_2CH_2O)_nH$. Due to the fact that the reaction conditions are ideal for ester interchange, moreover the final product contains free polyethylene glycol, it lead to the creation of the monoester and the diester $[RCOO(CH_2CH_2O)_nOCR]$ in the ratio 1:2:1. The Fatty Acid Ethoxylates can be obtained with the use of an alternative method of production of these compounds in the reaction in an ester kettle, where the polyethylene glycol with desired molecular weight is subjected to an esterification reaction with acid. Reaction temperatures and catalysts can be differentiated, however, in the region 100–200 °C. When fatty acid and polyethylene glycol are introduced in an equimolar ratio it results in a mixture similar to the product via ethoxylation, i. e. dominant in monoester. However, if high excess of polyethylene glycol is applied, monoester dominates.

Monoester products, even after purification can revert to the mixture on storage, which results in the adverse effect on wetting properties. The obtain esters in this procedure are hydrolyzed in both base and acid conditions, which results in that they are much less stable than alcohol ethoxylates. This is a limitation for them to be introduced in detergent formulations but they have many other mainly industrial uses. One of them is textile industry, in which they are used as emulsifiers, and substances that exhibit dispersive, lubricating as well as antistatic properties. They also found application in personal care, are used in institutional and industrial cleaning products, as crop protection agents, paints and coatings and adhesives [19].

8.4.4 Sorbitan esters/ethoxysorbitan esters

Sorbitan Esters and their ethoxylated derivatives form a very important group of nonionic materials and find many applications not only in foods, personal care products and cosmetics but also in pharmaceuticals. Fatty acid esters of sorbitan (the well-known as Spans) are produced by reacting sorbitol with a fatty acid at the temperature higher than 200 °C. First sorbitol is dehydrated to 1,4-sorbitan and afterwards esterification process takes place. In the aim to obtain a mono-ester one mole of fatty acid and one mole of sorbitol is introduced. There can be observed also the presence of by-product such as di-ester. The general formula of sorbitan monoester is presented in Figure 8.10.

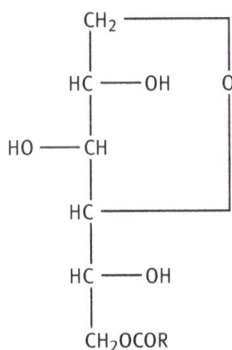

Figure 8.10: Sorbitan monoester.

Not only one hydroxyl group can be esterified, but also di- and tri-esters can be produces by esterification more than one OH group in the molecule. Depending on the nature of the alkyl group of the acid and whether the product is a mono-, di- or tri-ester several commercial products are available. Some examples of commercially available Spans are given below [1, 22]:

- Sorbitan monolaurate – Span 20
- Sorbitan monopalmitate – Span 40
- Sorbitan monostearate – Span 60
- Sorbitan mono-oleate – Span 80
- Sorbitan tristearate – Span 65
- Sorbitan trioleate – Span 85

Span^{TM} surfactants are lipophilic, thus they are generally soluble or dispersible in oil. They can form water in oil emulsions. Due to their very good emulsification properties they can find application in personal care and industrial cleaning products. They are used as agents in crop protection, water treatment, fiber finish, paints and coatings as well as lubricant. Sorbitan esters as well as their ethoxylated derivatives (the well-known as Tweens) are probably one of the most commonly used nonionic surfactants. Ethoxylated derivatives of Spans (Tweens) can be produced by the reaction in which ethylene oxide bound with any free hydroxyl group remaining on the sorbitan ester group. There is also a possibility to obtain Tweens in reaction where fist sorbitol is ethoxylated and then esterified. However, the final product exhibits different properties to the Tweens. Some examples of Tween surfactants are given below [1, 22]:

- Polyoxyethylene (20) sorbitan monolaurate – Tween 20
- Polyoxyethylene (20) sorbitan monopalmitate – Tween 40
- Polyoxyethylene (20) sorbitan monostearate – Tween 60
- Polyoxyethylene (20) sorbitan monooleate – Tween 80
- Polyoxyethylene (20) sorbitan tristearate – Tween 65
- Polyoxyethylene (20) sorbitan trioleate – Tween 85

The Spans are insoluble in water, but soluble in most organic solvents. They are characterized by low HLB values. The Tweens inversely to Spans are generally soluble in water and have relatively high HLB numbers. Sorbitan esters as well as their ethoxylated derivatives can be used as food additives, which is the main advantage of these compounds. In cosmetics and some pharmaceutical preparations they are used as inter alia emulsifiers [1].

Ethoxylated fats and oils is the class of nonionic surfactants which contain ethoxylated derivatives of castor oil and lanolin. Lanolin is also known as wool fat that contains a complex mixture of esters polyesters and alcohols with high-molecular-weight (aliphatic, triterpenoid and stroir) and fatty acids (unsaturated, saturated, hydroxylated and nonhydroxylated). Ethoxylation process is carried out on fractional products of lanolin, which consists of lower aliphatic alcohols and sterols. Castrol oil is a mixture of a triglyceride of fatty acids and is regarded as natural extract obtained from pressing of ricinus seeds. The fraction consists of ricinoleic, oleic, linoleic, palmitic, linolenic, stearic, arachidic and dihydroxystearic oils. Surfactants obtained by ethoxylation of lanolin and Castrol oil are good emulsifiers.

These products are also applied in pharmaceutical products, e. g. as solubilizers. They can be used in cosmetics industry [1, 21].

8.4.5 Alkanolamides

Alkanolamides is a group of nonionic surfactants which are derivatives of monoe-thanolamide and diethanolamide. One type of such surfactant can be obtained in the reaction between a fatty acid or ester with mono- or diethanolamide engaged in stechiometric proportions (type 1:1). The product of the synthesis is mono- or dia-lkanolamide. The other type of surfactants is obtained in the reaction of a ester or fatty acid with mono- or diethanolamide at molar proportion 2:1 (type 2:1). The product of the synthesis is a mixture of mono- or dialkanolamide. Type 1:1 has low water solubility at room temperature and can be used as viscosity modifiers in the presence of anionics. Type 1:1 alkanolamides found application in household deter-gent products and type 2:1 alkanolamides can used in formulations of heavy-duty surface cleaners. However the consumption of alkanolamides declined, which is a consequence of extensive development of alkyl ethoxylated compounds [21].

Reaction of alkanolamides with ethylene oxide leads to obtain ethoxylated al-kanolamides. Their properties are similar to ethoxylated alcohols. In addition they can improve the dispersibility of soap scum.

8.4.6 Amine ethoxylates

Amine ethoxylates belong to a small class of nonionic surfactants. They are much more applied in the industrial sector rather than the detergent industry. These com-pounds can be prepared by the reaction of ethylene oxide with primary or second-ary fatty amines. Their production are similar to alcohol ethoxylates and the same equipment can be used, however the first step is uncatalyzed. Initially, primary amine is dehydrated under vacuum, subsequently nitrogen is padded to the reactor and the required amount of ethylene oxide to produce diethanolamine is intro-duced. The addition of second mole of ethylene oxide is more rapid than it was made with the first one, however even the excess of ethylene oxide, the ethoxyla-tion process of the diethanolamine is little or none. The reaction is carried out at about 120 °C:

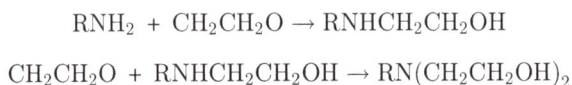

$$RNH_2 + CH_2CH_2O \rightarrow RNHCH_2CH_2OH$$

$$CH_2CH_2O + RNHCH_2CH_2OH \rightarrow RN(CH_2CH_2OH)_2$$

When the first step ethoxylation is achieved further ethoxylation is conducted with the base as a catalyst. This lead to reach the required degree of ethoxylation. The reaction is carried out in about 150 °C. There are two main compounds produced commercially. First diethanolamines, which are used in plastics as antistatic or anti-fog agents and second 15-mol ethoxylate, which is used as an adjuvant applied in herbicide formulations. The amines used are typically tallow or coco amines however other types can be used for specialities [1, 21].

8.4.7 Blockpolimer surfactants

Another class of nonionic surfactants is polyalkylene oxide copolymers. These compounds are the result of developed block polymerization techniques. Ethylene Oxide–Propylene Oxide Co-polymers (EO/PO) are produced with the method of sequential addition of ethylene oxide and propylene oxide to a reactive hydrogen compound with low molecular weight [19]. Two types can be distinguished: those prepared by reaction of poly(oxypropylene glycol) with EO or mixed EO/PO, giving block copolymers ((EO)n(PO)m(EO)n. The commercial products made by BASF is Pluronic^{TM} [19]. The second type of EO/PO copolymers are prepared by reaction of poly(ethylene glycol) with PO or mixed EO/PO. These will have the structure (PO)n (EO)m(PO)n and are called as reverse Pluronics [1].

The synthesis of the polyols can be carried out in the same reactors which are usually used for ethoxylation. In the first step of synthesis sodium hydroxide is dissolved in required amount of propylene glycol and heated up to 120 °C. The residue is dehydrated and padded with nitrogen. Then propylene oxide is applied and reaction is conducted until the required molecular weight is reached. Subsequently ethylene oxide is added. During the process maintains the temperature at 120 °C. All low boiling material is removed from the reaction mixture, neutralized, usually with phosphoric acid, then filtered and cooled. These process is more demanding, so the reaction conditions must be controlled carefully [19].

The polypropylene oxide part of the polymer is water insoluble, which means it acts as the hydrophobic part of the molecule, while the ethylene oxide is water soluble and acts as the hydrophilic part of the molecule.

Pluronics and reverse Pluronics exhibit many useful and interesting properties which have allowed them to be an important element in the surfactant formulation world. Their major property is low-foaming profile. Due to this property they are used in automatic dishwashing detergents or laundry detergents [21].

8.4.8 Alkyl glucoside

Alkyl glucosides are perhaps one of the oldest synthetic group of surfactants. They were first obtained by Fischer in 1893 but, only after the 1980s, the industrial produced has started. They are now interesting compound because they are considered to be obtained from natural materials and renewable resources, such as glucose and coconut- or palmoil-based alcohols [19]. Alkylpolyglycosides are long-chain acetals of polysaccharides. Currently available commercial products have relatively short (averaging 10 and 12.5 carbon atoms) alkyl chains. They show foaming, wetting properties, high detergency and biodegradation, which are similar to those of ethoxylates. However, alkylpolyglycosides have higher solubility not only in water but also in solutions which contain electrolytes. They are characterized by high solubility and stability in sodium hydroxide solutions. Although they are effective fatty soil removers with high detergency, they cause very low skin irritation, what recommends them as the ingredient of hand liquid dishwashing and hard surface cleaners [12].

In the production process of glycosides, a polyfunctional sugar component is reacted with a nucleophile such as a carbohydrate, an alcohol, or a protein [23]. Selective glycosylation reactions require special activation methods, the use of protective groups, or selective catalysis by enzymes. Examples of synthesis routes to glycosides are shown in Scheme (Figure 8.11) [23].

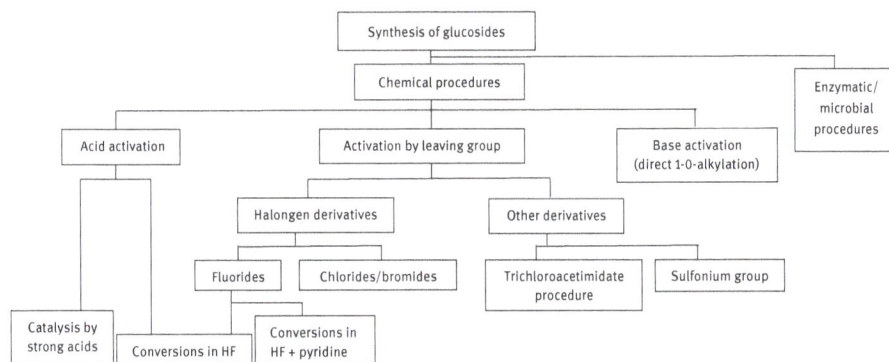

Figure 8.11: Glucosylation reactions [23].

The industrial process is based on the Fischer synthesis and commercial development work started some 25 years ago. The Fischer glycosylation can be described as a process in which, in a first step of relatively quickly reaction, where dextrose reacts to reach oligomer equilibrium. Next, slow degradation of the alkyl polyglycoside is observed. The reaction rate depends on both the acidity and the concentration of the acid in the alcohol. Secondary reactions are conducted with the use of acids as the catalyst. Afterwards, the acidic catalyst is neutralized by a suitable base, often used are sodium hydroxide or magnesium oxide. The reaction mixture contain less than 1% of residual glucose and 50% to 80% fatty alcohol, which is removed by continuous vacuum distillation to 1% or below.

During the process with decreasing fatty alcohol content causes increase in viscosity. This significantly worsens heat and mass transportation, especially in the last stage so thin-layer or short-path evaporators are preferred. The final product after distillation is almost pure alkyl polyglycoside. The main process steps for the synthesis of alkyl polyglycosides are presented in Figure 8.12 [23].

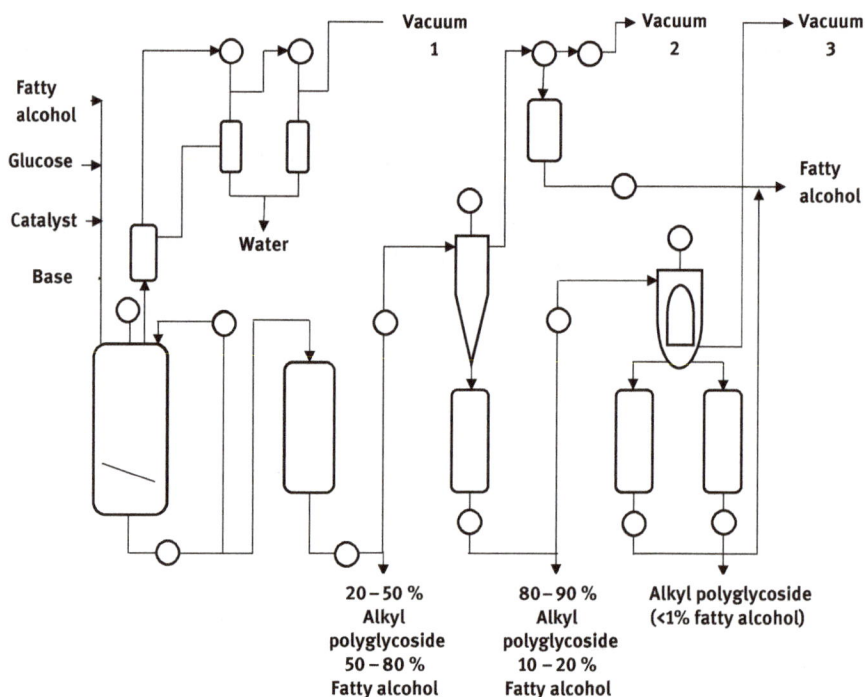

Figure 8.12: Industrial-scale glycosylation process for the reaction of glucose with fatty alcohol [23].

Quite a large group of nonionic surfactants are these that instead of alkyl chain possess in the molecule hydrophobic silicone moiety. These are the reaction products of a reactive silicone intermediate, with a capped allyl polyalkylene oxide, propylene oxide or a mixed ethylene oxide–propylene oxide copolymer. They are excellent wetting agents and present good lubricating properties especially of textile fibers. They are also good wetting agents for such polymers as polyester and polyethylene. They produced low to moderate foam in aqueous solution [12].

8.5 Biosurfactants

Biosurfactants are a group of surface-active compounds produced by microorganisms. Like the other groups of surfactants they exhibit the ability to reduce surface and interfacial tension. Over the last years, there has been an increase in interest in the topic of biosurfactants, because it is estimated that they have higher biodegradability than traditional surfactants. In addition, biosurfactants are compounds more stable at extreme pH, temperature and salinity values, it is also estimated that they have better foaming properties [24]. The properties of biosurfactants are diverse and result from their different structure. The biosurfactants in structural terms constitute a fairly broad group of compounds can be anionic, nonionic and also cationic. Essentially, the main classifications of biosurfactants are based on their chemical structure and microbiological origin, therefore, the glycolipids, phospholipids, fatty acids, polymeric biosurfactants and lipopeptides (surfactin) are distinguished [25, 26]. Previous research has made it possible to say that for the production of biosurfactants they can be responsible: yeasts, bacteria and some filamentous fungi. Table 8.1 below presents the microorganisms responsible for the production of biosurfactants groups [25]:

So far, biosurfactants have not been obtained on a wide industrial scale, however, if it manage to improve the yield of biosurfactants, this may change. In biotechnological processes, from 10% to 30% of the costs incurred are costs related to the raw material, therefore it is aimed at using cheap agro-raw materials in the biosurfactants production processes. Literature shows that in the production of biosurfactants can be used: rapeseed oil, babassu oil, sunflower and soybean oil, waste frying oils, oil refinery waste, curd whey and distillery wastes, potato process effluents, cassava flour wastewater [24, 27, 28]. The use of waste materials for the production of biosurfactants as a substrate is also important in ecological terms. In the case of environmental pollution, for example, hydrocarbons, biosurfactants may be used for remediation, they then influence the increase of bioavailability [29]. Biosurfactants have also found application in agriculture as biopesticides, in the production of food, cosmetics, detergents, petrochemical industry. Their properties: antibacterial, anticancer and antiviral are used in medicine [26].

Numerous beneficial properties of biosurfactants tend to search for technical solutions aimed at improving the process of obtaining them. Literature reports that biosurfactants can be effectively obtained in continuous fermenters with surfactant concentrators based on foam fractionation. The contents of the reactor are constantly mixed, and oxygen is also supplied. The air leaving the reactor generates foam at the top of the reactor. At the same time at the top of the fractionation column the foam is broken and the surfactant stream is discharged (Figure 8.13) [30].

Table 8.1: Main classes of biosurfactants and respective producing microorganisms [25].

		Biosurfactants class			
Glycolipids	Polymeric Surfactants	Lipopeptides	Fatty Acids	Particulate Surfactant	Phospholipids
		Producer microorganisms			
Acinetobacter calcoaceticus, Alcanivorax borkumensis, Arthrobacter paraffineus, Arthrobacter sp., Candida antartica, Candida apicola, Candida batistae, Candida bogoriensis, Candida bombicola, Candida ishiwadae, Candida lipolytica, Lactobacillus fermentum, Nocardia sp., Pseudomonas aeruginosa, Pseudomonas sp., Rhodococcus erythropolis, Rhodotorula glutinous, Rhodotorula graminus, Serratia marcescens, Tsukamurella sp., Ustilago maydis	Acinetobacter calcoaceticus, Acinetobacter calcoaceticus, Acinetobacter calcoaceticus, Acinetobacter calcoaceticus, Bacillus stearothermophilus, Candida lipolytica, Candida utilis, Halomonas eurihalina, Mycobacterium thermoautotrophium, Sphingomonas paucimobilis	Acinetobacter sp., Bacillus licheniformis, Bacillus pumilus, Bacillus subtilis, Candida lipolytica, Gluconobacter cerinus, Pseudomonas fluorescens, Serratia marcescens, Streptomyces sioyaensis, Thiobacillus thiooxidans	Arthrobacter paraffineus, Capnocytophaga sp., Corynebacterium insidibasseosum, Corynebacterium lepus, Nocardia erythropolis, Penicillium spiculisporum, Talaramyces trachyspermus	Acinetobacter calcoaceticus, Cyanobacteria, Pseudomonas marginalia,	Acinetobacter sp., Aspergillus, Corynebacterium lepus,

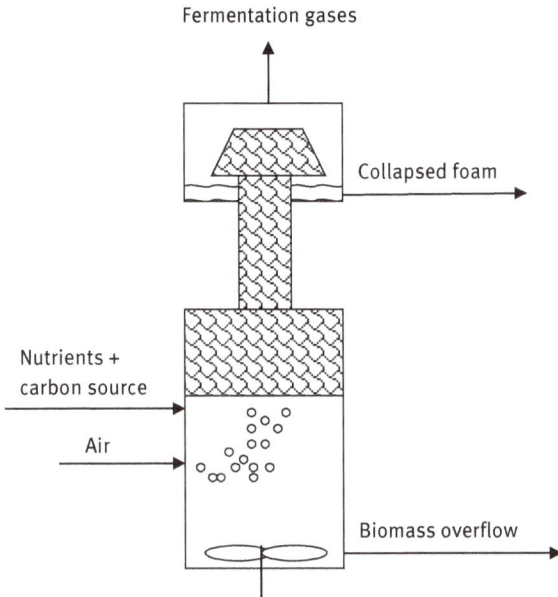

Figure 8.13: Schematic of a continuous/batch fermenter with a foam fraction unit [30].

References

[1]   Tadros TF. Applied surfactants. Weinheim: Wiley-VCH Verlag GmbH & Co. KGaA, 2005:19–51.
[2]   Salager JL. Surfactants types and uses. FIRP booklet, 300, 2002. https://www.alliedmarketre
       search.com/surfactant-markethttps://ihsmarkit.com/products/chemical-surfactants-scup.
       html.
[3]   Patel MK, Theiß A, Worrell E. Surfactant production and use in Germany: resource
       requirements and CO2 emissions. Resour Conserv Recycl. 1999;25:61–78.
[4]   Behler A, Biermann M, Hill K, Raths HC, Saint Victor ME, Uphues G. Industrial surfactant
       syntheses. Surfactant science series. New York: Marcel Dekker Inc, 2001.
[5]   Onyegbado CO, Iyagba ET, Offor OJ. Solid soap production using plantain peel ash as source
       of alkali. J Appl Sci Environ Manage. 2002;6:73–7.
[6]   He X, Guvench O, MacKerell AD, Jr, Klein ML. Atomistic simulation study of linear
       alkylbenzene sulfonates at the water/air interface. J Phys Chem B. 2010;114:9787–94.
[7]   Kocal JA, Vora BV, Imai T. Production of linear alkylbenzenes. Appl Catal A. 2001;221:295–301.
[8]   HERA linear alkylbenzene sulphonate (CAS No. 68411-30-3). Revised Environmental Aspect of
       the HERA Report 2013. Available at: http://www.heraproject.com/files/HERA-LAS%20revised
       %20February%202013.pdf. Accessed: 27 July 2019.
[9]   Zoller U, Sosis P. Handbook of detergents, part F: Production. CRC Press, 2008:2008.
[10]   In: Farn RJ, editor(s). Chemistry and technology of surfactants. Oxford: John Wiley & Sons, 2008.
[11]   Wieczorek D, Gwiazdowska D, Staszak K, Chen Y, Shen T. Surface and antimicrobial activity
       of sulfobetaines. J Surfactants Deterg. 2016;19:813–22.
[12]   Rosen MJ. Surfactants and interfacial phenomena. 3rd ed. Hoboken, New Jersey: John Wiley &
       Sons, Inc, 2004.

[13] Herrwerth S, Leidreiter H, Wenk H, Farwick M, Ulrich-Brehm I, Gruning G. Highly concentrated cocamidopropyl betaine – the latest developments for improved sustainability and enhanced skin care. Tenside Surfactants Deterg. 2008;45:304–8.

[14] Schnuch A, Lessmann H, Geier J, Uter W. Is cocamidopropyl betaine a contact allergen? Analysis of network data and short review of the literature. Contact Dermatitis. 2011;64:203–11.

[15] Hunter JE, Fowler JF. Safety to human skin of cocamidopropyl betaine: A mild surfactant for personal-care products. J Surf Det. 1998;1:235–9.

[16] Kwaśniewska D, Staszak K, Wieczorek D. Zieliński 2015 synthesis and interfacial activity of novel heterogemini sulfobetaines in aqueous solution. J Surfactants Deterg. 2015;18:477–86.

[17] Shoaib A, Fuller J. Amphoteric surfactants for household and I&I: widely used in the personal care industry, these materials offer many benefits to household manufacturers too. Househ Pers Prod Ind, 2002. Available at: http://www.highbeam.com/doc/1G1-85281422.html. Accessed 7 Oct 2013.

[18] Wieczorek D, Dobrowolski A, Staszak K, Kwaśniewska D, Dybik P. Synthesis, surface and antimicrobial activity of piperidine based sulfobetaines. J Surfactants Deterg. 2017;20:151–8.

[19] In: Farn RJ, editor(s). Chemistry and technology of surfactants. Oxford: Blackwell Publishing Ltd. *Hepworth 2006 P.*:Non-ionic Surfactants w 2006.

[20] Myers D. Surfactant science and technology. Haboken, New Jersey: John Wiley & Sons, Inc, 2005:29–79.

[21] Broze G. Surfactant science series, handbook of detergents, part a: properties. Boca Raton: CRC Press. De Guertechin, L. O. (1999). Surfactants: classification. w. 1999:7–46.

[22] Zieliński R. Surfaktanty. Budowa, właściwości, zastosowanie. Wydawnictwo Uniwersytetu Ekonomicznego w Poznaniu, 2013.

[23] von Rybinski W, Hill K. Alkyl polyglycosides—properties and applications of a new class of surfactants. Angew Chem Int Ed. 1998;37:1328–45.

[24] Mukherjee S, Das P, Sen R. Towards commercial production of microbial surfactants. Trends Biotechnol. 2006;24:509–15.

[25] Santos D, Rufino R, Luna J, Santos V, Sarubbo L. Biosurfactants: multifunctional biomolecules of the twenty-first century. Int J Mol Sci. 2016;17:401.

[26] Md F. Biosurfactant: production and application. J Pet Environ Biotechnol. 2012;3:2.

[27] Haba E, Espuny MJ, Busquets M, Manresa A. Screening and production of rhamnolipids by Pseudomonas aeruginosa 47T2 NCIB 40044 from waste frying oils. J Appl Microbiol. 2000;88:379–87.

[28] Dubey K, Juwarkar A. Distillery and curd whey wastes as viable alternative sources for biosurfactant production. World J Microbiol Biotechnol. 2001;17:61–9.

[29] Banat IM, Franzetti A, Gandolfi I, Bestetti G, Martinotti MG, Fracchia L, et al. Microbial biosurfactants production, applications and future potential. Appl Microbiol Biotechnol. 2010;87:427–44.

[30] In: Kjellin M, Johansson I, editor(s). Surfactants from renewable resources. Chichester, UK: John Wiley & Sons, 2010.

[31] Lomax EG. Amphoteric surfactants. New York: Marcel Dekker, 1996.

Maciej Staszak

9 Process simulation approach in computer aided industrial design

Abstract: The work presents the process of design using simulation approach. The fundamental unit operations were described including chemical reaction, separation in distillation units and heat exchange process together with thermodynamics considerations. Detailed description of the available methods and procedure is provided. The design process is based on the use of Chemcad software.

Keywords: process simulation, CHEMCAD, flowsheeting software, design

9.1 Introduction

The following text covers key elements of design process with the use of computer aided design (CAD) tools. This article is based on the use of Chemcad flowsheeting simulator but in case of any other CAD tools used the general concepts are valid. The methods of design in any chemical engineering CAD flowsheet software are in principle the same; therefore, they are also applicable for wide variety of other similar tools.

In general, the work steps required during the design process using CAD tools is analogous to the manual design stages in pre-CAD generation era. The act of design and procedure of simulation of industrial process are typically regarded to be on the opposite sides in the field of numerical approach to the problem. However, it is almost inevitable that the process of design is intertwined with process of numerical simulation at many stages of the designing course. Especially that many of the semiautomatic or automatic algorithms internally perform a simulation steps when converging to the required solution.

The CAD software, such as Process Systems Enterprise PSE, Chemcad, Aspen Plus, and many other available on commercial market or as free and open tools, typically represents the industrial installation as a graphical flowsheet composed of connected unit operations. The connections represent the material and energy flow paths and are understood as logical connectors between units. The process of solving such a flowsheet can be described as a process of finding the optimal unit operations settings, conditions and geometries for the process functioning with objectives desired. The unit operations and the entire flowsheet can be solved by taking a few general steps:

This article has previously been published in the journal Physical Sciences Reviews. Please cite as: Staszak, M. Process simulation approach in computer aided industrial design *Physical Sciences Reviews* [Online] 2020, 5. DOI: 10.1515/psr-2019-0038.

https://doi.org/10.1515/9783110656367-009

A. Defining and applying the process requirements and solving the mass and energy balances around unit operations and in the whole flowsheet.
B. Sizing the units to meet the requirements.
C. Rating the created model of unit operations and flowsheet.

The steps above are also subject of different levels of complexity. The first step (A) results in a general description of the process, namely actual flowrates, concentrations, temperatures and basic setting of units. The values obtained can be treated as the ideal ones in the sense that they fulfill the process requirements exactly as required by the designer in the ranges of feasibility. The second step (B) is performed on the basis of already estimated variables and is required to obtain sizes and geometries of unit operations involved in the process. The last step (C) is a verification step which can be understood as a process of simulation of created design. In this last step, a user/designer obtains realistic description of the process variables values, which differs from the idealization obtained at step (A). In some cases, it is enough to find optimal steady state of the process, but in some cases, like in a control system design, one must also perform the latter steps in transient mode. Transient mode or dynamics describes the process behavior in time and allows to check how the system behaves during startup or downtime and in any disturbance situation that can be applied to the system during its work. The dynamic approach enables the designer to find settings for units for which a time response analysis is crucial, for example in the case of PID controllers tuning.

The required level of complexity can essentially be applied at any of the steps mentioned. In general a designer may want to find optimal state for flowrate driven or for pressure driven description of the process. The first requires to apply flow data to the selected streams in the flowsheet, typically the feed streams, and then solving the balances according to the setting applied to the unit operations. The pressure has no direct impact on the flowrates in such a case so this is a simplified approach. The other method, more realistic and at higher level of complexity is to consider the pressures in the installation as driving forces for the flows. In this case, the problem is much more complex as it requires to perform a design of the hydraulics of the process. Many times such an approach results in complex flow situations like so called backflows or no flow at all, which in typical situations may not be intended but depict the realism of behavior of the flows for given units settings. Also the numerical difficulty is much higher in such a case which poses additional problems and demands larger knowledge of how to obtain converged solution. In this work, most often unit operations are described, their use and the concepts in the process of design.

The concept of the design is based on the knowledge of the nature of the process itself. The physical and chemical properties of the materials and resources, the chemical reactions kinetic data, thermodynamics of the systems involved understood as complete description of components and their mixtures, are the key elements when deciding on the directions of the design. A wide area of those aforementioned requirements is available through the process simulation software.

9.2 Physical and chemical properties

Typical way of property data storage is done in the form of specific correlations, which relate the required property with temperature at some given range. One of the most utilized commercial database is DIPPR (Design Institute for Physical Properties) [1] which in the half of year 2019 contained 2332 compounds in fully described database. Most of the compound are chemical individual but some are as mixture entity where an example is air. The typical commercial simulators that use this database are for example Chemcad and ProSim.

The DIPPR database stores the data in the form of values of coefficients (a, b, c ...) for appropriate temperature dependent equations (see Table 9.1). Those equations then are used by the software to estimate the properties of components.

Many software packages make use of other, similar correlations as well, where one of the examples is well known as Antoine equation for estimation of vapor pressure of pure components. The above data stored in the form of correlation equations are designed for pure components. On the other hand for the mixtures, the specific mixing rules must be applied to obtain realistic physicochemical description of the combinations of components used. The choice of proper mixing rules to the given mixture is performed by the CAD software itself, based on the property chosen and type of components taken into consideration and finally, on the thermodynamic model of calculation. The simplest of the mixing rules combine properties of components considered as ideal solution by assuming additivity and the use of molar fractions as a weighting factor. For ideal mixtures for which ratio of volumes of components V_i to the total volume of the mixture V are the same as molar fractions x_i, that is expressed by Amagat's law, see eqs. (9.1) and (9.2), for partial volumes:

$$V = \sum_i V_i \qquad (9.1)$$

and

$$\frac{V_i}{V} = x_i \qquad (9.2)$$

Different mixing rules apply to miscellaneous properties, especially when dealing with nonideal mixtures. Most popular, selected mixing rules are presented in the Table 9.2 below:

The important area of use of the mixing function, which need to be mentioned, are equations of state (EOS). Such equations relate the amount of matter within its

Table 9.1: Selected, most often used correlations for physical properties by DIPPR.

Property y	Recommended equation	DIPPR ID
Solid density, liquid thermal conductivity, liquid/solid heat capacity	$y = a + b \cdot T + c \cdot T^2 + d \cdot T^3 + e \cdot T^4$	100
Vapor pressure, liquid viscosity	$y = exp\left(a + \dfrac{b}{T} + c \cdot lnT + d \cdot T^e\right)$	101
Gas/liquid thermal conductivity, gas viscosity, surface tension	$y = \dfrac{a \cdot T^b}{1 + \frac{c}{T} + \frac{d}{T^2}}$	102
Second virial coefficients	$y = a + \dfrac{b}{T} + \dfrac{c}{T^3} + \dfrac{d}{T^8} + \dfrac{e}{T^9}$	104
Saturated vapor pressure, liquid density	$y = \dfrac{a}{b^{1 + \left(1 - \frac{T}{c}\right)^d}}$	105
Heat of vaporization, surface tension	$y = a(1 - T_r)^{b + c \cdot T_r + d \cdot T_r^2 + e \cdot T_r^3}$	106
Ideal heat capacity of gas	$y = a + b\left[\dfrac{c/T}{sinh(c/T)}\right]^2 + d\left[\dfrac{e/T}{cosh(e/T)}\right]^2$	107
Liquid heat capacity	$y = \dfrac{a^2}{\tau} + b - 2 \cdot a \cdot c \cdot \tau - a \cdot d \cdot \tau^2$ $- \dfrac{c^2\tau^2}{3} - \dfrac{c \cdot d \cdot \tau^4}{2} - \dfrac{d^2 \cdot \tau^5}{5}$	114
Alternative for liquid viscosity and vapor pressure.	$y = exp\left(a + \dfrac{b}{T} + c \cdot lnT + d \cdot T^2 + \dfrac{e}{T^2}\right)$	115
Liquid density	$y = a + b \cdot \tau^{0,35} + c \cdot \tau^{\frac{2}{3}} + d \cdot \tau + e \cdot \tau^{\frac{4}{3}}$	116
Alternative for ideal gas heat capacity	$y = a + b\left[\dfrac{(cT)^2 \exp\left(\frac{c}{T}\right)}{\left(\exp\left(\frac{c}{T}\right) - 1\right)^2}\right] + d\left[\dfrac{\left(\frac{e}{T}\right)^2 \exp\left(\frac{e}{T}\right)}{\left(\exp\left(\frac{e}{T}\right) - 1\right)^2}\right]$ $+ f\left[\dfrac{\left(\frac{g}{T}\right)^2 \exp\left(\frac{g}{T}\right)}{\left(\exp\left(\frac{g}{T}\right) - 1\right)^2}\right]$	127

volume and pressure, allowing to estimate various properties like density, phase, activity or fugacity. The equations of state are derived for single components in their fundamental form but their coefficients can also be a subject of mixing functions to give a description for mixtures.

From the user point of view, the estimation of mixture properties like its density, viscosity and other properties can be done by invoking the procedure for selected stream's composition and conditions. For a given stream at the flowsheet in Chemcad, which contains required definitions of temperatures, pressures, flows and compositions, a user can invoke a procedure contained in the menu "Plot/ Stream properties". That action will display a menu window for selected stream in which a user can make a choice of which property for the mixture to print in the form of a file and plot on the graph.

The proper selection of thermodynamic description is crucial to account for realism of design calculations. This is typically done on the basis of specific procedure,

Table 9.2: Most popular mixing rules for mixtures properties.

Property	Mixing rule	Description
Density	$\rho = \sum_i \rho_i x_i$	For mixtures in which excess volume is negligible
	$\rho = \sum_i \rho_i \dfrac{V_i}{\sum_i V_i + \sum_i V^E_{\,i}}$	For mixtures where excess volume $V^E_{\,i}$ is significant
Viscosity – gases	$\eta_{gmix} = \sum\limits_{i=1}^{N} \dfrac{y_i \eta_{gi}}{\sum\limits_{j=1}^{N} y_j \varphi_{ij}}$ where $\varphi_{ij} = \dfrac{\left[1 + \sqrt[2]{\dfrac{\eta_{oi}}{\eta_{oj}}} \cdot \sqrt[4]{\dfrac{M_j}{M_i}} \right]^2}{\dfrac{4}{\sqrt[2]{2}} \sqrt[2]{\dfrac{1+M_i}{M_j}}}$	Wilke mixing rule [2] for gas mixtures
Viscosity – liquids	$\ln \eta_{lmix} = \sum\limits_{i=1}^{N} x_i \ln \eta_{li}$	Grunberg-Nissan [3] classic mixing rule for liquids
Diffusion coefficients for liquids	$D_{12} = x_2 D^o_{12} + x_1 D^o_{21}$	Caldwell and Babb [4] concentrations-weighted arithmetic average, (binary mixtures)
	$D_{12} = \left(D^o_{12} \right)^{x_2} \left(D^o_{21} \right)^{x_1}$	Vignes [5] power-weighted average (binary mixtures)
	$D_{12} = \dfrac{\left(D^o_{12}\eta_2 \right)^{x_2} \left(D^o_{21}\eta_1 \right)^{x_1}}{\eta_{mix}}$	Leffler and Cullinan [6] viscosity-weighted average (binary mixtures)
	$D_{ij} = \prod\limits_{k=1}^{n} \left(\sqrt{D^o_{ij} D^o_{ji}} \right)^{x_k}$	Extension of Vignes to multicomponent mixtures [7].

so called software-wizard which helps the user to make proper choice regarding component selection, and temperature/pressure settings for the process to be designed.

9.3 Creating the flowsheet and streams definition

In the initial part of the work with the software a user makes a choice of components which take part in the process. It is important to mention that all the components, substrates, products, auxiliary substances must be selected for the given process. There is always a possibility to add additional components during the work, but the fundamental and correct way of work is to perform right choice of all the species existing in the process being designed. This becomes important and highly relevant in this initial moment of the work when the components selection impacts the algorithm of the thermodynamic wizard, which is a selector for the thermodynamics.

In some cases although, a full list of selected components might not be right choice for thermodynamic selector procedure. As an example suppose a user needs to perform a project in which there is a mixture of hydrocarbons and a utility (auxiliary) component, in this case sulfuric acid water solution. Suppose further that the acidic water solution, being a composition of two utility components, will never be mixed with the organic part and main design focus is devoted to the hydrocarbon part. In such a case a proper thermodynamic description should not take into account acidic water mixed together with the organic phase. The default behavior of the selector algorithm in such case will be to select the electrolyte package to perform accurate calculations for water and sulfuric acid solution. Instead, when the main focus is devoted to the hydrocarbon mixture, some equation of state should be used, like SRK (Soave–Redlich–Kwong) equation. For this reason the user is able to exclude the water and sulfuric acid from the decision list. The "Thermodynamic wizard" which is called automatically at initial components selection step or can be invoked manually from the "Thermophysical/Thermodynamics Wizard", allows making a selection of "Select components to ignore" which will command the software not to take such component into account during the thermodynamic model algorithm selector.

The selection of components and thermodynamic method are the first actions required to begin the project. The default way a CHEMCAD software works and calculates the flowsheet is called "Sequential modular". This specific option can be found and changed if required, in menu "Run/Convergence" which is a main place where a user can fine-tune global numerical settings. Sequential way of calculation consist on identifying the flowsheet and units order from inlets to outlets, and calculating consecutively each located unit separately. This means that the user needs to define feed streams and the minimum required settings for unit operations. Then the software will return an answer in the form of resultant compositions, pressures, temperatures and conditions for all the units in the flowsheet. Such way of performing the calculation has huge consequence on what can and what cannot be calculated or taken into account. Main advantage of such approach is that the flowsheets generally converge very fast to the solution required. The key drawback although is that the downstream units (located "after" in the sense of the flow direction), will not affect the upstream units (located "before" in the sense of the flow direction). There are some exceptions to that, for example when designing a control valve. If the user needs to set flowrate somewhere "backwards" in some upstream unit, the control valve can be set to do that. However, the general inability to affect upstream units and streams is the reason that such approach cannot account for many hydraulic effects. This is of course well known and accepted simplification, very useful at initial stages of working with the project. A user can always change this approach to "Simultaneous modular" in which the convergence can be much more difficult to obtain but the solver can take into account whole flowsheet "at once". This second approach enables for the interaction between units regardless of their position in the flowsheet. In such situation a unit located at outlet may interact or affect conditions at units located in the vicinity of inlet. This approach is required, and switched on automatically, when a user needs to take into account hydraulic effects in the flowsheet. Simultaneous modular approach generally takes all the equations that formulate the flowsheet problem and solves them as a single, large mathematical/numerical problem. A user should be aware that this approach presents higher degree of complexity and consequently is more difficult for the numerical engine to solve. It is advised then to use the default sequential modular approach at initial stages of design, and after obtaining converged solution, to switch to simultaneous modular if required. Also in such equation oriented (EO) approach not every thermodynamic model is available (state for Chemcad version 7). The example here is the electrolyte package available in sequential modular approach and not in the simultaneous mode.

The user is required at least to define the inlet streams conditions. The definition obeys the thermodynamic degree of freedom law, where user is required to fill in exactly as much information as needed. The stream thermodynamic conditions are temperature, pressure and vapor fraction (phase state). All three, given stream composition, are mutually related leaving two degree of freedom. A user is required to define two of these three thermodynamic variables. Typically, it is temperature and pressure leaving the vapor fraction to be calculated by the software. However, the selection is a free decision and if a user defines temperature and vapor fraction, the software will calculate corresponding pressure depending on the variables defined.

9.3.1 Flash calculations

This is important to explain the meaning and purpose of vapor fraction variable. The variable is defined on molar basis eq. (9.3) and describes the phase state of the mixture in the stream or unit operation:

$$VF = \frac{n_V}{n_L + n_V} \tag{9.3}$$

In other words, the vapor fraction defines the fraction of mixture that is vaporized. The VF equal zero means that the mixture is in the liquid state, the value one means it is in vapor state. Any fractional value between zero and one means the mixture is in both phase states and phases are in equilibrium – that is in most cases, but not always, a boiling condition. When a user defines a stream composition using VF variable, the zero value defines a saturated liquid, located at the bubble point line on the TP diagram. It means that any increase of temperature or decrease of pressure will start the liquid mixture to boil, a state when the first boiling bubble appears. On the other side if a user defines vapor fraction value to one, it means the mixture is vapor state, located at the dew point line on the TP diagram. It means that any decrease of temperature or increase of pressure will start vapor mixture to liquefy, a state when the first droplet appears.

The vapor fraction variable is essential element of so-called flash calculations, which are the calculations performed to estimate the state of the mixture. The flash calculation involves the thermodynamic model to describe the mixture and are based on solving the Rachford–Rice eq. (9.4):

$$\sum_{i=1}^{n} \frac{z_i\,(K_i - 1)}{1 + VF\,(K_i - 1)} = 0 \tag{9.4}$$

In the above the z_i is the mole fraction of the ith component in the n components mixture, understood as both liquid and vapor mixture, and K_i is the equilibrium constant of the ith component. The equilibrium constant is calculated on the basis of thermodynamic method chosen at the initial stages of the project.

Solving this equation for vapor fraction can be done only by numerical method, because no closed form solution exists. Depending on the complexity of the thermodynamics it can many times diverge, meaning that no physical solution was found. The problem of flash calculations is important due to its commonness, and its solving is required for many different unit operations. The first situation in which the flash calculation is performed takes place during the streams definition. Typically, the definition of temperature and pressure of a given mixture forms a mathematical problem which is numerically most stable and therefore easiest to solve for vapor fraction. On the other hand, a problem defined by the pressure and vapor fraction are harder to solve for unknown bubble or dew point temperature. Despite of the age of the Rachford–Rice equation and the flash problem, there is still ongoing scientific discussion and effort about a methods and algorithms to obtain realistic results. A user may expect a numerical problems or difficulties for high number of components in the mixture or for the mixtures, which involve complex thermodynamic description.

9.3.2 Dealing with flash numerical problems

A situation when the flowsheet solver returns a message about flash error may happen in different situations, for example: when defining a stream or when designing a heat exchangers or distillation unit. It is then important part of every flowsheet design calculation to manage properly the problematic state or condition to obtain meaningful, physically reliable results.

9.3.2.1 Flash tolerance

The key numerical parameter of a flash algorithm is the criteria of convergence. The acceptable numerical error of iterative algorithm searching for the solution is called tolerance. The default value of flash tolerance is set to 10^{-3} which means that the stepwise, iterative algorithms performs a numerical test which results in the relative difference between consecutive numerical calculation steps. A user can present this value as an amount of error that is accepted when looking for the solution. A higher value of tolerance is less rigorous and results in possible less accurate results but obtained faster, with smaller number of iterations. Setting this value to any lower will result in more rigorous approach, with smaller error acceptable and thus of higher accuracy. As a result of smaller tolerance value a longer calculation time should be expected. When a flowsheet calculations result in any flash numerical error message a user is encouraged to lower the flash tolerance, even several orders of magnitude. This can be done in menu "Run/Convergence" in which a proper input field is responsible for flash tolerance.

9.3.2.2 Local thermodynamics

For the situation when even very low flash tolerance does not allow to obtain correct results a user may change thermodynamic model for the unit operation of interest. The important fact from the flowsheet design procedure point of view is that it is better to obtain a solution with less accurate method than not to obtain the results at all. A less accurate thermodynamic method can many times cause less numerical flash problems. The typical example here is when the mixture is highly non-ideal and some of so called activity coefficient model was chosen like NRTL. This method being very accurate for highly non-ideal mixture may sometimes fail to converge when solving the flash problem. If the flash tolerance adjustment also failed then the user should switch the method to easier to converge, for example functional group method like UNIFAC could be used instead. UNIFAC method will usually be less accurate for highly non-ideal systems than NRTL but can still give reasonable good estimations.

The user may choose different than global method for selected unit operations by selecting unit at the flowsheet and navigating to menu "Thermophysical/ Thermodynamic settings" and selecting "Set local thermodynamics". The user will then be presented with the menu window of which "Local K-value model" option is crucial at this stage.

The proper selection of the thermodynamics requires a bit of practice and a knowledge about thermodynamic models themselves and their "behavior" during numerical simulations. It will not always be the case that the exemplary NRTL method should always be changed to UNIFAC. Some other choices might be better for specific cases. The description of thermodynamic modeling is however beyond the scope of this chapter.

9.4 Chemical reaction

One of the central aspects of the industrial process design is undoubtedly the chemical reaction and designing of chemical reactors. There are different ways how the chemical reaction can be described. The amount of matter being converted form one specie to another during chemical reaction can described by any proper mathematical model. It depends on the design objectives how complex the description needs to be. The available in the literature chemical kinetics data shows that many researchers use several mathematical models and formulations to accurately depicts how the reaction runs. Due to the fact of wide variety of available mathematical descriptions of chemical reactions it is not possible for any flowsheet simulator to account for all of them by typical user interface. The characteristic solution to this problem is to allow to use typical models of chemical reactions and possibly to share a method of more complex mathematical input, by means of some external programming language.

The four principal types of descriptions of chemical reaction exist and are widely used in teaching, academia and also in design software. Taking them into account in order of usefulness and complexity they are: chemical conversion, chemical equilibrium, thermodynamic approach, and kinetic description. These four types of description sometimes are referred as chemical reactions models. They found to be included into simulator software in the form of convenient graphical user interfaces allowing to input required chemical data. Based on the description of Chemcad software the four types are built in the form of adequate unit operations and are described below.

9.4.1 Thermal modes

Every of the four calculation approaches described below, allows also to define the thermodynamic conditions related to how the heat energy is to be processed. The three key modes are available, namely:

– Isothermal – software calculates the energy balance for the reactor and return amount of energy required to be exchanged to keep constant temperature of the process
– Adiabatic – software calculates the energy balance but in this case the temperature of the process in the reactor is being searched. The process is regarded as perfectly isolated system with no heat exchange.
– Heat duty – similar to the adiabatic in the sense of algorithm approach, but the user defines amount of heat energy exchanged during the process. The temperature is calculated. Process is neither adiabatic nor isothermal, except if the user leaves field energy empty, which is treated as zero heat exchange. This type can be referred to as diabatic process, exchanging heat together with temperature change.

From the numerical solution to the energy balance equation point of view, it is much easier to converge for isothermal case than for adiabatic. This is due to the formulation of the heat balance equation, for isothermal the unknown is heat energy, for adiabatic (and diabatic) the unknown is temperature. The very reason is that the temperature is entangled into the heat balance equation, opposite to the total heat exchanged, which is almost explicitly solvable. For complicated cases, when user experiences many numerical divergence related errors, the first suggestion is to start from isothermal settings. After successful isothermal recalculation then to switch to adiabatic case.

9.4.2 Chemical conversion approach

The chemical conversion description is built in the form of Stoichiometric reactor unit operation in Chemcad. The reaction conversion α is defined by well-known formulation eq. (9.5) for given reaction substrate:

$$\alpha = \frac{n_0 - n}{n_0} \tag{9.5}$$

In which the n correspond to amount of moles of given component in the mixture. The n_0 refers to initial number of moles. This simple approach allows user to make a definition of chemical reaction according to known stoichiometry and species conversion.

This is important to note that this calculation approach does not take into account chemical nature of the reactive species. It is not intended also to find the reactor volume nor residence time. If execution is successful it only presents output concentrations and temperature according to thermodynamic settings.

9.4.3 Equilibrium approach

State of chemical equilibrium is described by typical definition of equilibrium constant defined by activities a or molar concentrations c:

$$K_a = \prod_{i=1}^{N} a_i^{v_i}, \tag{9.6}$$

$$K_c = \prod_{i=1}^{N} c_i^{v_i}. \tag{9.7}$$

The N is the number of components and v_i are the stoichiometry coefficients. It is important to note that they are numbered according to thermodynamic convention, substrates are negative while products are positives values. The stoichiometry coefficients are convenient variables of calculation but sometimes they do not exactly reflect the nature of the reaction so in specific cases the arbitrary equilibrium constant exponentials values might be used. The Chemcad unit that provides the equilibrium model of chemical reaction to the designer is called Equilibrium reactor.

The name of equilibrium constant might be a bit confusing, in fact is not constant and depends on many factors. The most typical is temperature dependency. The relation of temperature to the equilibrium constant has in great range of temperatures exponential nature. This is typical to describe it by eq. (9.8) which contains specific constants denoted as $A, B, C \dots$:

$$\ln(K) = A + \frac{B}{T} + C\ln(T) + D \cdot T + E \cdot T^2 \tag{9.8}$$

When only a constant value of K is available then it is enough to specify only A constant by taking exponent of it, e^A. The more constants is specified the more accurate temperature dependency can be described. For the case when exactly to measurements of equilibrium constant $K1$ and $K2$ in two different temperatures $T1$ and $T2$ are available the closed source formula exist eq. (9.9) to estimate A and B constants from eq. (9.8):

$$A = -\frac{\ln(K2)T2 - \ln(K1)T1}{-T2 + T1},$$

$$\tag{9.9}$$

$$B = \frac{T1 \cdot T2(-\ln(K1) + \ln(K2))}{-T2 + T1}.$$

For three and more distinct measurements in different temperatures closed source formulation for the constants can also be found but they become impractical to use due to their lengthy appearance. For more numerous collections of data any nonlinear regression algorithm is suitable to find all the constants of eq. (9.8) (*Genfit* function of Mathcad or *NonlinearFit* of Maple software).

9.4.3.1 Approach fraction

The typical nature of equilibrium systems and chemical equilibrium in particular, is that when a reactive system approaches equilibrium state the rate of the process decreases. In many cases, it can take hours or days to reach almost exact chemical equilibrium state. In the light of the above, it is not practical and required in many industrial situations to wait until this state is reached. A designer may make a decision that the time enough for the process is when it reaches some specific percentage or fraction of the total equilibrium reach time. It depends on the exact reactive system at which point it is practical to stop the reaction and proceed further in the stream of reactive components. The approach fraction is useful to define such final point of the reaction. Initial point of chemical reaction, when the conversion is till zero, defines approach fraction also as zero. The equilibrium point, when conversion is at equilibrium, defines it as one. It means that when a designer uses exemplary value of 0.8 the assumption results in 0.8 (80%) of chemical approach to equilibrium state.

9.4.4 Thermodynamic approach

Main concept of finding the mixture composition after chemical reaction is to relate the concentrations and other mixture properties with total Gibbs free energy of the system. The Chemcad unit which provides this method to the user is Gibbs reactor. The formulation [8] eq. (9.10) relates amount of moles n_i, component fugacities at actual f_i and standard state f_{i0} with total Gibbs energy:

$$\Delta G_{tot} = \sum_{i=1}^{N} n_i G_i + RT \sum_{i=1}^{N} n_i \ln \frac{f_i}{f_{i0}} \qquad (9.10)$$

The problem is to estimate the values of n_i for all mixture components which minimize the ΔG_{tot}. To maintain physical results the minimization is bounded by the constraint of mass balance:

$$\sum_{i=1}^{N} n_i a_{ji} = N_j, \quad j = 1 \ldots K \qquad (9.11)$$

The above mass balance relation forms K equations set for every atom j in every component i in the mixture. The number of moles of given atom j in the molecule i is depicted by a_{ji}. The total amount of moles of atom j in the mixture is given by N_j. The K variable is the total number of atomic elements taking part in the reaction. Such problem is usually solved by the use of Lagrange multipliers algorithm which is suitable to find extrema bounded to specific constraint.

Setting the Gibbs model of chemical reaction in the flowsheet is probably the simplest of all approaches but it can sometimes pose numerical difficulties during calculations. Using Chemcad a user is required, in basic approach, to select thermal conditions. No stoichiometry coefficients are required. It is however very important to indicate (on separate tab of Gibbs reactor) the inert components, that is a substance that does not take part in reaction, so it is not involved into calculation.

9.4.4.1 Typical numerical problems
Characteristic problem with thermodynamic approach arise when designer selects adiabatic thermal conditions. When the procedure searches for compositions and temperature it may happen that the solution is not found. It is mainly caused by the initial, starting point of calculation which is far apart from the actual solution, minimization of Gibbs total energy. A stepwise manual method here can be helpful to find the adiabatic solution of the reactive process. Below is a proposed procedure to adiabatic case solving:
1   In the first step a user should calculate the problem using isothermal settings and observe the calculated amount of energy exchanged to keep that constant temperature.

2 Depending on the sign, negative or positive, a user should decrease or increase the isothermal temperature settings and repeat the calculation to find new value of exchanged energy.

3 When the exchanged energy value is located near zero (adiabatic) then user should switch back to adiabatic settings and find exact temperature of adiabatic process.

The only question is how close to zero energy exchanged is enough to start calculations with adiabatic thermal settings. Typically it is found when the energy starts to change rapidly with small isothermal temperature changes, especially when jumping to opposite sign is observed. Such behavior indicates that a user has located the vicinity of adiabatic process and from that point on the minimization algorithm should find the adiabatic result automatically.

9.4.5 Kinetic approach

The most versatile yet demanding model of chemical reaction is kinetic approach. Kinetics of chemical reaction describes, or in theory should explain, whole nature of a reaction. In most general description it gives an answer to how much matter is being transformed in the unit time in given volume. Many times it is formulated with temperature, pressure and/or other parameters dependent process. If properly measured the chemical kinetics may be used as accurate description in the field of chemical process design, taking into account any relevant process parameters. The most general formulations recognized in science or design field, refer to homo or heterogenic cases. Literature is rich of different type of kinetics formulations best suited to specific cases of chemical reactions. One of the most common and versatile expression for kinetics is the Arrhenius type exponential equation. It allows to account for the dependence of the conversion rate on the temperature, which is one of most important influence factor.

For the design purposes the chemical kinetics equation is formulated on the component basis, that is it describes the rate of consumption/creation of selected reactive component. For the component taking part in homogenous reaction the formulation is given by eq. (9.12):

$$r_i^{homo} = \sum_{j=1}^{N} \left(v_{ij} k_j \prod_{k=1}^{n_j} (C_k)^{a_{ki}} \right) \tag{9.12}$$

where

$$k_{ij} = A_j \exp\left(\frac{\Delta E_j}{RT}\right) \tag{9.13}$$

In the above A_j is frequency factor and ΔE_j is activation energy for j^{th} homogenous reaction. Number of homogenous reactions is given by N, number of reactants in j^{th} reaction is described by n.

Similarly, if the component takes part also in the heterogeneous reaction the description reads eqs. (9.14), (9.15):

$$r_i^{hetero} = \left(1 + \sum_{k=1}^{n_i} k_{ij}^{ads} \left(C_{kj}\right)^{b_{kj}}\right)^{-\beta_j} \tag{9.14}$$

where

$$k_{ij}^{ads} = \varphi_{kj} \exp\left(\frac{\Delta E_{kj}^{ads}}{RT}\right) \tag{9.15}$$

In the above φ_{kj} is the adsorption frequency factor and ΔE_{kj}^{ads} is adsorption activation energy. The coefficients b_{kj} and β_j are adsorption exponential factor for reactant k in reaction j and beta factor for reaction j, respectively. Both formulation reflect exponential temperature dependency of chemical reaction rate by Arrhenius type relations.

To obtain total rate of conversion for component i the above formulations are multiplied to yield:

$$r_i = r_i^{homo} r_i^{hetero} \tag{9.16}$$

Chemcad allows to specify all or selected part of the constants in the above formulations. Depending on the chemical reaction nature only a homogenous or heterogeneous part can be selected.

For any kinetic description, every reaction needs to be specified by its stoichiometric coefficients and at least by single rate constant. Most important fact is that kinetic approach is the only approach that allows to determine required reaction volume for different reactor geometries given the conversion desired.

Using Chemcad designer may specify required number of chemical reactions, the geometry of the reactor, and five different thermal setting. It is also very important to note that kinetic specification obeys separate unit selection, available at the second tab of the kinetic module. Every reaction parameters, corresponding frequency factors, activation energies etc. are to be specified on separate windows.

9.4.5.1 Reactor geometry

Two types of geometries can be selected for calculations using kinetic approach: tank and tubular geometry. According to a specification a user makes some of the options available changes. The most important although is that a good practice is to initialize the reactor using CSTR (continuous stirred tank reactor) geometry even if tubular is required. The reactor initialization allows afterwards to perform next steps by using any of the two geometries available (CSTR or PFR), but in case of tubular geometry selection it makes numerical calculations much easier to converge. The selection of CSTR mode allows designer to switch into design mode (given reactor was initialized) and make a definition of required reaction conversion to obtain an answer what is the volume needed. Such option, only available for CSTR, is very handy when searching for the optimal volume of the reactor. Because the PFR has more complex geometry specification, at least diameter and number of tubes and the length, the design options are not available. The design mode can be selected in the field area "Specify calculation mode" by using one of the two available options: "Specify volume, Calculate conversion" and "Specify conversion, Calculate volume", where the first is default one.

9.4.5.2 Thermal modes

Thermal modes available in Chemcad are geometry dependent. Isothermal, adiabatic and heat duty modes refer only to continuous stirred tank reactor and reflect design requirements exactly as stated in 4.1. The two more available are intended for tubular reactor geometry and are explained below.

9.4.5.2.1 Temperature profile

Specification of temperature profile allows designer to apply a conditions when the reactor wall temperature is known variable. The temperature may change along the reactor and can be specified in the form of appropriate table Chemcad displays when user makes this selection. The case is in most cases non-isothermal and non-adiabatic because the temperature of reaction changes according to the local temperature values. The exchanged heat energy is calculated for every geometrical position in the reactor and summated to obtain total heat duty.

9.4.5.2.2 Utility conditions

When setting the PFR thermal "Utility conditions" mode and conditions it is very important to keep in mind the flow configurations. At the very first a user is required to select proper iconic representation of the module on the flowsheet, which allows to attach additional feed and additional output side streams, similarly to Figure 9.1.

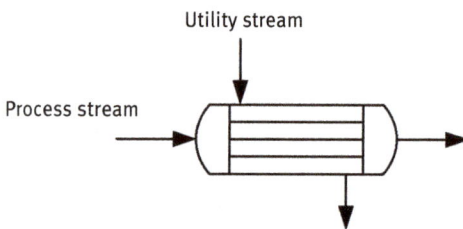

Figure 9.1: Tubular reactor with utility streams (co-current setup).

Those side streams are utility streams and define the flow of the heating or cooling media. Note to users of Chemcad version 7, unfortunately this icon is not available and without access to older version this icon cannot be accessed. This specific iconic module of kinetic reactor must allow to attach one process stream and one utility stream and corresponding outputs. As a workaround to this problem a user may export (save project with icon of kinetic reactor with properly connected streams) it from older version (e. g. Chemcad version 6) and import into Chemcad version 7.

From the heat exchange point of view two key setups are available: co-current and counter-current flows. Both setups can be selected at the second tab of the kinetic module. The setups differ obviously by the directions of reactive mixture and utility stream flows. The utility condition of the reactor must also be specified, namely heat exchange coefficient (area of heat exchange is calculated by the software based on the geometry sizes provided).

9.4.5.2.2.1 Numerical considerations

There is one very important thing of the numerical nature to keep in mind when selecting the flow setup. The tubular reactor geometry type is longitudinal in its nature. That means it has to be integrated in mathematical sense to obtain overall values of the process variables. The integration is internally defined by the numerical algorithm to start from the inlet process stream. Such a numerical algorithm has important effect on the limitations during the calculations.

When we take into consideration longitudinal dimension of the tubular reactor then for the co-current flow both utility and process stream enter in the same starting location. A user is required to define the temperatures at this starting point, namely in both reactor input streams. In effect a user needs to specify both inlet temperatures of the stream and outlet temperatures will be calculated accordingly.

The different, more problematic case is for counter-current setup. Both streams enter from opposite sides, process stream enters in inlet (starting) location and utility stream enters at the outlet (ending) position of the reactor. The numerical integration algorithm starts the integration from the inlet (starting) reactor location. As a result, a user needs to input inlet temperature of the process stream and outlet temperature of the utility stream. This means that the software back-calculates the inlet temperature of the utility counter-current stream. Such approach might sometimes pose some difficulties, also of numerical nature, because the user is required to enter the resultant, outlet temperature, which in fact is the variable which is sought for. The typical method is to recalculate flowsheet a few times to find reasonable, required temperatures at utility inlet. A general tool "Sensitivity study" might be very helpful in such repetitive recalculations.

9.4.5.3 Case studies

9.4.5.3.1 Single rate constant
This is simplest situation when reaction kinetics was measured for constant temperature results in single, constant value of rate coefficient. It is easy to note that eq. (9.13) reduces to A_j (frequency factor). Frequency factor is obligatory parameter to be specified in the field corresponding to chemical reaction kinetics.

Example: Direct chlorination of ethene reaction kinetics [9]

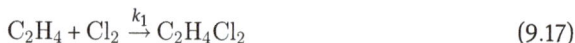

$$C_2H_4 + Cl_2 \xrightarrow{k_1} C_2H_4Cl_2 \qquad (9.17)$$

The temperature range investigated was 30–75 °C, the pressure range was up to 2 atm. The kinetics of the above reaction is described by typical formulation:

$$r = k_1 C_{C_2H_4} C_{Cl_2} \qquad (9.18)$$

The rate constant is given by single value:

$$k_1 = 0.132 \frac{m^3}{mol \cdot s} \qquad (9.19)$$

The user needs to specify only frequency factor field in the corresponding position of chemical kinetics description, paying close attention to the unit selection on the kinetic module tab.

9.4.5.3.2 Activation energy and frequency factor
Formulation of kinetics temperature dependency in the form of Arrhenius type equation is typical approach. Such formulation allows designer to make a corresponding specifications using Chemcad input fields for kinetics of selected reaction. Although the Arrhenius formulation hides the underlying reaction mechanism it is a good representation of temperature dependency of the rate of chemical reaction due to is practicality and simplicity.

Example: Benzene nitration by nitric acid [10]

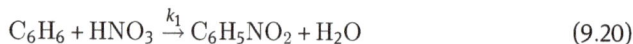

$$C_6H_6 + HNO_3 \xrightarrow{k_1} C_6H_5NO_2 + H_2O \qquad (9.20)$$

The benzene nitration was conducted by an excess of 45–70 weight % nitric acid in a homogeneous liquid phase. The temperature range of measurements was 60–90 °C. Estimated kinetic parameters for the kinetic formulation eq. (9.21), where rate constant k_1 is given by Arrhenus type equation:

$$r = k_1 C_{C_6H_6} \qquad (9.21)$$

were given by the tabulated data gathered for temperature range and selected nitric acid concentrations in Table 9.3:

Table 9.3: Frequency factors and energies for benzene nitration, frequency factor unit 1/s.

Concentration of HNO$_3$, weight %	60.0 °C	70.5 °C	80.1 °C	88.1 °C	Activation energy, kJ/mol
44.35	4.00·10^4	1.30·10^5	2.70·10^5	5.20·10^5	90 ± 2
56.7	1.14·10^6	3.33·10^6	7.34·10^6	1.45·10^7	89 ± 2
60.51	4.63·10^6	1.20·10^7	2.65·10^7	–	85 ± 2
62.28	8.61·10^6	–	–	–	–
63.8	2.02·10^7	–	–	–	–
66.69	6.39·10^7	–	–	–	–

Authors estimated temperature dependency for different ranges of nitric acid concentrations that are given by frequency factors and corresponding activation energies. Depending on the designer choice of the nitric acid solution concentration the required parameters can be directly specified in the corresponding tab of kinetic module, in a fields that correspond to the frequency factor and activation energy. Again proper attention must be paid to unit selection which is unique for reactor module (not units of whole flowsheet). There is no simple possibility to account for different kinetics according to different concentrations, it can only be done on the programming basis using for example Visual Basic for Applications macros.

9.4.5.3.3 Reaction catalyzed
Many chemical reactions kinetics cannot fit to the Arrhenius rate equations. Such example is reaction catalyzed when the catalyst amount or concentration is explicitly defined into rate expression.

Example
methyl acetate synthesis catalyzed by sulfuric acid [11]

$$CH_3OH + CH_3COOH \xleftrightarrow[K_2]{k_1, \ H_2SO_4} CH_3COOCH_3 + H_2O \tag{9.22}$$

The reaction was examined at 1:1 mole ratio of acetic acid to methanol, with the temperature starting from 32 °C to 60 °C and the catalyst concentration was in the range from 0.0633 mol/L to 0.3268 mol/L. The kinetic expression for the rate of reaction reads:

$$r = -k_1 C_{CH_3COOH} C_{CH_3OH} + k_2 C_{CH_3COOCH_3} C_{H_2O} \tag{9.23}$$

where
$$k_2 = \frac{k_1}{K_{eq}} \tag{9.24}$$

The rate constant is given by the modified Arrhenius equation by introducing additional variable, catalyst concentration $C_{H_2SO_4}$ [mol/L], which cannot be entered by Chemcad default fields:

$$k_1 = 3 \cdot 10^8 C_{H_2SO_4} + 8 \cdot 10^7, \quad \left[\frac{L}{mol \cdot min}\right] \tag{9.25}$$

The equilibrium constant is determined to be almost independent of temperature in the range investigated:

$$K_{eq} = 5.07 \tag{9.26}$$

The forward activation energy ΔE_f is equal to 62.721 J/mol and for backward reaction ΔE_b is equal to 62.670 J/mol. Final formulation of the kinetic expression must be rearranged to explicit component conversion rates:

$$r_f = -k_1 \exp\left(\frac{\Delta E_f}{RT}\right) C_{CH_3COOH} C_{CH_3OH}$$

$$r_b = \frac{k_1}{K_{eq}} \exp\left(\frac{\Delta E_b}{RT}\right) C_{CH_3COOCH_3} C_{H_2O} \tag{9.27}$$

In order to define individual, non-Arrhenius kinetics, Visual Basic for Applications (VBA) macro can be used. When defining general parameters for the kinetic unit, a user must select "Each reaction" for the expression type. This allows later on to select predefined template macro for chemical conversion rates. The equilibrium reaction in this example is converted into two distinct equations for forward r_f and backward r_b reaction. Such approach is expected by the predefined VBA macro. For both reactions Chemcad displays a specific window for kinetic parameters input. Compulsory data are stoichiometric coefficients, frequency factor (which will not be used so user should enter dummy variable e. g. 1). The kinetic expression list allows to select how user needs to enter the kinetic data – a VBA macro should be selected, which enables shows additional, second list which contains available macros. The RxnTemplate is right choice because this is already prepared Arrhenius type, Langmuir–Hinshelwood equation in the form of a VBA macro. The next step is to prepare actual macro for the kinetic calculations. Menu "Tools/Visual Basic" allows to create a user defined code. In the "Project" tool window a user should select and open a "Reactions". A macro RxnTemplate should be opened and be ready to edit. Programming rules are beyond scope of this text but there exist huge amount of materials available for reading in the form of internet resources or books. The subroutine RxnTemplate contains in its header section a comma separated list of variables that are passed form Chemcad core to macro, and one variable that should be calculated by the macro and returned to Chemcad core. The variables required for exemplary macro are fTemperature and C(). The f prefix in the name of temperature variable means the number is of floating type (real number). The brackets in the concentrations means it is a vector of concentration data for components which is ordered according to user selection made at the start of the project. This is very important however to note that numbering starts from 0th index. The pRateForm is the return variable which is to be calculated, it must contain actual value of reaction rate in units specified at the kinetic module window. The nRxnID is the number of actual chemical reaction for which Chemcad call this macro. Unfortunately and opposite to the definition of C(), the reactions are indexed starting from 1st value. For the sake of this example suppose a user selected the components in following order: acetic acid, methanol, methyl acetate, water, sulfuric acid.

The actual content of the macro should be removed, the part starting after the instruction On error ... to the line containing pRateForm = ..., the part is responsible for the default calculation scheme, which is not suitable for actual example. According to equations set eq. (9.27) the content of the macro should read:

```
If nRxnID = 1 Then
    pRateForm = -(3e+8*C(4) + 8e+7) * Exp(62.721/(8.314 * fTemperature)) * C(0) *
C(1)
    End If
    If nRxnID = 2 Then
```

pRateForm = (3e+8*C(4) + 8e+7) * Exp(62.721/(8.314 * fTemperature)) * C(2) * C(3)/5.07
   End If

The macro above calculates distinct values of reaction rates depending on the Chemcad requirements. When Chemcad invokes the macro, it passes the nRxnID value for which rate must be calculated. This is assured by the If-Then clause. This macro definition allows to account for catalyst concentration for the esterification example reaction.

Table 9.4: Chemical reactors summary, given input streams contain all required substrates.

Approach	Data required	Achievable result
Reaction conversion	– Stoichiometry – Reaction conversion α – Energy regime – isothermal, – adiabatic, – heat duty.	Output species concentration. Heat exchanged or output and reactor temperature.
Equilibrium	– Stoichiometry – Chemical equilibrium – Energy regime – isothermal, – adiabatic, – heat duty.	Output species concentration. Heat exchanged or output and reactor temperature.
Thermodynamic	–Energy regime – isothermal, – adiabatic, – heat duty.	Output species concentration. Heat exchanged or output and reactor temperature.
Kinetic	– Stoichiometry – Reaction kinetic data – Arrhenius or Hinshelwood type data – Unrestricted possibility to use any formulation by VBA macro – Energy regime – isothermal, – adiabatic, – heat duty. – Geometry – Tank (CSTR) – Tubular – Volume – Design mode possible for required conversion of reaction	Output species concentration. Heat exchanged or output and reactor temperature. Volume if design mode used for CSTR type Temperature and concentration profile for tubular reactor type

The Table 9.4 contains summary of the chemical reactor models available in Chemcad software, together with the key features.

9.5 Separation by distillation

One of the most common industrial process for fluid mixtures separation is distillation in column apparatus. Chemcad contains five units available for distillation in columns, of which the most complex yet versatile will be discussed here, that is SCDS unit (simultaneous correction distillation simulation). Calculation or simulation of this process requires proper description of vapor-liquid equilibrium. In fact this is an element that is key for the accuracy of the results that can be obtained from calculations. The choice of proper thermodynamics is performed at the very beginning of the flowsheet setup, when the user selects the components.

Although setting the options for distillation column in Chemcad is rather straightforward, the calculation of distillation unit is one of the most difficult from the numerical point of view. Regardless of the type of the column, packed or trayed, in mathematical terms it is described by huge equations set, which size is dependent on the size of the column and number of components involved. It is typical that mathematical model of the distillation column consist of several hundreds of algebraic equations in steady state and differential ones in transient mode. This poses high degree of complexity, especially visible when applying numerical method to solve such mathematical problem. For detailed description of distillation column mathematical models a reader should refer to literature [12, 13].

9.5.1 General column setup

A mixture being processed in the distillation unit is composed of components that are characterized by different boiling points. This is important to recognize a situation when there exist a component which in the process temperature becomes gas, typically for the process that runs above its critical temperature, and thus such component cannot be liquefied. Default behavior of distillation unit is to liquefy whole mixture at the top of the column to obtain liquid distillate stream (Total condenser option). However if the mixture contains at least one gaseous specie then partial condensation (Partial condenser) should be selected. This ensures that no gaseous specie will be liquefied and in effect no, unusual in many cases, cryogenic temperatures at the top will be obtained.

The compulsory data which must be entered by the user are height of the column counted by theoretical, balance segments in the default VLE (vapor liquid equilibrium) approach, and location of feed stream or streams. It must be noted that segments are numbered in top-down method. When user by the choice of calculation model, decides that the column is trayed, the segments are understood as trays. If the decision is to packed column type then in such a case the segment stay as arbitrary segments into which packing is mathematically divided and balanced.

Best approach to calculate and design a distillation column is to first use the VLE model, then size the column by the use of sizing procedure (menu "Sizing/ Distillation") and then to test the results by the use of one of the detailed, technical models: tray or packed column. The VLE model is default, general one which gives overall description of the process without delving into actual types and sizes of the equipment. VLE approach solves material and energy balance thus it gives in the results the flows, concentrations and temperatures at output streams and inside, along the column. The results obtained by VLE model are excellent starting point for detailed calculations using one of the technical, mass transfer models. The latter have additional functionality paid by larger workload, however. Due to including additional equations for mass transfer for each column segment they become much more difficult to solve by the numerical algorithm. On the other hand a user obtains the information about hydraulic regime inside the column, which highly affect efficiency of the process. The software enables indication of the flood or weeping situation which might happen for too high or too low flowrates in the column.

9.5.2 Process parameters

Main process parameters are settings for the condenser and for the reboiler. They are available on the second tab of the SCDS unit. The two lists correspond to the way by which the two key distillation elements work. Keeping in mind the difficulties that might appear during the calculations it is a good practice is to start the calculation from numerically easy settings. The most often utilized and very useful options are presented below.

9.5.2.1 R/D or V/B

These are typically numerically easiest options and they indicate exactly how they are named. The R/D option is the reflux ratio, namely ratio of reflux flow R to the distillate flow D. On the other side the V/B is the reboiler ratio, in this case it is ratio of vapor flow V that enters the column from the reboiler to the bottom stream B drawn from the column. Both values needs to be greater than zero. This is also typical behavior that altering their value, for example from 1 to 2 has much greater impact on the process than altering from 20 to 40, as indicated by the definition of the ratios.

9.5.2.2 Condenser and reboiler duty

These fields are required if the user has some knowledge of how much energy to draw (condenser) or to input by heating (reboiler) into the distillation system. This value is rarely known apriori but the option is useful if a designer finds optimal condition for the column in the design mode (next option) and wants to the condenser or reboiler remain exactly at the conditions found. For condenser the entered value must be negative, for reboiler it must be positive.

9.5.2.3 Mole/Mass fraction of component in distillate or bottom stream

This is a design mode for the column. Its function is to find process conditions for condenser or reboiler to satisfy the designer input, required concentration of selected component at selected stream. Designer specifies component and demanded concentration dependent on the concentration measure selected. By repetitive, stepwise numerical algorithm the software looks for the condenser/reboiler specification to match the concentration at output. Once the results are found the user may want to check the values of heat duty found on the displayed third numerical tab and use them for further calculation. This action is however not required but it saves calculation time because every time flowsheet is recalculated the distillation module will again start searching for already found data.

9.5.3 Notes

- Successful column calculation is highly dependent on the numerical properties of the mathematical problem being solved. To avoid many numerical difficulties the typical way of solving the column model should follow from easier setup to the more difficult one in stepwise manner. The numerical difficulty or easiness is quantified by type of the problem; number of equations involved into it; nature of the equations in the sense of the numerical conditions (stiff, ill-conditioned equations sets are more difficult to solve).
- Following the steps below is a suggested method of performing calculation which is not converged at first run.
    1 Finding initial, preliminary solution for "easy" settings. The easiness of the problem means a user sets condenser option for R/D and reboiler for V//B using rather small values up to, but sometimes larger could be required.
        - If the solution still cannot be found a user should refer to numerical options and define the value of "Interations" field to some large value, for example 1000. This sets the limiting number of iterations a which numerical engine stops large enough for many problems to converge.
        - If the solution still cannot be found a user may temporarily remove a component of lowest boiling point from the distillation mixture. If successfully calculated then later should return it to original quantity in stepwise manner (see next point no. 2).
        - If the solution still cannot be found a user may temporarily lower the pressure because high values of pressure typically poses higher degree of numerical difficulty. Then later if the solution is found a user should

revert back to the original pressure values, using stepwise manner (see next point no. 2).

– In some very complex cases, when the solution cannot be found a user may change local unit thermodynamics to simpler one. This can be done by selecting the unit at the flowsheet and using menu "Thermophysical/ Thermodynamics settings" and checking "Set local K-model/BIPs".

2 Having found the solution a user should select "Reload column profile" option form numerical settings, which allows the algorithm to start from advantaged initial position in the next calculations. Having selected the numerical setting, the more complex or difficult settings can be applied, for example the design mode mentioned in 5.2.3.

3 The numerically difficult properties or options should be specified in stepwise manner. That means that after every change a user should recalculate the unit to find new solution for actual settings. Reader is referred here to the logic of homotopy approach [14] which is a way of continuous or stepwise solving from easier to more difficult, more realistic cases. The idea is that it is easier to find solution to very complicated problem by first solving easier ones, and approach consecutively to the more difficult case.

4 The difficult cases might involve: distillation of mixture of components with high boiling temperatures differences, high pressure distillation, mixture described by complex (highly nonlinear) thermodynamics, columns with a big number of stages, mass transfer models.

– It is important to note that the default settings of "No condenser" and "No reboiler" mean that in this case the calculated column must be prepared with an external top liquid input stream and bottom vapor/gas input stream, respectively. In other words the default settings correspond to absorption column rather than the distillation. In this case at least two inlet streams must be present, where one is a top and the other is bottom inlet. Both streams locations must be set for the top (liquid) and bottom (gas). Running such column with distillation type inlet stream setup (only one feed) will result in error of no flows inside the column.

– The third tab contains numerical settings specific to the distillation unit. It is worth noting that for such complex mathematical equations set the starting, initial values for the algorithm to begin are crucial. Having found some preliminary results for the column, the "Reload column profile" is very useful option to consider. It allows the user to set current found solution, namely the profiles of concentrations, flows and temperatures as starting point for succeeding numerical calculations. In most cases this largely speeds up the calculation process and what is very important, it allows to minimize risks of numerical errors.

– Chemcad allows to make a design of the distillation column twofold. The first approach is to use integrated approach where whole distillation process, condenser, reboiler, reflux drum and the column itself is integrated into one unit operation. This makes the calculations easier from the designer point of view. The possibility to use a condenser and reboiler setup inside the unit options simplifies the whole process of calculating the distillation column. The other way of solving is distributed approach in which a user prepares flowsheet of all units placed and connected explicitly. The settings for condenser and reboiler are left as default, and the user is responsible to put into the flowsheet the heat exchangers in proper positions, the reflux drum and other required units. Although this approach poses higher degree of difficulty it provides larger flexibility to extend the design, for example into more complex control systems using dynamic options.

9.6 Heat exchangers

The heat exchanger design process includes three stages: general mass and heat balance, detailed technical design and rating sized equipment. In general the first step is fundamental for successful heat exchanger design and thus of key importance, and will be discussed here. Important terminology involves naming of the streams in terms of their purpose. Main purpose of heat exchange is to alter the thermal conditions of some selected stream. This stream is called process stream. This is a stream which is given or calculated by upstream flowsheet unit and its flowrate should not be altered. Only the thermal conditions should change. In order to be able to perform this task a second source or sink of heat energy must be available. This is provided by the other stream – utility stream, which is used to heat up or cool down the process stream, depending on the design requirements.

The heat transfer in the heat exchanger unit is described by the general equation:

$$\dot{Q} = K \cdot A \cdot (T_2 - T_1) \tag{9.28}$$

In which $\dot{Q}$ is heat stream transferred through surface area A when the temperature difference across A is given by $(T_2 - T_1)$. The variable K is heat transfer coefficient. Having in mind the heat exchanger construction the surface A is the dividing wall area which separates two sides of the heat exchanger.

Different thermal settings can be defined for the outlet streams depending on the design requirements or data available. Following is the list of thermal conditions accessible in Chemcad heat exchanger unit:

- Temperature – most obvious case when a designer defines required outlet temperature. The mass and energy balance will be calculated accordingly. A user must pay close attention to the feasibility of the heat exchange process when defining its value.
- Vapor fraction – Chemcad handles this input as a designer decision that the exiting fluid at outlet is in the boiling state. Values between 0 and 1 indicate liquid and vapor at boiling state, respectively. Any fractional value indicates

the fluid is a mixture of liquid and vapor in given by vapor fraction proportions at boiling state. Even more attention is required when defining this value, because the boiling points of the mixture might not always be known before calculations (it can of course be examined for example by the use of stream definition table). Unknown resultant boiling temperature might turn out to be outside of feasibility range for the heat exchange process.

– Subcooling – the difference between boiling point and actual temperature of the fluid. Selecting this definition results in that the fluid at outlet is in subcooled liquid state, that means it is below bubble point line on the TPXY equilibrium graph. In other words the liquid is not in boiling state and will not start boiling with the small temperature increase.

– Superheating – the difference between actual temperature and boiling point of the fluid. Selecting this definition results in that the fluid at outlet is in superheated vapor state, that means it is above dew point line on the TPXY equilibrium graph. In other words the vapor is not in boiling state and will not start to liquefy with the small temperature decrease.

– Heat duty – the amount of heat energy exchanged during the process. The outlet temperatures are calculated accordingly to the heat change for both streams.

– Heat transfer coefficient and area specification – the two variables serves as one thermal input definition. According to eq. (9.28) given the K and A, and flowrates and temperatures from the flowsheet inlet streams, Chemcad calculates the resultant temperatures at outlets.

Depending on the position of the heat exchanger in the flowsheet and on the design objectives, the process may have different number of degree of freedom. Several possible situations are presented in next parts of the text.

9.6.1 Given process and utility streams, one thermal degree of freedom case

The situation when both inlet streams, process and utility, are fully defined and designer requires one single outlet temperature to be calculated. The outlet temperature typically refers to the outlet process stream. The other stream temperature is calculated to satisfy the mass and energy balance of the heat exchanger. With one degree of freedom a user may define only one thermal condition.

9.6.2 Given one inlet stream and other to be calculated, two thermal degrees of freedom

The situation when process stream is given at inlet and a user needs to adjust the flow of the utility stream to obtain required temperature change. The designer indicates two thermal conditions both referring to both streams at outlet. The process stream is given by the design objectives or calculated by upstream unit. Thus its flowrate cannot be changed. By the use of "Utility option" a user must indicate the stream ID which is the utility stream. The flow of the indicated utility stream will be back calculated by the heat exchanger unit to satisfy mass and energy balance. This option is useful when a user needs to calculate the utility stream flow requirements. With two degrees of freedom two thermal conditions must be specified paying close attention they refer to process and utility outlet streams. The additional, second, utility stream thermal data refers to outlet utility stream conditions and must be provided upon the designers decision.

9.6.3 Given heat transfer coefficient and area of exchange, no thermal degree of freedom

Defining the heat transfer coefficient and area of exchange for heat exchanger unit locks and disables the possibility to define any thermal outlet data (although the fields are still available to input). Given the data together with mass and heat balance equations is enough to calculate both outlet streams temperatures leaving no thermal degree of freedom definition for the user. This case is useful when both inlet streams flowrates are given and cannot be changed. Such situation arises in the case when a heat exchanger unit is located internally in the flowsheet, when both inlet streams are calculated by other upstream units. This happens for example when a heat exchanger is used to recover or recycle some heat energy in the flowsheet.

9.6.4 Given heat duty, no thermal degree of freedom

The heat duty field available in heat exchanger unit is very useful when a user needs to calculate the heat effect of heat source or sink on the stream. One of the examples can be the design of the reboiler for the distillation column in the form of explicit unit definition on the flowsheet. The column enables to calculate the heat duty required at the reboiler for the distillation to work properly. To be able to perform a detailed design of heat exchanger reboiler unit, one may copy the column calculated heat duty into the heat exchanger, copy stream data from the column internal specification (column profiles) and perform a heat exchanger calculations. This enables the user to proceed to sizing heat exchanger procedure, using typical TEMA reboiler specifications (not covered by this text).

Funding: This work was supported by the Polish Ministry of Science and Higher Education [grant number 0912/SBAD/2007].

References

[1]   https://dippr.aiche.org/ Accessed: 02 Aug 2019.
[2]   Wilke CR. A viscosity equation for gas mixtures. J Chem Phys. 1950;18:517–19. DOI: 10.1063/1.1747673.
[3]   Grunberg L, Nissan AH. Mixture law for viscosity. Nature. 1949;164:799–800. DOI: 10.1038/164799b0.
[4]   Caldwell CS, Babb AL. Diffusion in ideal binary liquid mixtures. J Phys Chem. 1956. DOI: 10.1021/j150535a014.
[5]   Vignes A. Diffusion in binary solutions. Variation of diffusion coefficient with composition. Ind Eng Chem Fund. 1966. DOI: 10.1021/i160018a007.
[6]   Leffler J, Cullinan HT. Variation of liquid diffusion coefficients with composition. Binary systems. Ind Eng Chem Fund. 1970. DOI: 10.1021/i160033a013.
[7]   Kooijman HA, Taylor R. On the estimation of diffusion coefficients in multicomponent liquid systems. Ind Eng Chem Res. 1991. DOI: 10.1021/ie00054a023.
[8]   Lwin Y. Chemical equilibrium by gibbs energy minimization on spreadsheets. Int J Engng Ed. 2000;16:335–9.
[9]   Lakshmanan A, Rooney WC, Biegler LT. A case study for reactor network synthesis: the vinyl chloride process. Comput Chem Eng. 1999;23:479–95. DOI: 10.1016/S0098-1354(98)00287-7.
[10]   Danov SM, Kolesnikov VA, Esipovich AL. Kinetics of benzene nitration by nitric acid. Russ J Appl Chem. 2010;83. DOI: 10.1134/S1070427210010337.
[11]   Mekala M, Goli VR. Kinetics of esterification of methanol and acetic acid with mineral homogeneous acid catalyst. Chinese J Chem Eng. 2015;23:100–5. DOI: 10.1016/j.cjche.2013.08.002 ISSN 1004–9541.
[12]   Luyben W, Cheng-Ching Y. Reactive distillation design and control. Hoboken, New Jersey, USA: Wiley-AIChE. ISBN-13: 978–0470226124. 2008.
[13]   Kister H. Distillation design. Michigan: McGraw-Hill Education, 1992. ISBN-13: 978–0070349094.
[14]   Liao S. Homotopy analysis method in nonlinear differential equations. Beijing: Springer, 2012. ISBN-13: 978–3642251313.

Xavier Montané, Josep M. Montornes, Adrianna Nogalska,
Magdalena Olkiewicz, Marta Giamberini, Ricard Garcia-Valls,
Marina Badia-Fabregat, Irene Jubany and Bartosz Tylkowski

10 Synthesis and synthetic mechanism of Polylactic acid

Abstract: At present, Polylactic acid (PLA) is one of the most used biodegradable polyesters. The good properties and its biodegradability make that PLA can replace the fossil fuel derived polymers in different applications. PLA can be synthesized by using different methodologies. Among them, the most widely used forms on an industrial scale are the direct polycondensation of Lactic acid and the ring-opening polymerization of cyclic Lactide. The final properties of the obtained PLA are dependent on the used stereoisomers of the raw materials (Lactic acid and/or Lactide) and the conditions employed to polymerize them. Therefore, the comprehension of the synthetic mechanism of PLA is crucial to control the stereoregularity of PLA, which in turn results in an improvement of the polymer properties. So, distinct mechanisms for the synthesis of PLA by ring-opening polymerization using different catalysts systems (organometallic catalysts, cationic catalyst, organic catalyst, bifunctional catalysts) are examined in this review.

10.1 Introduction

Plastics or commodities, which are produced from fossil fuels, are used in many different applications due to their good properties and their versatility: packaging, encapsulation, electrical equipment, construction, medical applications, etc. However, the fact that these materials are not biodegradable cause environmental problems. So, a considerable research effort has been devoted to the replacement of commodities by biodegradable polymers to reduce the environmental impact in the recent years [1–3].

Depending on the used raw materials and the synthetic methods, biodegradable polymers can be classified in three different categories [4]:

- Polymers generated from biomass: agro-biopolymers from agro-resources (polysaccharides from starches, lignin, cellulose or polymers based on proteins and lipids) [5].

This article has previously been published in the journal Physical Sciences Reviews. Please cite as: Montané, X., Montornes, J. M., Nogalska, A., Olkiewicz, M., Giamberini, M., Garcia-Valls, R., Badia-Fabregat, M., Jubany, I., Tylkowski, B. Synthesis and synthetic mechanism of Polylactic acid *Physical Sciences Reviews* [Online] 2020, 5. DOI: 10.1515/psr-2019-0102.

https://doi.org/10.1515/9783110656367-010

- Polymers made by microbial production routes (polyhydroxyalkanoates, PHAs, are aliphatic polyesters naturally produced via a microbial process in a sugar-based medium) [6].
- Synthetic polymers using monomers obtained from agro-resources: Polylactic acid (PLA) from Lactic acid (LA) or Polyglycolic acid (PGA) from glycolic acid [3].

10.2 Polylactic acid

PLA is the most widely used bio-based polymer used in many application fields due to its low production cost combined with its properties (biocompatibility, stiffness, high strength). The global PLA market is expected to reach USD 2.169,6 million by 2020 [7]. PLA is a thermoplastic aliphatic polyester composed of LA monomers. The production of this LA generally derives from agricultural resources: LA is usually obtained by microbial fermentation of carbohydrates derived from biomass feed-stocks [2].

The main cost for LA production through fermentation process is the substrate, that nowadays is produced mainly using starch-rich crops such as maize (*Zea mays*) [8]. With the aim of reducing costs, avoiding substrates that can compete with the food chain as well as valorizing agro-industrial wastes, alternative substrates are being studied: cassava bagasse, cellulosic biosludge, wheat straw, paper sludge, apple pomace and waste sugarcane bagasse among others [9]. Production of Lactic acid from lignocellulosic waste requires of a pre-treatment to facilitate the accessibility of fermentative microorganisms to the nutrients, which can increase the costs of LA production.

Among all the studied organic wastes for LA fermentation, cheese whey is one of the most promising because it contains lactose, which is easy to convert to sugar monomers through biological fermentation, and other nutrients such as proteins, fat, vitamins and mineral salts. The main drawbacks are the costs of whey transportation that needs to be refrigerated to avoid its spoilage. However, it can be solved by means of filtration technologies (reverse osmosis or ultrafiltration) to concentrate the whey [10].

The process of fermentation to Lactic acid is done by microorganisms from different classes: Lactic acid bacteria (LAB; *Lactobacillus, Bacillus, Corynebacterium,* etc.), fungi (*Rhizopus*), yeasts, microalgae or cyanobacteria. Nonetheless, the most commonly used microorganisms are *Lactobacillus* sp. or engineered strains of *Escherichia coli* to improve the production yields or the optical purity by deletion of L- or D-lactate dehydrogenase [9].

Operational conditions of fermentation and further purification of Lactic acid can influence also in the production yields and purity of the final product. During fermentation, for example, pH should be regulated between 5 and 7 by means of the addition of a neutralizing agent (calcium carbonate or calcium hydroxide) to

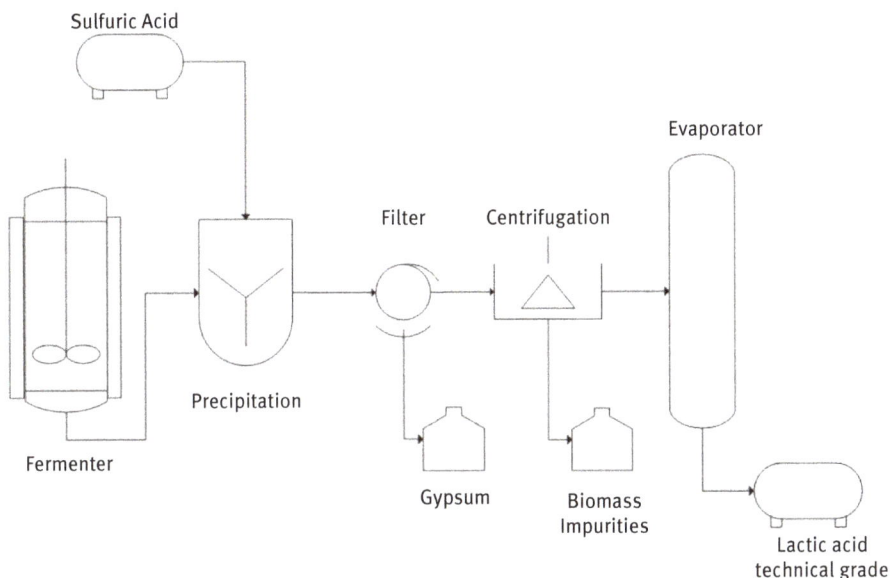

Figure 10.1: Flow diagram of the conventional process of fermentation and purification of Lactic acid [11].

avoid bacteria inhibition due to acidic condition. Then, down-stream processes (Figure 10.1) include a first step of precipitation of lactate salts and dissolution by addition of sulfuric acid. This purification process generates a huge amount of gypsum that requires of landfill disposal. Therefore, alternative LA purification systems are under study, such as electrodyalisis or other membrane processes [11].

Another advantage of PLA is that is not only a bio-based polymer but it is also easily degraded by simple hydrolysis of the ester bond in biological processes such as composting [12].

The possibility to synthesize PLA from different stereoisomers (D-LA, L-LA, a mixture of both or D-Lactide, L-Lactide or meso-Lactide) can influence the properties of the PLA polymer during its polymerization. Therefore, thermal, mechanical and biodegradation characteristics of Lactic acid polymers depend on the distribution of stereoisomers within the polymer chains. It is reported that high-purity L- and D-Lactides form stereo regular isotactic poly(L-Lactic) (PLLA) and poly(D-Lactic) (PDLA), respectively. The two stereoisomers of Lactic acid and the three stereoisomers of Lactide are depicted in Figure 10.2.

The microstructure of the four different PLA stereoisomers depending on their tacticity is depicted in Figure 10.3. Heterotactic PLA is usually obtained from a rac-Lactide mixture (mixture of D-Lactide and L-Lactide), while syndiotactic PLA is obtained when meso-Lactide was used as raw material in the polymerization reaction.

a)

OH

OH

O

D-(-)-Lactic acid

b)

OH

OH

O

L-(+)-Lactic acid

c)

O

O

O

O

Meso-lactide

d)

O

O

O

O

D-(-)-Lactide

e)

O

O

O

O

L-(+)-Lactide

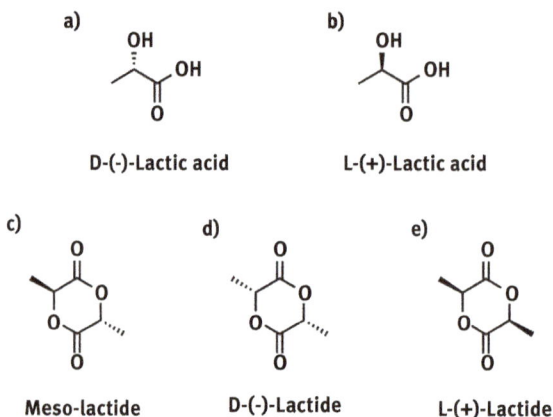

Figure 10.2: Chemical structure of: (a) D-(-)-Lactic acid, (b) L-(+)-Lactic acid, (c) meso-Lactide, (d) D-(-)-Lactide and (e) L-(+)-Lactide monomers.

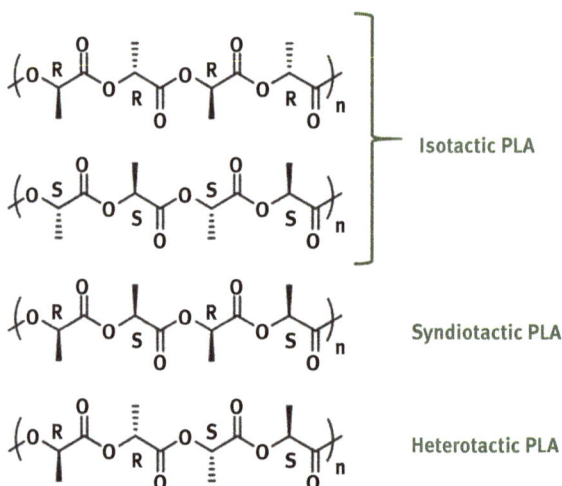

Isotactic PLA

Syndiotactic PLA

Heterotactic PLA

Figure 10.3: Chemical microstructure of: (a) isotactic PLA, (b) syndiotactic PLA and (c) heterotactic PLA.

Table 10.1 presents some properties of different PLA depending on the stereo-chemistry of the synthesized PLA. For example, the meso- and D,L-Lactide form atactic PDLLA which is amorphous. On the other hand, the stereoregular isotactic PLLA are semicrystalline polymers. From all of them, PLLA shows the best thermal properties with a melting point around 180 °C and a glass transition temperature between 55 °C and 60 °C.

As deduced from the Table 10.1, the stereoregularity of the PLA clearly improves its thermal and mechanical properties, since the PLA with the best

Table 10.1: Properties of different PLA [13–15].

Lactic acid polymers	Glass transition temperature T_g (°C)	Melting temperature T_m (°C)
PLLA	55–80	173–180
PDLLA	43–53	120-170
PDLA	40–50	120–150

properties is the isotactic PLLA. Therefore, many efforts to understand the polymerization mechanism of Lactic acid and Lactide isomers focus on control and improving the stereoregularity of the growing PLA polymeric chains.

10.3 The synthesis of PLA

The properties of PLA-derived polymers make this material highly attractive for different applications in which the biodegradability of the polymer plays a crucial role. For this reason, the improvement of the synthesis of PLA has been deeply studied in order to obtain materials with enhanced properties (mechanical, thermal or optical properties) [14, 16, 17].

PLA can be synthesized by using different polymerization methods (Figure 10.4) [18]:
- Polycondensation of Lactic acid monomer.
- Ring-opening polymerization of Lactide.
- Azeotropic dehydration condensation reaction from Lactic acid.

From the three methods presented above, the two most common used in industry were the direct polycondensation of Lactic acid and the ring-opening polymerization of Lactide.

10.3.1 Polycondensation of Lactic acid

The presence of the hydroxyl in the α-position to a carboxylic acid group in the chemical structure of Lactic acid facilitates the polycondensation of LA molecules. In this case, the polycondensation reaction of LA monomer connects the hydroxyl groups with the carboxylic acid groups. At the same time, water is produced as byproduct.

Although the polycondensation process is the cheapest way to obtain PLA, this process shows different limitations that cannot allow the obtention of PLA with high molecular weights:

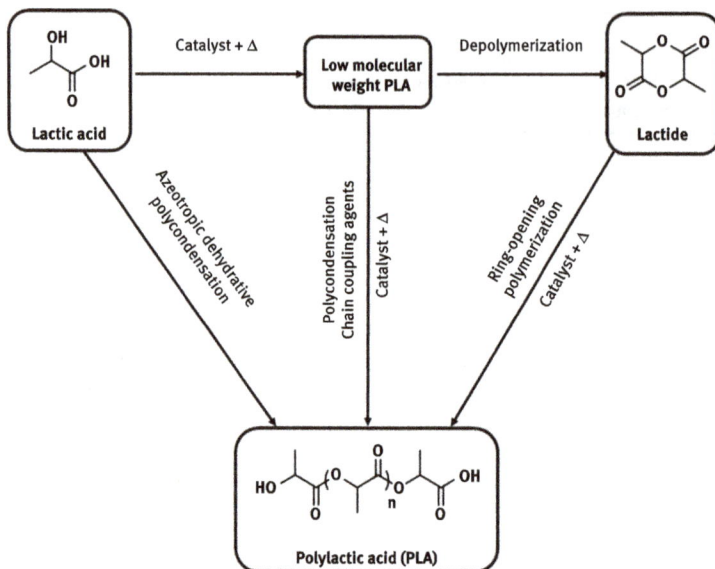

Figure 10.4: Synthetic routes for the obtention of Polylactic acid.

- High temperatures will be required to eliminate the generated water during the polymerization reaction. Moreover, a temperature above the melting of its polymer is required to carry out the polycondensation without solvent.
- On the other hand, direct polycondensation from Lactic acid should be carried out at less than 200 °C, because the formation of Lactide monomer is favored above that temperature.

The direct polycondensation route is an equilibrium reaction, which difficulties removing trace amounts of water in the later stages of the polymerization and generally limit the ultimate molecular weight that is achievable by this approach.

Concerning the difficulty in removing water from a high viscous medium, the exploration and modification of the melt polycondensation. The addition of a solvent was proposed by *Ajioka et al.* [19]. In this work, the authors described the synthesis of PLA with an average molecular weight up to 300.000 g/mol by using different organic solvents in the solution media. The water was removed by a Dean Stark apparatus. They report that the rate of polymerization is proportional to the boiling point of the solvents.

10.3.2 Ring-opening polymerization (ROP) of Lactide

The most common route for the industrial production of PLA with high molecular weight is the ring-opening polymerization of Lactide, a cyclic di-ester (Figure 10.4).

This method was firstly reported by Carothers in 1932 [20]. As shown in Figure 10.4, the first step of this route consist in the synthesis of a low molecular weight PLA, which is then catalytically depolymerized though internal transesterifications to Lactide, that is the real monomer for the obtention of PLA [21–23]. Finally, the ring of Lactide opens to form high molecular weight PLA.

By this process, PLA with very high molecular weight within a short polymerization time can be produced. An advantage vs. the direct polycondensation reaction of Lactic acid is that there is no need for the removal of water because the ROP of Lactide is not a condensation polymerization. Nevertheless, the Lactide monomer obtained by depolymerization of Lactic acid oligomers is not too much pure, which requires an additional purification step that lead to an increase of the final price of the process comparing to the petrol based thermoplastics [24].

Different metal catalysts can be used in this reaction, which can take place in solution, in the melt or as a suspension. Depending on the catalyst, isotactic, syndiotactic or heterotactic PLA can be isolated [25].

10.3.3 Azeotropic distillation of Lactic acid

One of the alternatives to the direct polycondensation of LA is the azeotropic distillation. By this method, the water generated during the polycondensation is removed efficiently by using an appropriate azeotropic solvent. In the year 1994, Mitsui Toatsu Chemicals patented an azeotropic distillation process using a high-boiling point solvent to drive the removal of water in the direct esterification process to obtain high-molecular weight PLA [26]. The control of the equilibrium between monomer and polymer can be tuned in the selected organic solvent to produce PLA with relatively high molecular weight in one-step.

By the azeotropic distillation, the temperature of the polymerization is lower than the polymer melting point, which avoids the presence of impurities caused by depolymerization and racemization [27]. In this process, the chosen solvent is critical for the polymerization conditions and the final properties of the obtained PLA.

10.4 Catalysts

The study of different methodologies developed for the synthesis of PLA is a necessity to control the stereochemistry of PLA during the polymerization reactions and subsequently their final properties. One of the parameters that influence the mechanism of the polymeric chain growth is the catalyst. Different catalysts were used for the synthesis of PLA: protonic acids, metals, metal-oxides, metal-halides and organic salts. They are specially used in the ROP of Lactide.

10.4.1 Organometallic catalysts

From all of them, Stannous (II) 2-ethylhexanoate (Sn(Oct)$_2$) is the most widely used because of high reaction rates, the solubility in the monomer melt, and the ability to produce high molecular weights. Moreover, Sn(Oct)$_2$ starts the ROP of various lactones and Lactides.

A coordination-insertion mechanism is proposed for the reaction between the Lactide and Sn(Oct)$_2$. An alcohol molecule, which act as an initiator such as MeOH or the propagating hydrolyzed Lactide, exchange with the octoate ligands, followed by the coordination of Lactide to the metal center. In the third step, the activated nucleophilic alkoxide proceed by the opening of the Lactide. The generated linear monomer starts the propagation, which occurs by subsequent Lactide coordination and alkoxide insertion until the metal-alkoxide bond is cleaved by termination reactions. By this method, the obtained PLA contains an ester end group derived from the initiator. The mechanism is depicted in Figure 10.5 [28, 29].

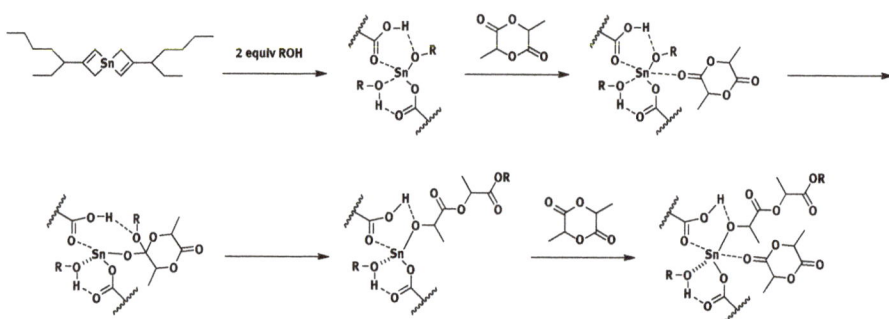

Figure 10.5: The coordination-insertion mechanism of the Lactide ROP catalyzed by Sn(Oct)$_2$.

By this method, the obtention of PLA with molecular weights up to 10^6 g/mol at 140–180 °C with an Sn(Oct)$_2$ concentration between 100–1000 ppm in 2–5 h is reported by *Drumright et al.* [30]. The polymerizations that follow a coordination-insertion mechanism always show an excellent control over polymer molecular weight due to the living character of the Lactide ROP with metal alkoxide complexes [31].

The replacement of Tin catalysts by other metals that are less toxic than Tin is an important factor considering the possible applications of the synthesized biopolymers in fields like medicine [32]. Different Zinc [33, 34], Magnesium [34, 35], Iron [36] or Aluminum [37] complexes were used as catalyst for the ROP of Lactide. Moreover, a better control of the stereoregularity of the PLA growing chains could be achieved with them [19, 32].

The mechanism of the Lactide ROP catalyzed by Zn complex is shown in Figure 10.6. As observed in Figure 10.6, in the first step of the ROP of the Lactide,

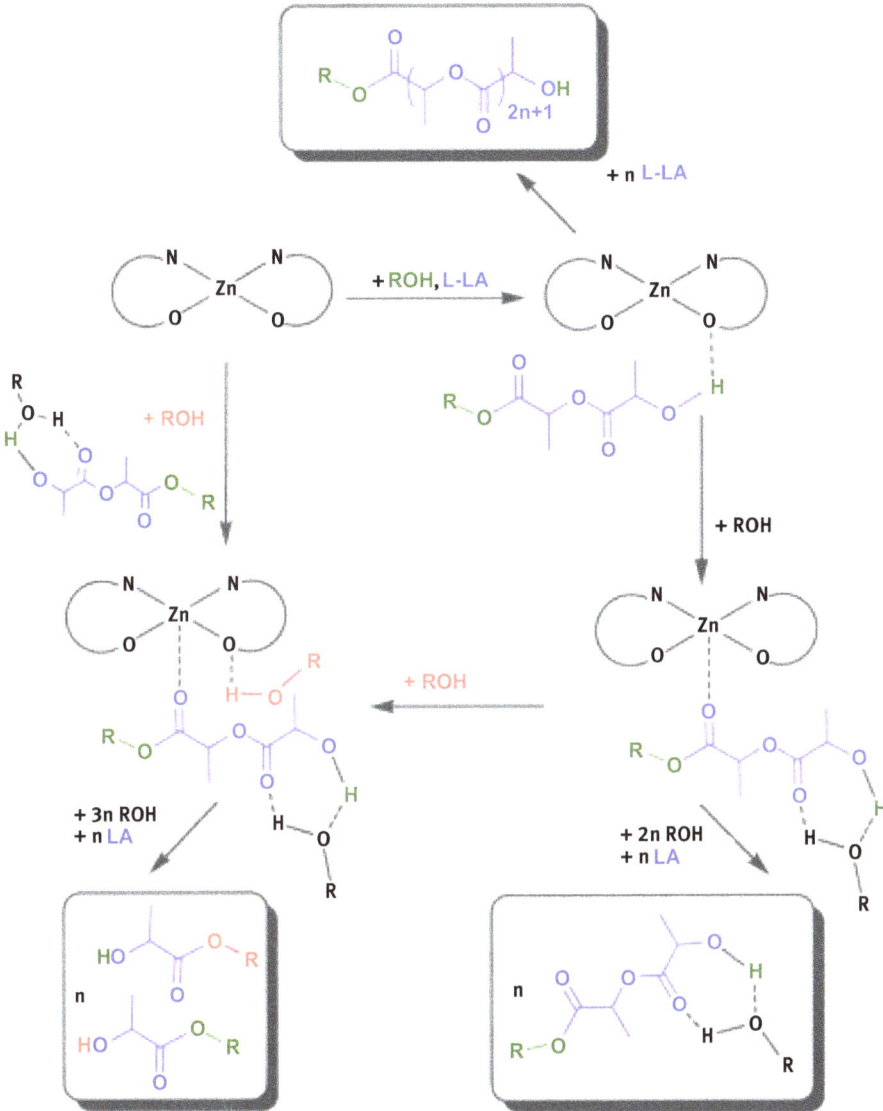

Figure 10.6: Proposed mechanism for the ROP and the alcoholysis of Lactide catalyzed by ZnL$_2$ complex [38].

the Zn metal center coordinates to one molecule of Lactide by a carbonyl group. In the second step, one alcohol molecule bonds to the oxygen donor atom of the ligand by hydrogen bond. After that, the alkoxy group generated from the alcohol molecule initiates the ROP of the Lactide. The catalytic cycle of PLA production continues until complete consumption of Lactide monomer when a molar ratio of

alcohol to zinc complex of 1:1 is used, because the formation of the complex between alcohol and the first open Lactide molecule is the crucial stage during ROP
in the presence of the Zinc catalyst.

10.4.2 Cationic catalysts

The major drawback of the metal catalysts is the incorporation of the toxic metals
on the polymer chain end, which lead to a toxicity risk for biomedical applications.
    Therefore, other catalysts were explored for the ROP of Lactide. Apart from Tin
compounds, protonic acids were found to be the best catalyst to obtain high-molecular weight PLA at relatively low temperatures (around 130 °C) [19]. In this manner,
strong acids such as sulfuric acid, phosphoric acid or sulfonic acids were used as
catalysts in the ROP of Lactide.
    The cationic mechanism of this polymerization is shown in Figure 10.7. In the
initial step, the Lactide ring is activated by a proton of the acid. In the second step,

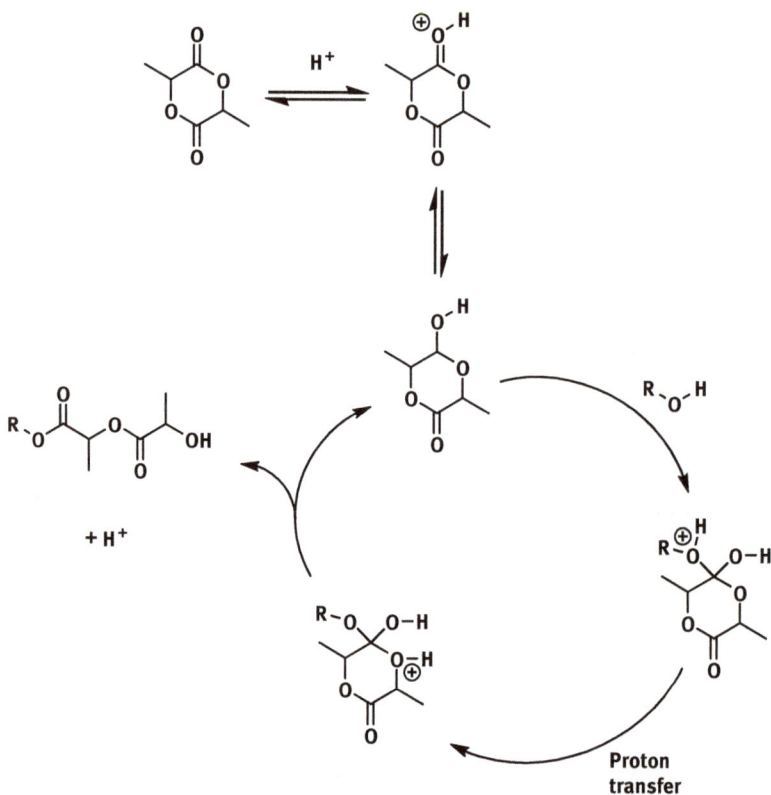

Figure 10.7: Proposed mechanism for the activated monomer pathway for the cationic ROP of Lactide.

the activated ring of Lactide is opened by the attack of the hydroxyl of the initiator, generating the linear monomer that undergoes chain growth.

Using sulfonic acids, molecular weights up to 100.000 g/mol were obtained in the ROP of Lactide carried out in solution.

10.4.3 Organic catalysts

Regarding to the principles that concern the green chemistry, the development of biocompatible methods with organic catalysts that not include metals for the synthesis of PLA that will be applied to biomedical area will reduce the environmental pollution.

There are different organic catalysts for the ROP of cyclic esters. From them, nucleophilic catalysts, such as tertiary amines, phosphines, and stabilized singlet carbenes, were effective polymerization catalysts for strained cyclic esters like Lactide. From all the organic catalysts, 4-(dimethylamino)pyridine (DMAP) was the first one described in literature [39]. The catalytic scheme using DMAP in the ROP of Lactide is depicted in Figure 10.8. In the proposed mechanism, the nitrogen bond of the heterocyclic amine attacks the carboxylic carbon. After that, a hydroxyl group arising from the initiator reacts with the Lactide-DMAP catalyst complex, leaving a terminal ω-hydroxyl group to act as a nucleophile to react with additional Lactide monomer.

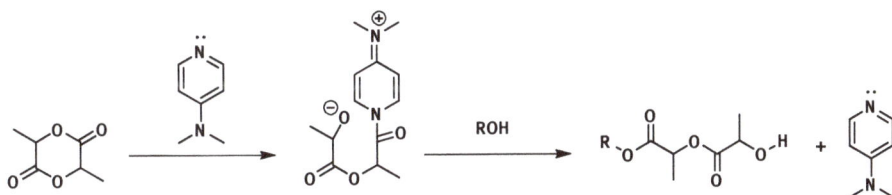

Figure 10.8: Proposed mechanism for the polymerization of Lactide using DMAP as catalyst.

In the same manner, N-Heterocyclic Carbenes (NHC) are used as a nucleophilic organic catalyst [40].

The catalyzed ROP of Lactide with NHC follows the same mechanism described before in the case of DMAP (Figure 10.9). Again, the formation of a zwitterion between the carbene and Lactide is formed, which act as a nucleophile to the hydroxyl group of the initiator. The linear monomer formed by the esterification in the last step of the cycle will start the chain-growth of PLA polymer.

10.4.4 Bifunctional catalysts

Another route that has been explored during the last years for the development of more controlled Lactide polymerization systems is the use of bifunctional catalysts.

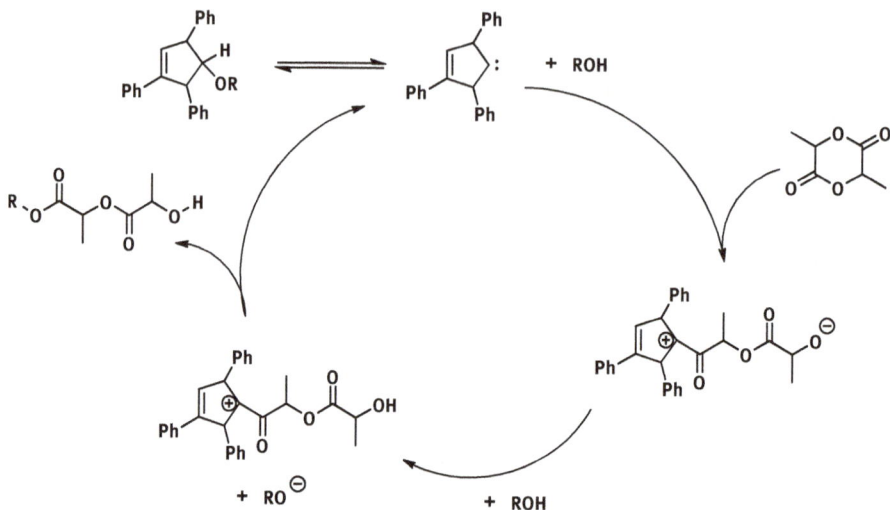

Figure 10.9: Proposed mechanism for the polymerization of Lactide using Triazole Carbenes.

A bifunctional catalysts system takes advantage of two different functional groups. One of these functional groups activate the monomer and the second functional group activates the alcohol of the initiator. In a bifunctional catalyst, both functionalities can be present in a single catalyst or two different catalysts can be used to activate the polymerization of Lactide.

The bifunctional catalysts used in the ROP of Lactide are organic molecules. These molecules often contain a hydrogen bond donor and acceptor moieties to activate the Lactide and the alcohol that acts as initiator, respectively. In these cases, an efficient activation of the Lactide monomer is achieved at room temperature. However, often high catalyst loadings will be required to achieve convenient rates.

As an example of a bifunctional catalyst, a thiourea-amine catalyst was also explored for the synthesis of PLA. The different hydrogen bonding possibilities of this catalyst by different amines are required for the activation of both the alcohol from the initiator and the Lactide to catalyze the ROP of cyclic esters. In this polymerization, the use of a solvent that are not able to do hydrogen bonding is crucial [41]. The authors of this work demonstrated that the presence of both functionalities: thiourea and amine, are fundamental for the activation of the Lactide ROP, although they do not necessarily have to be present in the same molecule.

As observed in Figure 10.10, the carbonyl oxygen of the Lactide is connected to the thiourea group by hydrogen bonding and the initiator attack the acyl-oxygen bond by the tertiary amine, resulting a linear monomer, which starts the ROP.

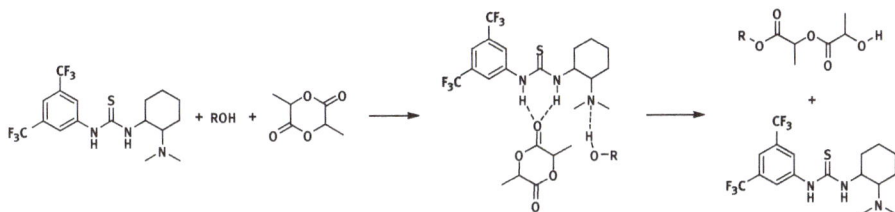

Figure 10.10: Proposed mechanism for the polymerization of Lactide using Thiourea-amine catalyst.

Dove et al. reported a high conversion for these systems at RT in solvents such as CH_2Cl_2, $CHCl_3$ or toluene, which can hardly lead to hydrogen bonding. They also remark that well-defined PLA architectures with narrow polydispersity were obtained adjusting the amount of catalyst and the monomer/initiator ratio. Nevertheless, 5 % of thiourea-amine catalyst is needed to obtain PLA.

10.4.5 Stereoselective Lactide polymerization

As mentioned above, different polymer microstructures are obtainable, depending on the Lactide stereoisomer that is used in the polymerization reaction and the way in which it is incorporated into the polymer. The microstructure of final PLA is directly correlated to the thermal and mechanical properties of the biopolymer. Different catalytic systems demonstrate high selectivity for Lactide polymerization and lead to a stereoregular ROP [42].

Due to the high cost of the purification of Lactide isomers, the ROP of (rac)-Lactide was carried out using different catalysts, preferentially organometallic compounds, in order to obtain isotactic PLA. The synthesis of isotactic PLA from (rac)-Lactide was firstly reported by Spassky and coworkers by using a chiral aluminum alkoxide complex [43]. Furthermore, Coates and Ovitt reported that by using an enantiopure aluminum complex [Al(SalBinap)(OR)], meso-Lactide is converted into crystalline syndiotactic PLA ($T_m = 150$ °C) [44].

The mechanism suggested by Coates and Ovitt was based on a site control at a chiral site. The coordinated ligands to the Aluminum confer a stereorigid chiral sphere, which strongly influence the Lactide monomer that will coordinate to the metal complex. The proposed mechanism is depicted in Figure 10.11.

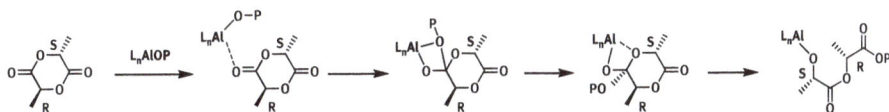

Figure 10.11: Proposed mechanism of ROP of meso-Lactide [44].

10.5 Conclusions

To sum up, PLA is a biodegradable polymer that has recently attracted much attention for the large-scale replacement of the non-biodegradable synthetic polymers produced from fossil fuels due to its optical and mechanical properties.

The main synthetic procedures to obtain PLA are stablished. The first one comprises the direct polycondensation of Lactic acid. The second one involves the ring-opening polymerization of cyclic Lactide monomer. However, different factors can determine the final properties of the obtained PLA in all the synthetic methods used (catalyst, temperature, solvent, etc.). Different improvements were done in the distinct synthesis of PLA. Specially, several catalysts have been developed for the ROP of Lactide. Mainly, there are two different groups of catalysts: metal-based and organic catalysts. Nowadays, the development of different organocatalysts that presents the same reactivity than the organometallic complexes in PLA polymerization due to their low toxicity is an interesting topic due to the potential applications of PLA in biomedical applications. Moreover, these organic catalysts are relatively inexpensive and highly active.

Regarding polymer stereoregularity and final properties, the selection of the catalysts have an important role in the produced PLA. Nowadays, the obtained PLA presents a low PDI because these ROP are living polymerizations. To improve the properties of PLA and expand their possible applications, significant efforts were carried out to study and understand the mechanism and the synthesis of PLA.

Acknowledgements: We acknowledge EURECAT internal research program: Plan of Research and Innovation 2019 (PRI 2019) and ACCIÓ - Regional Agency for the Business Competitiveness of the Generalitat de Catalunya for financial support of Bioplace project.

References

[1]   Bastioli C. Starch-based technology. In: Bastioli Catia, editor(s). Handbook of biodegradable polymers. Shawbury, Shrewsbury, Shropshire, SY4 4NR, UK: Rapra Technology Limited, 2005:257–86.

[2]   Zhu Y, Romain C, Williams CK. Sustainable polymers from renewable resources. Nature. 2016;540:354–62.

[3]   Haider TP, Völker C, Kramm J, Landfester K, Wurm FR. Plastics of the future? The impact of biodegradable polymers on the environment and on society. Angew Chem Int Ed. 2019;58:50–62.

[4]   Lee C, Sungyeap H. Ano overview of the synthesis and synthetic mechanism of poly (Lactic acid). Mod Chem Appl. 2014;2:1–5.

[5]   Bertan LC, Tanada-Palmu PS, Siani AC, Grosso CR. Effect of fatty acids and "Barzilian elemi" on composite films base don gelatin. Food Hydrocoll. 2005;19:73–82.

[6]   Sabbagh F, Idayu I. Production of poly-hydroxyalkanoate as secondary metabolite with main focus on sustainable energy. Renew Sustain Energy Rev. 2017;72:95–104.

[7]   Kumar S, Krishnan S, Mohanty S, Nayak SK. Synthesis and characterization of petroleum and biobased epoxy resins: a review. Polym Int. 2018;67:815–39.

[8]   Ghaffar T, Irshad M, Anwar Z, Aqil T, Zulifqar Z, Tariq A, et al. Recent trends in Lactic acid biotechnology: a brief review on production to purification. J Radiat Res Appl Sc. 2014;7:222–9.

[9]   Abdel-Rahman MA, Tashiro Y, Sonomoto K. Recent advances in Lactic acid production by microbial fermentation processes. Biotechnol Adv. 2013;31:877–902.

[10]   Panesar PS, Kennedy JF, Gandhi DN, Bunko K. Bioutilisation of whey for Lactic acid production. Food Chem. 2007;105:1–14.

[11]   Komesu A, Wolf Maciel MR, Maciel Filho R. Separation and purification technologies for Lactic acid – a brief review. Bioresources. 2017;12:6885–901.

[12]   Gorrasi G, Pantani R. Hydrolysis and biodegradation of Poly(lactic acid). Adv Polym Sci. 2018;279:119–52.

[13]   Thakur KA, Kean RT, Zupfer JM, Buehler NU, Doscotch MA, Munson EJ. Solid state 13C CP-MAS NMR studies of the crystallinity and morphology of poly-(L-lactide). Macromolecules. 1996;29:8841–51.

[14]   Madhavan Nampoothiri K, Nair NR, John RP. An overview of the recent developments in polylactide (PLA) research. Bioresour Technol. 2010;101:8493–501.

[15]   Södergard A, Stolt M. Properties of Lactic acid based polymers and their correlation with composition. Pro Polym Sci. 2002;27:1123–63.

[16]   Auras R, Harte B, Selke S. An overview of polylactides as packaging materials. Macromol Biosci. 2004;4:835–64.

[17]   Inkinen S, Hakkarainen M, Albertsson AC, Södergård A. From Lactic acid to poly(lactic acid) (PLA): characterization and analysis of PLA and its precursors. Biomacromolecules. 2011;12:523–32.

[18]   Hamad K, Kaseem M, Yang HW, Deri F, Ko YG. Properties and medical applications of Polylactic acid: a review. Express Polym Lett. 2015;9:435–55.

[19]   Ajioka M, Enemoto K, Suzuki K, Yamaguchi A. Basic properties of Polylactic acid produced by the direct condensation polymerization of Lactic acid. Bull Chem Soc Jpn. 1995;68:2125–31.

[20]   Carothers WH, Dorough G, Natta FV. Studies of polymerization and ring formation. X. The reversible polymerization of six-membered cyclic esters. J Am Chem Soc. 1932;54:761–72.

[21]   Kricheldorf HR, Berl M, Scharnagl N. Poly(lactones). 9. Polymerization mechanism of metal alkoxide initiated polymerizations of lactide and various lactones. Macromolecules. 1988;21:286–93.

[22]   Dechy-Cabaret O, Martin-Vaca B, Bourissou D. Controlled ring-opening polymerization of lactide and glycolide. Chem Rev. 2004;104:6147–76.

[23]   Dove AP. Controlled ring-opening polymerisation of cyclic esters: polymer blocks in self-assembled nanostructures. Chem Commun. 2008;48:6446–70.

[24]   Penczek S, Szymanski R, Duda A, Baran J. Living polymerization of cyclic esters – a route to (bio)degradable polymers. Influence of chain transfer to polymer on livingness. Macromol Symp. 2003;201:261–9.

[25]   Nuyken O, Pask SD. Ring-opening polymerization – an introductory review. Polymers. 2013;5:361–403.

[26]   Enemoto K, Ajioka M, Yamaguchi A. Polyhydroxycarboxylic acid and preparation process thereof. US Patent 5.310.865, 1994.

[27]   Masutani K, Kimura Y. PLA synthesis. From the monomer to the polymer. In: Jimenez A, Peltzer M, Ruseckaite R, editor(s). Poly(lactic acid) science and technology: processing, properties, additives and applications. Cambridge: Royal Society of Chemistry, 2015:1–36.

[28] Kowalski A, Duda A, Penczek S. Mechanism of cyclic ester polymerization initiated with tin(II) octoate 2. Macromolecules fitted with tin(II) alkoxide species observed directly in MALDI-TOF spectra. Macromolecules. 2000;33:689–95.

[29] Kowalski A, Duda A, Penczek S. Kinetics and mechanism of cyclic esters polymerization initiated with tin(II) octanoate. 3. Polymerizaiton of L,L-dilactide. Macromolecules. 2000;33:7359–70.

[30] Drumright RE, Gruber PR. Polylactic acid technology. Adv Mater. 2000;12:1841–6.

[31] Byers JA, Biernesser AB, Delle Chiaie KR, Kaur A, Kehl JA. Catalytic systems for the production of Poly(lactic acid). Adv Polym Sci. 2018;279:67–118.

[32] Wu J, Yu T-L, Chen C-T, Lin C-C. Recent developments in main group metal complexes catalyzed/initiated polymerization of lactides and related cyclic esters. Coordin Chem Rev. 2006;250:602–62.

[33] Williams CK, Breyfogle LE, Choi SK, Nam W, Young Jr VG, Hillmyer MA, et al. A highly active Zinc catalyst for the controlled polymerization of Lactide J Am Chem Soc. 2003;125:11350–9.

[34] McKeown P, McCormick SN, Mahona MF, Jones MD. Highly active Mg(II) and Zn(II) complexes for the ring opening polymerisation of lactide. Polym Chem. 2018;9:5339–47.

[35] Chamberlain BM, Cheng M, Moore DR, Ovitt TM, Lobkovsky EM, Coates GW. Polymerization of lactide with Zinc and magnesium b-Diiminate complexes: stereocontrol and mechanism. J Am Chem Soc. 2001;123:3229–38.

[36] O'Keefe BJ, Monnier SM, Hillmyer MA, Tolman WB. Rapid and controlled polymerization of lactide by structurally characterized ferric alkoxides. J Am Chem Soc. 2000;123:339–40.

[37] Hormnirun P, Marshall EL, Gibson VC, White AJ, Williams DJ. Remarkable stereocontrol in the polymerization of racemic lactide using aluminum initiators supported by tetradentate aminophenoxide ligands. J Am Chem Soc. 2004;126:2688–9.

[38] Jędrzkiewicz D, Czeluśniak I, Wierzejewska M, Szafert S, Ejfler J. Well-controlled, zinc-catalyzed synthesis of low molecular weight oligolactides by ring opening reaction. J Mol Catal A-Chem. 2015;396:155–63.

[39] Nederberg F, Connor EF, Möller M, Glauser T, Hedrick JL. New paradigms for organic catalysts: the first organocatalytic living polymerization. Angew Chem Int Edit. 2001;40:2712–15.

[40] Conner EF, Nyce GW, Myers M, Mock A, Hedrick JL. First example of N-heterocyclic carbenes as catalysts for living polymerization: organocatalytic ring-opening polymerization of cyclic esters. J Am Chem Soc. 2002;124:914–15.

[41] Dove AP, Pratt RC, Lohmeijer BG, Waymouth RM, Hedrick JL. Thiourea-based bifunctional organocatalysis: supramolecular recognition for living polymerization. J Am Chem Soc. 2005;127:13798–9.

[42] Dijkstra PJ, Du H, Feijen J. Single site catalysts for stereoselective ring-opening polymerization of lactides. Polym Chem. 2011;2:520–7.

[43] Spassky N, Wisniewski M, Pluta C, Le Borgne A. Highly stereoelective polymerization of rac-(D,L)-lactide with a chiral schiff's base/aluminium alkoxide initiator. Macromol Chem Phys. 1996;197:2627–37.

[44] Ovitt TM, Coates GW. Stereochemistry of lactide polymerization with chiral catalysts: new opportunities for stereocontrol using polymer exchange mechanisms. J Am Chem Soc. 2002;124:1316–26.

Index

https://doi.org/10.1515/9783110656367-011

www.ingramcontent.com/pod-product-compliance
Lightning Source LLC
Chambersburg PA
CBHW080931220326

41598CB00034B/5752